How to Prepare for the

FIREFIGHTER EXAMINATIONS

Third Edition

by
James J. Murtagh
Assistant Chief,
New York City Fire Department
Adjunct Professor,
John Jay College of Criminal Justice,
Western State Oregon College

Sundry
S

All inquiries should be addressed to:
Barron's Educational Series, Inc.
250 Wireless Boulevard
Hauppauge, New York 11788

Library of Congress Catalog No. 94-46425

International Standard Book No. 0-8120-9086-1

Library of Congress Cataloging-in-Publication Data

Murtagh, James J.
Barron's how to prepare for firefighter examinations / by James J. Murtagh.—3rd ed.
p. cm.
ISBN 0-8120-9086-1
1. Fire extinction—United States—Examinations, questions, etc.
I. Title. II. Title: How to prepare for fire fighter examinations.
III. Title: Fire fighter examinations.
TH9157.M57 1995
628.9'2'076—dc20 94-46425
CIP

PRINTED IN THE UNITED STATES OF AMERICA
5678 100 9

CONTENTS

ACKNOWLEDGMENTS

Thanks are due to the following for permission to reprint materials:

Cathy Briner, Employment Manager of the City of Eugene, Oregon for *Assessment Centers for the Selection of Entry-Level Firefighters.* 1981

City of Los Angeles Police Officer and Firefighter Selection Unit for "Doing Your Best on the Firefighter Interview." Revised 1977

Palo Alto Fire Department for information about its Assessment Lab.

Donald J. Schroeder and Frank A. Lombardo for excerpts from *How to Prepare for the Police Officer Examination.* Copyright © 1982 by Barron's Educational Series, Inc.

Joseph P. Spinnato, Fire Commissioner, City of New York, for use of materials published in Uniformed Training and Operations *Manuals* and in Fire Prevention *Directives.*

John Wiley & Sons, Inc. for definitions from *Fire Science Dictionary* by Boris Kuvshinoff. Copyright © 1977 by Boris Kuvshinoff. Reprinted by permission of John Wiley & Son's, Inc.

We would also like to thank the following for permitting us to use test questions:

City of Los Angeles Police Officer and Firefighter Selection Unit
Louisville Civil Service Board
City of Madison (WI) Fire Department
City of New York Department of Personnel

Acknowledgment is also made to the U.S. Government Printing Office, for the use of materials from government publications.

We are grateful to the City of New York Department of Personnel for granting us permission to reprint examinations numbers 3040 and 1162. The Diagnostic Procedures and Answer Explanations that accompany the tests are the work of the author and not of the City of New York.

Certain questions selected from *Pre-Examination Booklet for the Philadelphia Fire Department Fire Fighter Examination,* (1984). Reprinted by permission of Educational Testing Service. Permission to reprint the above material does not constitute review or endorsement by Educational Testing Service of this publication as a whole or of any other testing information it may contain.

The preparation of this text is not the endeavor of a single person; it is the work of many. I would like to thank Battalion Chief Jim Moran, Rae Monaco, and Jim Beirne for their assistance. I offer a special thanks to Liza Burby and Jane O'Sullivan for their outstanding efforts in editing and insuring that the manuscript has been properly prepared. I give special thanks to my daughter Kim and my son James for reading, analyzing, and then constructively criticizing the entire text, and I give an extra special and loving thanks to my wife, Fran, without whose help this book would not have been possible. I thank her for her Herculean efforts in assisting me in the preparation of the manuscript and for her love, faith, and patience. Finally, I would like to thank the members of the fire service across the country who have actively helped and supported me, and the fire service as a whole for what it has given me and for its dedication to SERVICE.

James J. Murtagh

Welcome to the Fire Service

You have taken a most important first step—from thinking to action, from planning to doing. You've decided to study, to prepare, and to be ready. The fire service is an action-oriented organization, with the actions determined by previous training, study, and other preparation. Your actions in preparing for your upcoming examination demonstrate that you have the qualities necessary to be a successful firefighter.

This test preparation guide has your success as its goal. It requires, as will the fire service, study, training, and commitment. It contains not only information about the role of the fire service and the career of the firefighter, but also sample Firefighter Examinations with clear explanations of all answers, analyses of the various types of questions and helpful review material, practice exercises to determine your weaknesses, and tips for improvement.

The questions on which you will practice are based on actual test questions. The techniques and strategies you will learn to answer these questions are those that have proved to be consistently successful.

PART ONE

Becoming a Firefighter

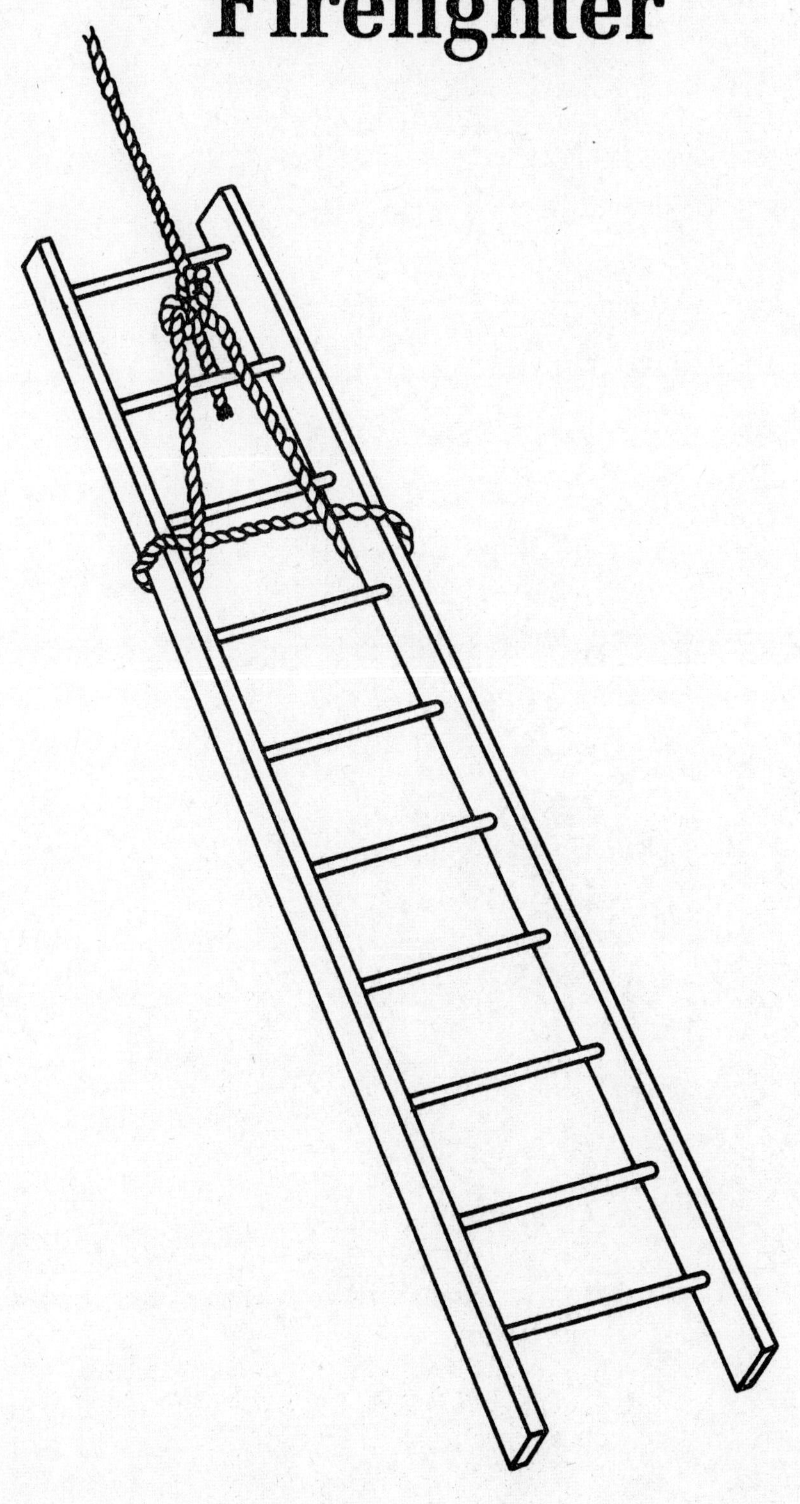

CHAPTER 1

Overview of the Firefighter

Job and Role of the Firefighter

Fire fighting has traditionally been a team-based occupation, with four primary goals:

1. The prevention of unwanted fires.
2. The prevention of loss of life and property from an unwanted fire.
3. The confinement of a fire to its place of origin.
4. The control and extinguishment of the fire.

Although these four activities are still prominent for the role of the firefighter, the occupation of fire fighting has become a multi-dimensional occupation. Firefighters are routinely called to major disasters like earthquakes, plane crashes, and building collapses; minor emergencies like gas leaks, sparking electrical wires, water leaks, and people trapped or stuck in elevators; or other unusual situations like hazardous materials incidents, leaks, spills, and gas releases; and firefighters play a major part in the nation's emergency medical response system.

After approximately 16 weeks of training, which includes firefighting and emergency medical procedures, the recruit firefighters are assigned to a fire fighting unit, where they work with experienced firefighters under the direction and control of a fire officer. During duty hours they must maintain a state of preparedness that will allow them to drive to respond quickly to a fire, other emergency, or emergency medical call.

At a fire, they connect hose lines to hydrants, carry these hose lines to the fire, and operate them to protect the occupants of the building and extinguish the fire. Firefighters position and use portable and aerial ladder equipment to make rescues and to gain entrance to buildings to search and ventilate smoke-filled areas. Other activities include salvage (the protection of property from water and fire damage), first aid, and the operation of specialized emergency equipment such as rescue tools, fire pumps, fire boats, and communication equipment. When not responding to calls for help, firefighters are involved in many other activities such as fire prevention inspections and enforcement, fire education and fire investigation classes, and the promotion of good community relations.

While fire fighting is the most visible aspect of a firefighter's role, emergency medical care, fire prevention, fire education, and community relations occupy a major part of the time on the job. Firefighters are trained to inspect buildings for hazardous conditions that may result in a fire. Armed with their knowledge of fire hazards and local fire and building laws, they check exit routes, the storage and use of flammable and combustible materials, overcrowding of public places, and improper use of equipment or materials in all kinds of buildings. Firefighters are also trained to administer emergency first aid. There are several levels of training: Certified Pulmonary Resuscitation (CPR), Certified First Responder (CFR), Emergency Medical Technician (EMT), and Paramedic. The training and skill requirements are standardized, however the level of training varies for each community.

Another important task of a firefighter is public education—the chance to meet with people and discuss the hazards of fire, and the opportunity to interact with a class of school children, to teach them the dangers of fire, and show them the correct reaction to such an emergency.

Fire fighting is a diversified job requiring a commitment to learn and to put into operation what has been learned. It requires that large amounts of time be spent in drilling and in practicing fire fighting operations. It also requires that time be spent in learning fire safety practices and laws. Finally, it requires an active interest in learning about people.

Employment Opportunities and Conditions

The job of a firefighter is extremely dangerous and demands knowledge, physical strength, agility, stamina, and courage, as well as compassion, understanding, and adaptability. It requires working in hostile environments, being a member of a team, and interacting with fellow firefighters and the public.

Fire departments generally have a staff that works hard to give prospective candidates the essential help they often need. Fire departments are equal opportunity employers and strive to have a diverse work force. Many minorities and women have successfully competed for these positions and are now firefighters and fire officers. Minorities and women are encouraged to apply and compete for these positions.

Some fire departments, working with local high schools and colleges, have established Fire Cadet programs. These programs prepare the cadet to be a firefighter and to meet and pass all the requirements for entry to the fire department. This training may be accepted as experience for those departments that have an experience requirement, or may qualify for some preference credit.

Fire departments must comply with the Americans with Disabilities Act (ADA). This means they must make a conditional offer of employment before any testing relating to physical or medical conditions occurs. This conditional offer of employment can be withdrawn by the fire department if, after a reasonable accommodation, the candidate cannot perform the essential tasks firefighters must perform.

Firefighting is a 24-hour-a-day, 365-day-a-year occupation. To meet this demand, firefighters work in shifts. Each fire department establishes its own shift schedule and manning requirements. Although there are no absolute rules governing how shifts are worked, the two most prevalent shifts are 24-hour tours and the split shift. The first requires that firefighters work 24 hours of duty followed by either 48 or 72 hours off duty. The second has two different schedules. One requires that firefighters work 9-hour day and 15-hour night tours, or 10-hour day and 14-hour night tours. After each set of day tours, the firefighter is given approximately 48 hours off, and after each set of night tours 48 to 72 hours off. The other breaks the day into three 8-hour shifts, with systematic rotation through each of the 8-hour tours over a period ranging from 1 to 3 months. The work week can vary from 40 to 56 hours, depending on local conditions.

While on duty, firefighters live in a close relationship, preparing and eating meals together, and sleeping in dormitories. When called to a fire or emergency, they are expected to respond immediately, and to put into operation all that they have learned with an almost zealous sense of commitment. Firefighters are the guardians of safety for society and must be devoted to that end.

SALARY AND BENEFITS

The salaries of most firefighters are determined by labor and management negotiations. Salaries vary from region to region and often reflect the local cost of living and the size of the community being serviced.

As a rule, the larger cities offer higher salaries and greater benefits. It must be kept in mind, however, that these urban centers often represent greater fire and emergency activity and greater exposure to danger. Salaries are lowest in the South and highest in the West. To determine the salary scale of your local fire department, read the notice of examination or consult the department's personnel office. If you wish information about salaries and benefits of other fire departments, you may refer to *The Municipal Year Book,* published by the International City Management Association. This should be available at a local library. You may also write to the International Association of Firefighters, 1750 New York Avenue NW, Washington, D.C. 20006, or to the International Association of Fire Chiefs, 4025 Fair Ridge Drive, Fairfax, Virginia 22033-2868.

Most fire departments provide medical and health care coverage, paid sick days, and pension and retirement benefits, which include disability or service retirement. Service retirement is generally at half-pay after 20 or 25 years of service. Vacation periods and personal days are provided, and almost all fire departments furnish appropriate uniforms and safety equipment.

CAREER OUTLOOK

Firefighter recruits will be trained for approximately 16 weeks at their local fire academy. Generally these recruits will serve a one-year probation during which time their abilities, demeanor, and character will be evaluated. Traditionally, promotion in the fire service has been from firefighter to lieutenant or captain, to battalion chief, to assistant chief or deputy chief, and, finally, to chief of department. These promotions almost always come from within the department. At the present time, however, some lateral entry into high-level positions is possible. Because of a nationwide program to standardize firefighter qualifications, opportunities to enter the fire service at all levels should increase. This process (*lateral entry*) will be enhanced further by the Apprenticeship Program, developed jointly by the International Association of Fire Fighters and the International Association of Fire Chiefs, with cooperation from the Joint Council of Fire Service Organizations, the National Professional Qualifications Board, the National Fire Protection Association Qualifications Committee, and the Bureau of Apprenticeship and Training, U.S. Department of Labor.

Some departments have multiple career tracks specializing in areas such as fire investigation, fire inspector, or emergency medical service. The National Fire Protection Standard, NFPA 1001, Fire Fighting Professional Qualifications, is generally accepted as the standard for entry-level training. There are other NFPA standards, such as Fire Officer Professional Qualifications, Fire Inspector Professional Qualifications, Fire Investigator Professional Qualifications, Fire Service Instructor Qualifications, Airport Fire Fighter Qualifications, Aircraft Rescue and Fire Fighting Qualifications, and Fire Department Safety Officer. These national standards are not mandatory but are generally accepted as guides to what is expected of a person in the position.

In most fire departments, candidates for promotion are selected from a civil service list, developed in a manner similar to that used for the Fire Fighter Examinations. This list, however, reflects eligibility requirements as well as the specific knowledge, skills, and abilities required of a particular rank. As might be expected, the higher the rank, the more difficult and challenging the examination process will be. The successful candidate can also expect a higher salary, more diversified assignments, and greater self-satisfaction.

General Requirements for Applicants

Although no absolute requirements can be given, the following are examples of typical requirements to be a firefighter:

- AGE: Generally the age entry range is 18 to 21.
- CITIZENSHIP: Most departments require the candidate to be a citizen to be appointed to the job, but not to take the entrance examination.
- RESIDENCE: Many departments require the candidate to be a resident of the community at the time of the examination and to maintain such residence thereafter. Some fire departments give additional credit, 5 to 10 points added to the final score, to candidates who reside in the community. If you are not a resident of the community, you may be required to become a resident as a condition of employment after you are hired.
- DRIVER'S LICENSE: This is normally a requirement to be appointed, but often is not needed to take the test.
- EDUCATION: Almost all fire departments require a high school diploma or its equivalent. Some fire departments require college education.
- CERTIFICATION: Certification as an EMT may be required. Many fire departments will require that this be obtained during the first year of employment.

- EXPERIENCE: Although this is not generally required, some departments do require it and others give additional credit, or allow it to be substituted for other requirements such as education.
- GROOMING: This is generally not a consideration for employment. However, many fire departments do have strict hair length and facial hair standards.
- VETERAN PREFERENCE: Qualified armed forces veterans can request that preference points be added to their final score.
- NONSMOKING REQUIREMENT: A number of fire departments require that the candidate be a nonsmoker. Upon appointment, firefighters are required to sign a pledge that they will not smoke during their employment; otherwise, they face dismissal. This is to ensure that heart and lung disability claims are in fact job-related.
- CHARACTER INVESTIGATION: After an eligibility rating in the examination has been established, an investigation may be conducted to secure additional evidence of the candidate's qualifications and fitness with particular reference to integrity, reliability, and general suitability.

Applying for the Examination

To apply for the test, begin by finding out when the examination will be held and then obtain an appropriate application. Information about the examination and application process can usually be obtained from your city's department of personnel and from your local fire department. Sometimes applications and information can be obtained from your school guidance office or your local library.

The most effective method of locating these agencies is to look them up in your local telephone directory. If you cannot find them, call the operator and ask for assistance. When you telephone the agency, have a pencil and paper ready to jot down instructions and directions. Explain clearly and politely that you need:

1. Information about the Firefighter Examination.
2. The necessary application.
3. The correct procedure for filling out the application.

If there is a test in the near future, the agency will give you the necessary information. However, if no examination is currently scheduled, you must ask the agency how the examination notice is made public: where it will be made available, and in which newspaper it will be published. Be persistent!

Most jurisdictions require applicants to file for the examination by filling out and submitting an application by a certain date. If you mail your application, it should be postmarked by midnight on the closing date indicated in the job announcement. After you file, you will be notified by mail as to where and when you should appear for the examination. In some communities, "walk-in exams" are given which do not require prior filing. Know what is required in your community. If there is a filing date cutoff, make sure you file before that date.

GETTING AND CHECKING YOUR SCORE

You will be told, either in the examination booklet or by the monitor, when and where the official answers will be posted and/or published. When they become available, carefully check your own answers to determine how well you have done.

Unless you get a perfect score, you will probably not agree with all the official answers. If you feel strongly that an answer you chose is as good as the official answer, or better, you should file a protest. The protest procedure should be indicated on the examination instructions or be available at the office of the examiner. Usually, protests must be submitted to the examiner within a limited time.

A protest should include an explanation of why you believe your choice is as good as the official answer, or better, and should list any books or other materials that will support your claim. If you plan to protest more than one question, submit each protest on a separate sheet of paper. On the bottom of each sheet, sign your name and give your address. After the examining agency has reviewed all of the protests, a final determination will be made and a final official answer key will be published.

KEEPING IN CONTACT WITH THE AGENCY

Several months may elapse from the date of the examination to the posting of the final examination key and the establishment of the official list. When the list is established, you will be notified by mail of your score and list number. The time period until the first appointment may be even longer.

It is your responsibility to keep abreast of what is going on during these waiting periods and to notify the examining agency of any change of address or status that may affect its ability to get in touch with you. Often you can get pertinent information from your local civil service newspaper or newsletter. Some agencies have a special telephone number for giving such information.

GETTING APPOINTED

Once the list is established, the agency will begin to process the candidates in list-number order. It will ask for background employment information, while setting up appointments for medical and psychological examinations. It is important that you complete and return all forms as soon as possible. Follow instructions carefully, and ask questions when you don't know what to do.

CHAPTER 2

Firefighter Examinations

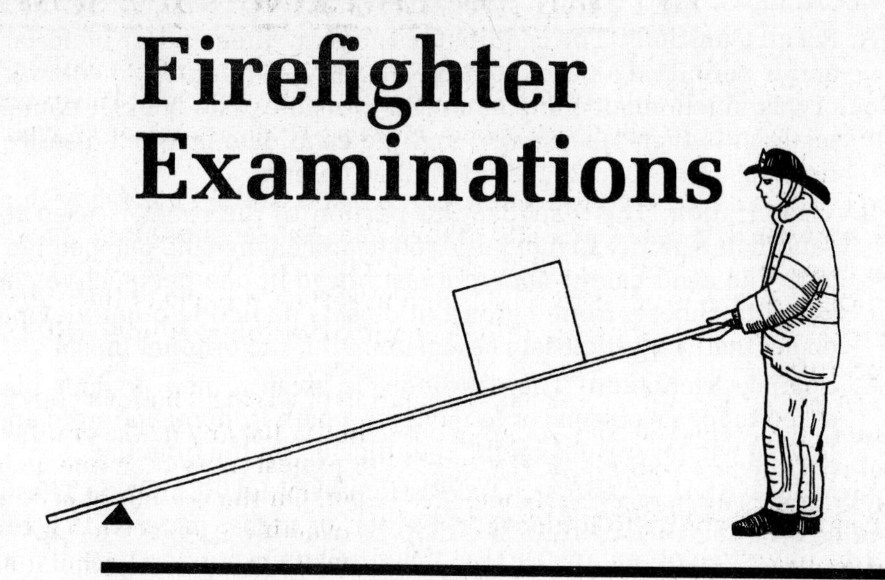

Firefighter Examinations across the nation have been challenged in the courts and as a result have been improved substantially. Numerous court decisions have resulted in examinations designed to measure accurately skills directly related to the occupation of fire fighting. Today's examiners strive to ensure that the examinations they prepare are nondiscriminatory, correspond to the fire fighting position in question, and measure the physical and mental ability required of a firefighter. Questions that require the candidate to have prior knowledge of fire fighting, fire prevention laws, or rules and regulations of fire departments should not be asked. You will not be tested on actual job knowledge, unless some clearly stated prerequisite is required. You will be tested for the kinds of knowledge, skills, and aptitudes needed to become a successful firefighter.

This guide has been designed to identify your weaknesses, sharpen your test-taking skills, and reduce the stress and anxiety of the examination process—in short, to help you become a member of the fire service.

Purpose of the Examinations

The central purpose of Firefighter Examinations is to develop a list of qualified candidates who have expressed a desire to become firefighters. This list is used as a guide by civil service agencies in hiring new firefighter recruits. Candidates who score the highest are offered positions first; therefore, it's in your best interest to place as high on the list as you can. Careful, diligent work with this guide will lead to the achievement of this goal.

Written Examination

EXAMINATION FORMATS

Analyses of testing practices for the position of firefighter throughout the nation indicate that the typical 3½-hour Firefighter Examination may contain the following types of questions:

1. RECALLING, VISUALIZING, AND SPATIAL ORIENTATION QUESTIONS.
   A. Recall Questions. The candidate is given written and/or pictorial material and is permitted a period of time, usually 20 minutes, to commit to memory as much about the contents as possible. The booklet containing the material is then taken away, and the candidate is asked a series of questions, usually 10 to 20, based on the contents.
   B. Visualization. The visualization portion of the exam is used to test the candidate's ability to mentally rotate and change the perspective of an object. The candidate is shown a test image in one perspective and is then asked to choose, from a group of objects depicting other viewpoints, the object that is most closely associated with the original image.
   C. Spatial Orientation. The candidate is given a map or floor plan and is asked to locate objects, or to navigate a path from one point to another following whatever rules might be indicated on the drawing.
2. READING AND VERBAL COMPREHENSION QUESTIONS. These test the candidate's ability to understand written or verbal material that reflects the kind of written or oral instruction that firefighters must be able to understand. This material may be entitled "Understanding Job Information."
3. QUESTIONS ON UNDERSTANDING AND APPLYING BASIC MATHEMATICS AND SCIENCE. The candidate is given a series of basic math and science questions, and is tested on the ability to understand them and to apply the information to typical firefighter situations.
4. QUESTIONS RELATING TO TOOLS AND EQUIPMENT. These are designed to evaluate the candidate's knowledge and understanding of tools and their proper use.
5. QUESTIONS ABOUT DEALING WITH PEOPLE. This type of question tests the candidate's ability to interact with peers, supervisors, and the general public. Both emergency and nonemergency situations are considered.
6. QUESTIONS RELATING TO MECHANICAL DEVICES. The prospective firefighter is asked to demonstrate knowledge and understanding of general mechanical devices. Pictures are shown, and questions are asked about the mechanical devices shown.
7. QUESTIONS THAT TEST JUDGMENT AND REASONING. This type of question tests the candidate's ability to interpret and handle situations that a firefighter may encounter on the job.

Don't be concerned if the description of the test format given above seems difficult. Each of the areas is thoroughly explained in Part Three of this book. For each area, helpful hints and strategies are offered. Finally, you will be given many sample questions in each area, as well as fully explained answers to each question.

Oral Interview/Assessment Centers

ORAL INTERVIEW

The oral examination or oral interview is, as its name implies, a nonwritten examination; it has been used in the fire service for many years. There are two types of oral interviews: the unstructured and the structured interview. The unstructured interview makes use of open-ended questions, that is, questions that do not require specific answers. In the structured interview a set of predetermined questions and appropriate responses to them are developed by the testing agency before the interview. The questions are normally hypothetical and job related and require solutions that measure the ability to think clearly and creatively while demonstrating knowledge of the subject area.

The city of Los Angeles has prepared material on the oral interview as an aid to fire fighter candidates. For your benefit it is reprinted here by permission. Although all oral examinations will not be exactly as outlined on the following pages, you will do well to follow the directions and recommendations given.

ABOUT THE INTERVIEW*

There are three parts to every job interview. Each part is important.

1. **THE JOB** is the duties and responsibilities that need to be done.
2. **THE INTERVIEW BOARD** is the men and women who make a judgment on how well your qualifications match the requirements of the job.
3. **YOUR QUALIFICATIONS** are your education, experience, knowledge, abilities and/or personal qualities. You must show the interview board how well your qualifications fit the job.

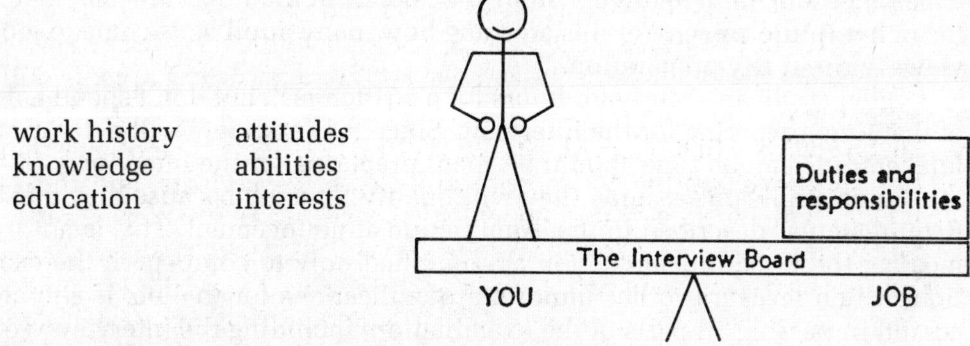

The interview board compares you with the job.

The interview board which conducts your Firefighter interview will probably be composed of three or four members. They will include a Fire Captain, a Personnel Department staff member, and men and women from business, government and community organizations. They are chosen because of their familiarity with the Firefighter program and are experienced in hiring employees for their own organizations.

All interviews conducted may be tape recorded.

Before the interview, the interviewers will be briefed by Personnel Department staff and a member of the Fire Department. The briefers will describe the department's operation and the qualifications which this department believes necessary to do the job.

The briefer's job is to make the interview board aware of the duties of the job which the applicants who are eventually hired will be doing. The briefer is **not** permitted to discuss individual candidates.

The Personnel Department briefer will provide the interviewers with rating sheets which describe the qualifications to be evaluated in the interviews and the standards to be applied in these evaluations.

The briefer also makes the interview board aware of areas they should not consider in the interview—areas that are not related to job performance—such as race, religious creed, color, national origin, sex, age, political affiliations, or sexual orientation.

The Personnel Department tries to avoid using interviewers who are likely to know some of the candidates. This is to avoid any bias, favorable or otherwise, toward any candidate. Interviewers are instructed to disqualify themselves from interviewing any candidates about whom they believe they cannot rate objectively because of prior knowledge about the candidates.

If an interviewer does not disqualify himself or herself from examining a candidate whom the interviewer knows, **it is the candidate's right,** if he or she wishes, to have the interviewer remove himself or herself from the board or to have an interview with another board if one is available.

The names, titles and affiliations of interviewers on all interview boards will be posted in the waiting room area. Check the list to see who your interviewers are. If you recognize an interviewer and believe that you could not receive a fair interview from him or her, tell the receptionist so that the interviewer can be taken off the board for your interview.

BEFORE THE INTERVIEW

The job interview is one of the most important events in the average person's experience since the relatively short time spent in the interview may determine his or her future career. Yet it is amazing how many applicants come to job interviews without any preparation.

Often applicants study for hours for a written test, but don't spend a fraction of that time preparing for the interview. Since the interview will determine your final grade, as much time should be spent preparing for the interview.

Some applicants assume they will qualify for a job because they meet "The Requirements" described in the examination announcement. This is not true. By meeting these requirements, you are qualified only to compete in the examination, which measures other important qualifications for the job. If you are successful in passing all parts of the examination, including the interview, you may then be considered for a job.

Getting Ready

Be aware of the exact date, time and place of your interview. This may sound almost too basic to mention, but it's an unfortunate candidate who assumes that the interview is to be held in a certain place, and then discovers two minutes before the interview that the appointment is somewhere else. Equally unfortunate is the candidate who arrives at the right place and time, only to find that the appointment is tomorrow or, worse, yesterday. Keep the interview notice with you. Don't rely on your memory.

Plan to arrive for your interview at least 15 minutes early. A few extra minutes will help to take care of unexpected emergencies. It is frequently difficult to find a parking place quickly. . . . Late arrival for an interview is seldom excusable.

You should present a neat, businesslike appearance for your job interview. It is usually appropriate for you to dress as you would for an office job.

DURING THE INTERVIEW

Just as you are about to enter the interview room, the receptionist will tell you the name of the chairperson, who is selected randomly. The chairperson will introduce you to the other board members and ask you to sit down.

Your Conduct

Much of the impression you make upon the interview board will be made by your conduct.

Your courtesy, alertness and self-confidence are important; so, you should try to speak in a self-assured tone of voice; smile occasionally; look the interviewers in the eye as you listen and talk. Sit erect, but be relaxed; and be prepared to answer the questions that are likely to be asked of you.

The board members realize that it is normal for people to feel nervous in this situation. Experienced interviewers will discount a certain amount of nervousness. But you should try to avoid doing obvious things such as drumming your fingers or twisting a handkerchief. If you are prepared to answer the questions that will be asked of you, you will probably find that you will not be as nervous as when you are unprepared. It is usually best not to smoke or chew gum.

The Questions

Remember that the interview board will be trying to measure your qualifications based on the information about job requirements given to them by the departmental briefer and the Personnel Department briefer.

The interview board will be exploring and evaluating those qualifications which have not been measured by prior parts of the examination. These qualifications include such things as personal qualities, oral communication, attitudes, goals and interests. This does **not** mean, however, that material covered on the prior tests you have taken will not be explored further.

The interview board will be given your application to review prior to the time that you enter the room. Be ready for at least one question at the start, such as:

1. Tell us something about yourself.
2. Why are you applying for a Firefighter position?
3. Why do you want to work for the City?

These are not easy questions to answer without some previous thought. Try them before deciding for yourself. You should be able to answer these kinds of questions without hesitation. Your preparation will help get you off to a good start.

You should also be prepared to answer questions about your abilities, training, and experience, such as:

1. Tell us how your previous work experience or training has prepared you for this job.
2. What are your major assets in regard to this job?
3. In what areas related to the job you are applying for do you need to improve yourself the most? How have you compensated for this weakness or deficiency?

The specific areas to be measured in the interview for Firefighter fall in five general areas. All candidates should expect to be asked questions relating to the subjects listed below:

A. **Work History**—This area will include your present job, jobs you have held in the past and your experience in the military, if any.

B. **Educational History**—This area of questioning may deal with your experiences in high school as well as any schooling or training you have undertaken since your graduation.

C. **Interpersonal Relations History**—This area will include questions about how you deal or have dealt with the people around you in everyday life.

D. **Interview Behavior and Communication Skills**—This area of measurement refers to the way in which you handle yourself in the interview. Raters will be looking at your behavior and manner as well as your communication skills.

E. **Reasoning and Problem-Solving in Answering Situation Questions**—Questions in this area will be geared to find out about your ability to think out reasonable solutions to everyday problems that can arise in a Firefighter's work. Answering them calls for good practical judgment, not special knowledge of firefighting.

You are **not** expected to have had previous training or experience in firefighting. The interviewers will be concerned mainly with how you have responded to whatever jobs and educational opportunities you have had. The Fire Department expects to train you in firefighting, so the interviewers will not be looking for any direct connection between the kinds of work you have done and firefighting.

By discussing the areas listed above, your interview board will be able to measure **your** suitability for Firefighter work. Keep in mind that you will be rated competitively with all other candidates on these factors.

A review of a copy of your application and the examination announcement should help you to answer these and other questions related to the job you are applying for. You should review them immediately before the interview to make sure they are fresh in your mind.

Your Answers

Most interviews follow a simple question-and-answer formula. Your ability to answer quickly and accurately is very important, but don't rush yourself if it will hurt your ability to answer questions well. If your answers are confused and contradictory, you will not do well.

The greatest preventive against contradictory answers is the plain truth. A frank answer, even if it seems a little unfavorable to you, is better than an exaggeration which may confuse you in the next question. Being friendly, honest and sincere is always the best policy.

Don't answer just "yes" or "no" to any question. Expand on your answer at least a little. Volunteering information is often helpful in showing how you qualify for the position, but be completely honest, because you will almost always be asked more about your answer. It is also important to know when to stop answering a question. You should try to avoid repeating yourself, giving information that is unrelated to the question, or talking too much on any one point.

Ask the interviewers to repeat or explain any questions you do not understand. This may be embarrassing, but it is better than answering the wrong question.

Be certain your employment application is complete and accurate in all respects before presentation to the interview board. False or incomplete statements made during the selection process may be cause for disqualification or dismissal at a later date.

If something went wrong on a previous job, explain the circumstances and accept the blame if it was your own. If you have been fired and you are asked about it, admit it, and explain what you have learned from this experience. Negative experiences can be turned into an asset for you, if you can show how you have changed or improved yourself after recognizing your mistakes.

Make sure that your good points get across to the interviewers, but try to be factual and sincere, not conceited. If you are describing your best qualities, be concrete. Give examples of how these qualities have helped you and your previous employers. This is where your preparation will pay off.

Remember that you have known yourself all of your life. The interviewers, however, have only a short period of time to try to get to know you and to recognize your capabilities. Be sure to help them all you can by giving them the information they need to properly evaluate you.

The End of the Interview

Toward the end of the interview, you will be asked if you would like to add anything. If you believe that there is something in your background the interviewers should know that hasn't been mentioned, this is your chance. This is also a good time to briefly sum up what you believe makes you a good candidate for the job.

Sometimes candidates protest their interviews after receiving their scores, with the claim that the interview board did not ask them about experience, training or other background which the candidate believes is important. Interviewers don't have time to ask enough questions to bring out all the qualifications a candidate may possess. That is the reason we instruct interview boards to ask all candidates if they have anything they would like to add; and such protests as the one mentioned above are not valid if you do not take advantage of the opportunity to answer the last question fully.

Try, however, to make your final statements or your answer to the above question concise because the interview board has a schedule to keep and there are other candidates waiting.

The chairperson will indicate the end of the interview by thanking you for coming in. Thank the interviewers for their time and consideration.

AFTER THE INTERVIEW

The results of your examination will usually be mailed to you within two weeks after your interview.

The Personnel Department makes every effort to assure that all candidates receive a fair interview and that the persons who serve on interview boards are competent.

The score that a candidate receives in an interview depends to a large extent on the presentation of his or her qualifications to the interview board. Candidates who receive low scores frequently disagree with the judgment of the interview board. The Civil Service Commission, however, does not consider a difference of opinion between the candidate and the interviewers regarding the candidate's qualifications as a valid grounds for protest.

INTERVIEW APPEALS

If you believe that any of the persons on your interview board were prejudiced, or that there was fraud involved, or that the interview was not properly conducted, you should file a written protest within two working days after you complete your interview. The reasons for your protest and the facts supporting your charges must be submitted in writing.

All too frequently when we fail, we blame someone else for our failure. The wise thing is to try to determine why we failed so that the next time we can succeed. You can find out why you failed the interview by coming to the Firefighter Selection Unit during the review period.

CHECKLIST FOR PREPARING FOR THE INTERVIEW

- ☐ Read this section carefully.
- ☐ Make a list of your good points and think of concrete examples that demonstrate them.
- ☐ Practice answering the six questions on page 13.
- ☐ Review a copy of your application and the examination announcement.
- ☐ Make sure you know the exact date, time, and place of your interview.
- ☐ Dress neatly.
- ☐ Bring the interview notice with you.
- ☐ Have enough money for parking.
- ☐ Leave in time to arrive at least 15 minutes before the appointed time.

ASSESSMENT CENTERS

The use of the assessment center for the selection of firefighters is relatively new. It is seen as an effective alternative to the oral examination and can be expected to find greater use in the future. The assessment center is an interactive process allowing a group of firefighter candidates to participate in simulations of job-related functions and processes under controlled conditions. The candidates are rated on how well they respond to the activities and how they handle each assignment. As the candidates go through the simulations, they are systematically observed and rated by a panel of assessors.

The concept of the assessment center derives its strength from the fact that fire fighting tasks can be identified and that these tasks have certain underlying qualities which can be measured. The assessment exercise is developed to measure specific behaviors in an atmosphere that approximates the real thing.

The assessment center (assessment exercise) is much more comprehensive than the traditional examination. It generally takes a full day or more to complete the tasks. The exercise consists of five to eight separate components, each designed to evaluate a different set of skills, abilities, and knowledge. The assessors are a group of knowledgeable people who have been specifically trained to conduct the assessment center. This special training is necessary because each assessment center is developed specifically for a particular community and its Firefighter Examination.

The assessors will measure your ability to perform; they will do this both as individuals and as a group. Their composite score will be your score. In other words, your overall score will not be determined by just one assessor.

The assessment exercise is made up of several job-related tasks. Each component is scored individually. Here are examples of typical exercise components that have been used:

1. Triad—candidates are divided into groups of three and allowed 2 minutes to get acquainted with each other as much as possible. The candidates are then regrouped until all participants get to meet each other.
2. In Basket—candidates are allowed up to an hour to complete a series of administrative tasks that would be required of a firefighter. These include reports, memorandums, notifications, complaints, and decision-making tasks.
3. Manipulative Tasks—a job-sampling procedure in which participants connect hoses, fittings, and hydrants so they are operational.

4. P.T.A. Presentation—participants are allowed 5 minutes to devise a solution to fire prevention problems, using materials provided. They then present their solution to a panel, simulating a P.T.A. audience.
5. Written Reports—given the description of a fire, the participants must transfer the information to a standardized reporting format.
6. Dilemma Sources—an individual selected as the leader is allowed to study a Tinker Toy model and then must direct his/her "team" in the construction of a duplicate structure.
7. Fire House—an individual must react to a hypothetical problem he/she may face as a fire fighter when not actually fighting fires, such as establishing and coordinating a cooking schedule.

There are eight dimensions that are typically used in assessment centers:

- Work standards
- Mechanical/manipulative aptitude and skill
- Oral communication skill
- Written communication skill
- Interpersonal relations
- Problem analysis
- Learning ability
- Leadership

These are explained below.

FIREFIGHTER ASSESSMENT CENTER DIMENSIONS

Work standards
Does a good job in a safe manner and stays with a task until it is completed; displays initiative by taking action beyond what is required and by modifying behavior appropriately to maintain effectiveness in changing situations; meets organizational norms of punctuality and reliability and maintains effectiveness in pressure situations.

Mechanical/manipulative aptitude and skill
Correlates common tool usage with mechanical/manipulative task performance; discerns pipe and hose dimensions and matches related thread types; "makes things work" with or without the aid of tools or other equipment; understands interrelationship of moving parts in mechanical apparatus.

Oral communication skill
Listens effectively and comprehends verbal instructions; makes a clear, effective, and persuasive presentation of ideas or facts in individual or group situation; speaks in clear, distinct, and understandable manner; uses appropriate tone, grammar, and vocabulary; exhibits effective nonverbal communication skills, such as eye contact, gestures, and listening practices.

Written communication skill
Reads with good comprehension and clearly expresses ideas in writing using correct spelling, vocabulary, grammatical form, and legible penmanship.

Interpersonal relations
Identifies and reacts to the needs of others and analyzes the impact of self on others; displays empathy as appropriate and deals with others nondefensively; uses tact, diplomacy, and discretion as appropriate.

Problem analysis

Perceives relations between various items, gathers relevant information, and takes appropriate action based on the information at hand.

Learning ability

Comprehends and assimilates, retains, and recalls large amounts of factual information required for fire protection and emergency medical task performance.

Leadership

Guides individuals and/or a group toward successful task completion or goal attainment by soliciting participation from all members; gets effective ideas accepted without being domineering; motivates others to action; and mediates in troublesome situations.

HOW TO PREPARE FOR THE ASSESSMENT CENTER

Is Preparation Possible?

It has been said by a number of prominent examiners that a candidate cannot prepare for an assessment center. This statement is both correct and incorrect.

It is correct in the sense that there is no text to study, nor are there sample test questions that can be practiced or rehearsed. The assessment center is an interactive exercise derived from the work of the firefighters in a particular community. There is no way to predict what specific activities the examiners will choose for the exercise.

The statement is incorrect, on the other hand, because prior knowledge of what kinds of questions or tasks will be encountered and how they will be scored is significantly helpful.

Knowing What to Expect

Knowing what to expect reduces anxiety and stress while eliminating the chance of being caught totally unprepared. The following description of each of the typical components listed on pages 16 and 17 will give you an idea of what to expect.

- *Triad.* You will be required to interact with several strangers, not just once but several times. The purpose is to see how you fit into a group and how comfortable you are with strangers. This component relates to the dimension of *interpersonal relations.* You should make an effort to put others at ease, say "please" and "thank you," display concern and fairness, and be sensitive to the needs of others. You can improve your interpersonal relations skills by trying to be more outgoing, for example, by starting a conversation with another person while waiting on lines. The more relaxed and familiar you are with this exercise component, the better you will do.
- *In Basket.* This is a paper and pencil test of your administrative abilities. You will be required to fill out forms, write short memos, record a telephone message, and/or take a complaint from a citizen. This component relates to the *written communication skill* dimension. You must write legibly, clearly, and concisely. Use appropriate vocabulary, grammar, punctuation, spelling, and sentence structure. You can improve your penmanship by practice, and you can gain confidence in other skills by developing the habit of writing down some of your telephone conversations and by paying attention to directions when filling out forms such as credit card, bank loan, and driver's license applications.
- *Manipulative Skills.* Here you are required to do a job-related task, and your performance is assessed either as to how well you did the process or as to

how well it turned out. This component relates to the *mechanical/manipulative aptitude and skill*, the *learning ability*, and the *work standards* dimensions. Follow all directions to the best of your ability. Ensure that only safe practices are used, and do not become rattled or upset if things do not go the way you expected. Keep trying, and stay calm. Since this exercise requires a familiarity with basic tools and mechanical devices, a review of Chapters 9 and 10 will be helpful.

- *P.T.A. Presentation.* In this exercise you will be given time to prepare or rehearse a short presentation to a group. Some public speaking ability is required. This component relates to the *leadership* and the *oral communication skill* dimensions. Your challenge is to organize quickly and then deliver a reasonably realistic presentation. To organize quickly, get the known facts and put them in order. Discuss them quickly with the other members of your group; take notes as the discussion proceeds. Prepare a short outline that addresses the questions *who, what, when, where, why,* and *how.* Your talk will be informal. You will not be expected to make a speech, but you should be able to express yourself audibly, clearly, and concisely, maintain eye contact, and utilize proper body and hand gestures, while stating the problem, offering some solutions, and answering questions. To prepare for this section, study the materials in this guide, discuss them in a group setting, and keep in mind that nervous tension is both natural and good for you. Above all, don't allow yourself to be intimidated at the idea of "making a speech."

- *Written Reports. See In Basket.*

- *Dilemma Sources.* This is also known as decision making. In this section you will be given a problem for you and the rest of the group to solve. It may be a Tinker Toy model, a game problem, or a written problem. This component relates to the *problem analysis* dimension. Be prepared to express your opinions, making specific recommendations, deriving a possible course of action, discarding irrelevant factors, and finally making a decision. You must decide what information is necessary to solve the problem. You can prepare for this section by reviewing Chapter 8. Keep in mind that the ability to size up a problem, develop possible alternatives, put a solution into operation, adapt to setbacks, and then revise your plan is the key element to this exercise.

- *Firehouse.* This component requires you to be a self-starter, to be an organizer, and to act as an informal group leader. It can measure many dimensions: *leadership, interpersonal relations, work standards,* and *problem analysis.* To be prepared, you must do what has been outlined in the sections you have just read, and you must practice and follow the instructions detailed in this guide.

Because of the vast number of individual problems and the great variations in what can be expected in the exercise components, it is impossible to define each component specifically.

We have attempted to show you how an assessment center may be set up and conducted. Remember that it will be testing for general knowledge and abilities. Knowing what to expect and doing some thinking beforehand about the topic areas is the best way to prepare.

Physical Ability Test

The physical ability test is a major part of the examination for the position of firefighter. Understanding this part of the examination and being prepared for it are critical factors for your success.

In physical terms, fire fighting is an extremely demanding occupation. It requires agility, strength, and stamina. The fire fighting environment is normally hazardous and constantly changing. Fire fighting calls for the wearing of special protective clothing and breathing equipment and the use of tools that are often heavy. Because of the extremes encountered in the fire fighting environment—hot and cold, wet and dry, night and day, clean and contaminated air—the protective clothing and equipment must be durable and effective. The need for these qualities has led to the development of equipment that is often heavy and cumbersome.

Given the demands of the occupation—saving life and property, the challenges of the environment, and the weight and constraint of the protective equipment, the need to ensure that firefighter recruits are physically capable of learning and performing the tasks required is obvious.

The physical ability examination is designed to evaluate the candidate's ability to perform fire fighting activities. In the recent past the courts have held that the physical ability examination must be related to the tasks that are actually performed by firefighters. Studies reveal that the firefighter must have a high level of aerobic energy, strength, and a significant ability to resist fatigue.

The tasks that firefighters perform require a person who can run, jump, and bend while lifting, pulling, or carrying heavy weights.

WHAT TO EXPECT

There is a great deal of similarity in fire fighting from community to community. The courts have ruled, however, that each community must analyze its own needs and then develop a test to meet these needs.

The physical portion of the examination is designed to measure your ability to perform typical fire fighting tasks. Proper exercise and a familiarity with the typical events will help boost your score significantly.

The illustrations presented in this section are representations of typical test events and may not meet exact test specifications. Each community varies the number of events, the requirements for each event, and the manner in which the event is set up or carried out. However, these test illustrations are accurate representations of the type of activities firefighters perform on a daily basis.

Before you begin each event, it is extremely important to find out from the testing agency, fire department, or personnel department what restrictions may be imposed for each event. Some questions you should try to get answered are:

- What kinds of personal protective equipment are required or excluded from being used during the testing process?
- What constitutes the completion of each test event?

Each event can have a different criteria that signifies the completion of the exercise. You may be required to cross a designated finish line, place a hose, a tool, or a ladder in a specified area, ring a bell, hit a completion button, or tell the monitor you have completed the exercise. You do not want to lose time or points by having to backtrack to execute the "event completion component."

When practicing these events, make sure the event completion component is part of your practice routine. If it becomes a routine part of your style, there is less chance that you will forget or skip it on the day of the test.

HOSE CARRY

The hose carry activity tests your ability to lift a 50-pound length of fire hose from either an elevated position or from the floor and carry the hose on the shoulder for a distance of approximately 75 to 150 feet. You may be required to climb stairs while carrying the hose and this may be a timed event. (see stair climb)

You may be required to wear an air pack or weighted pack simulating an air pack while performing this test event.

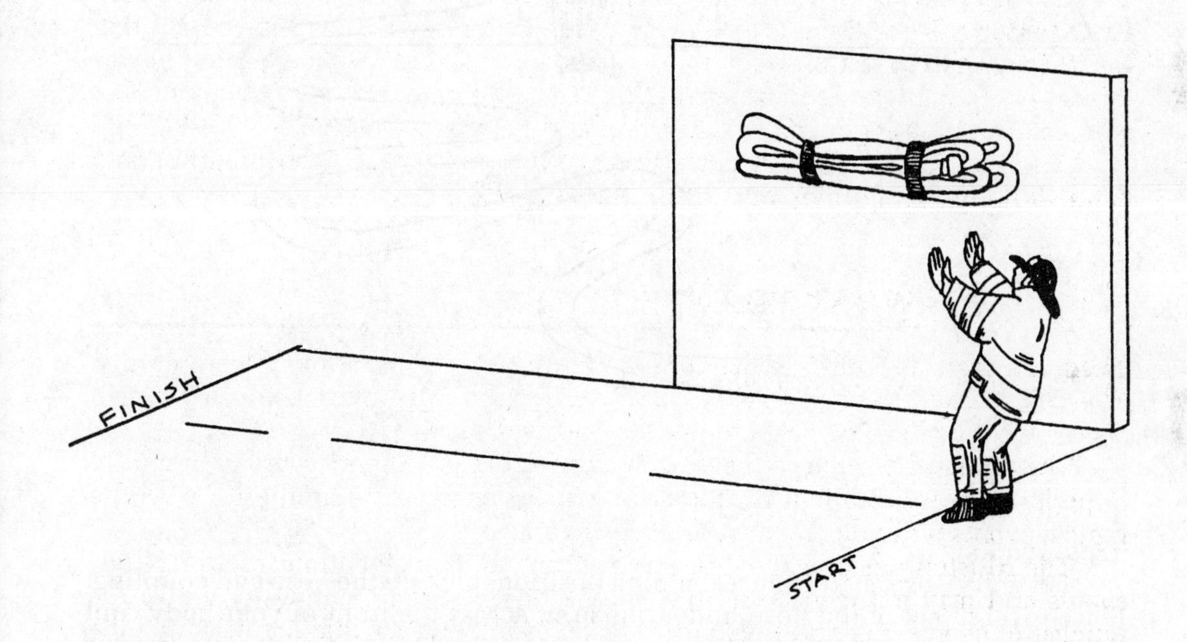

Questions

Before you attempt this event, find out if there are any restrictions about how the hose must be carried.

Tips/Tricks

When carrying the hose, keep the load centered on your shoulder to prevent it from shifting forward or backward. Walk or jog in a position that is comfortable for you. Leaning slightly forward with your upper body may be helpful.

HOSE DRAG

The hose drag is used to measure your ability to drag (move) hose a distance of approximately 50 to 200 feet. This exercise simulates a firefighter moving a hose line closer to a fire.

You will be required to pick up one end of several connected lengths of fire hose and then drag them a specified distance. The hose may be filled with water. You may be required to wear an air pack while performing this test event.

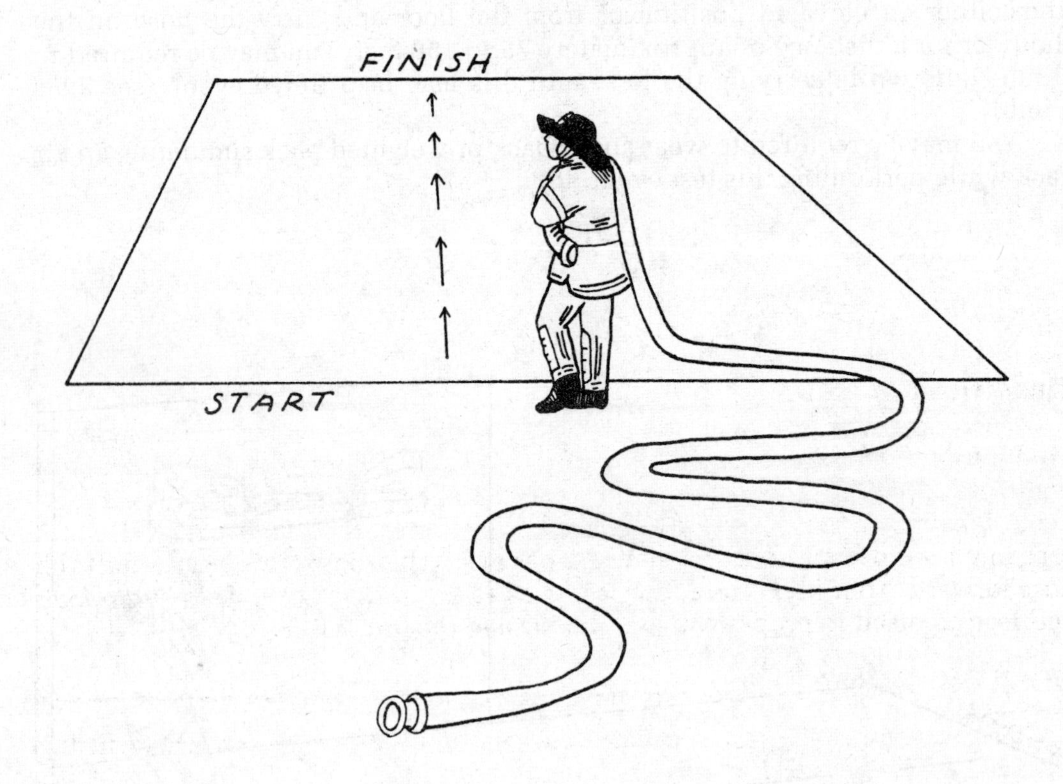

Tips/Tricks

For a dry hose, start from a standing position, pick up the hose and coupling. Place several feet of the hose under one arm, across the front of your body, and then up and over your other shoulder, so that the nozzle or hose butt rests on your back at hip level. For a water-filled hose, drape the hose over your shoulder so that both hands can comfortably hold the hose down in front of your body, about waist high. Lean into the direction you will be dragging the hose and use the strength of your legs to push you and the hose forward.

HOSE ADVANCEMENT

The hose advancement test is used to measure your ability to work and drag a fire hose in a confined space for approximately 50 feet. This activity simulates moving a hose into a fire area in preparation to fight the fire. The hose may be filled with water or other substance to simulate the weight of a water-filled hose line.

From a crouching or kneeling position, you will drag a hose line into and through a covered course. This simulates taking the hose through a hallway and to the seat of the fire. You may be required to wear an air pack while performing this test event.

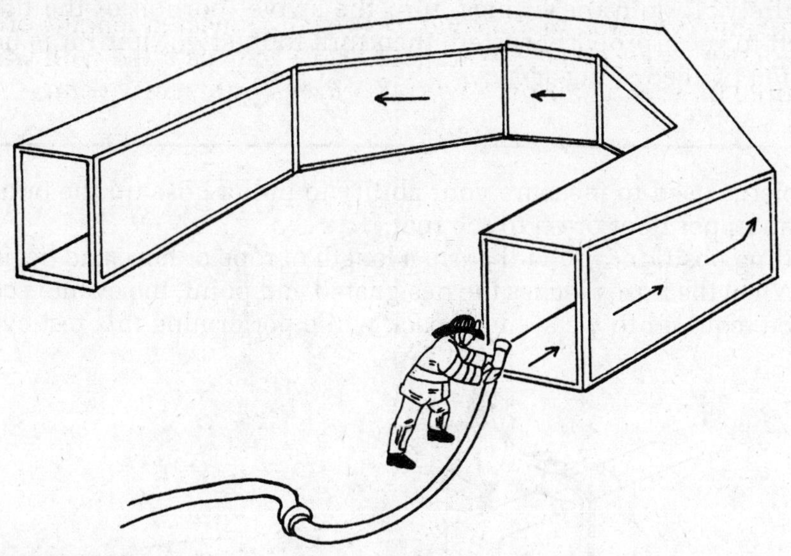

Tips/Tricks

When dragging hose in a confined space, grip the hose with one hand just behind the hose butt and extend that arm fully forward. With your other hand and arm bent at a 90 degree angle, grip the hose 15 to 18 inches in back of the first hand. Clasp your hands tightly over the hose. Keep your hands well apart while dragging the hose and use the strength of your upper body and legs to pull the hose forward. Your hands and arms are used to anchor the hose, while your body and legs are used to propel you and the hose to the finish line.

HOSE COUPLING

The hose coupling test is used to measure your ability to join a hose fitting with a fire hydrant fitting or with another hose fitting.

From a standing position, you will attach a female hose coupling to a male coupling on a fire hydrant or on another hose line. You may be required to repeat this action several times and to wear an air pack while performing this test event.

Tips/Tricks

Hold the female coupling with one hand. About two feet back from the coupling, tuck some hose under your arm and use your arm and body to keep the hose stationary while you work with the fittings. At least one of the fittings will be mounted securely. To join the fittings, turn the swivel portion of the fitting slightly to the left to get it properly seated, then turn to the right until it is hand tight, or as directed by the testing agency.

HOSE HOIST

The hose hoist is used to measure your ability to pull a hose up the outside of a building to an upper floor or on to the roof.

From a standing position, you will grasp a length of rope or hose and begin to pull it upward. When the hose reaches the designated end point, the event is complete. You may be required to wear an air pack while performing this test event.

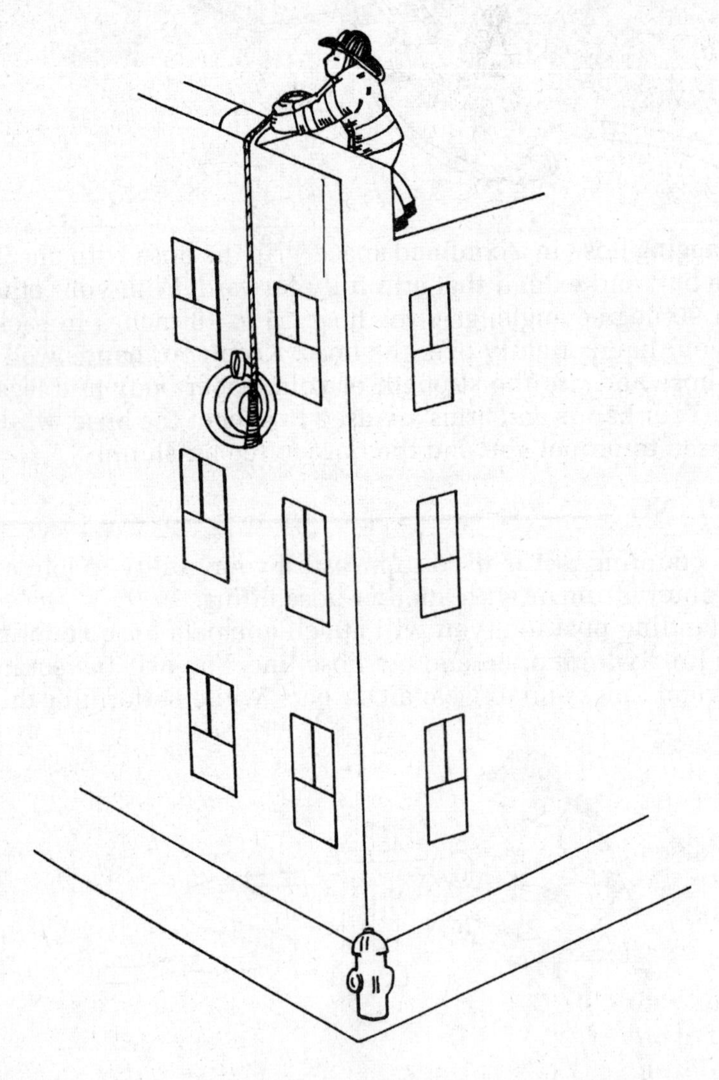

Tips/Tricks

When hoisting a rope or a hose, use a hand-over-hand technique. Try to develop a steady rhythm and avoid jerking movements. You may be required to pull the hose or rope approximately 50 to 100 feet to get it up to the roof or on to the platform.

STAIR CLIMB

The stair climb is used to measure your ability to climb stairs while carrying fire fighting hand tools or their equivalents. This test may include the carrying of hand tools, a length of folded hose, or a spare air pack cylinder.

From a standing position, you will pick up the tools, or similar test equipment weighing approximately 15 to 40 pounds, and then climb approximately three to six flights of stairs to the designated stop point. You may be required to repeat this event two or three times and to wear an air pack while performing this test event.

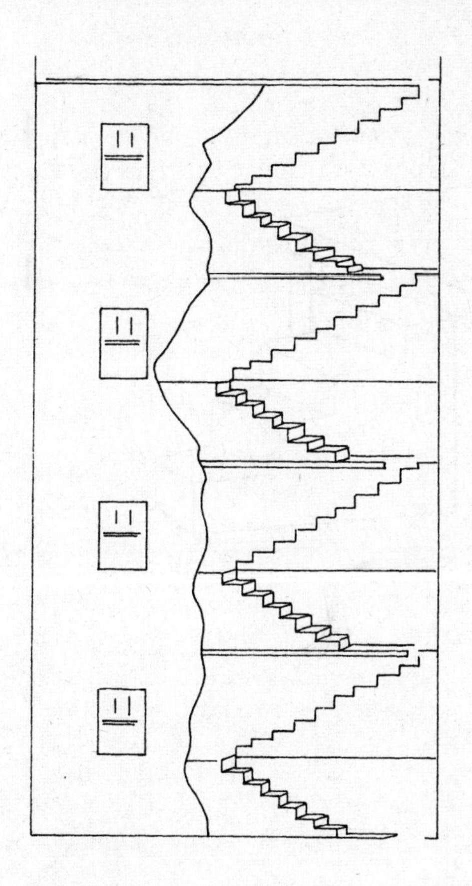

Questions

Before beginning this event, make sure you understand the requirements and restrictions.

- Do you have to step on each step, or can you skip steps?
- Can you use the hand rail if one is present?

Tips/Tricks

Use a steady stair-climbing pattern. Don't exhaust yourself on the first flights and then be out of breath on the upper flights. This is an event that is easy to practice and helps develop good cardiovascular and physical fitness.

Note: You may be required to simulate climbing stairs on a mechanical stair machine. While this has all of the cardiovascular and physical fitness benefits of a standard set of stairs, the technique is slightly different. When working on the mechanical stair, it is important to keep your balance by leaning slightly forward.

LADDER CLIMB

The ladder climb is used to measure your ability to ascend a 20- to 24-foot ladder. You may be required to dismount the ladder at the upper height, walk around the ladder, remount the ladder, and climb down the ladder. You may be required to wear an air pack, or to carry a tool while performing this test event.

From a standing position, you will climb the ladder and dismount at the top. If the event requires it, you may also be asked to climb back down the ladder.

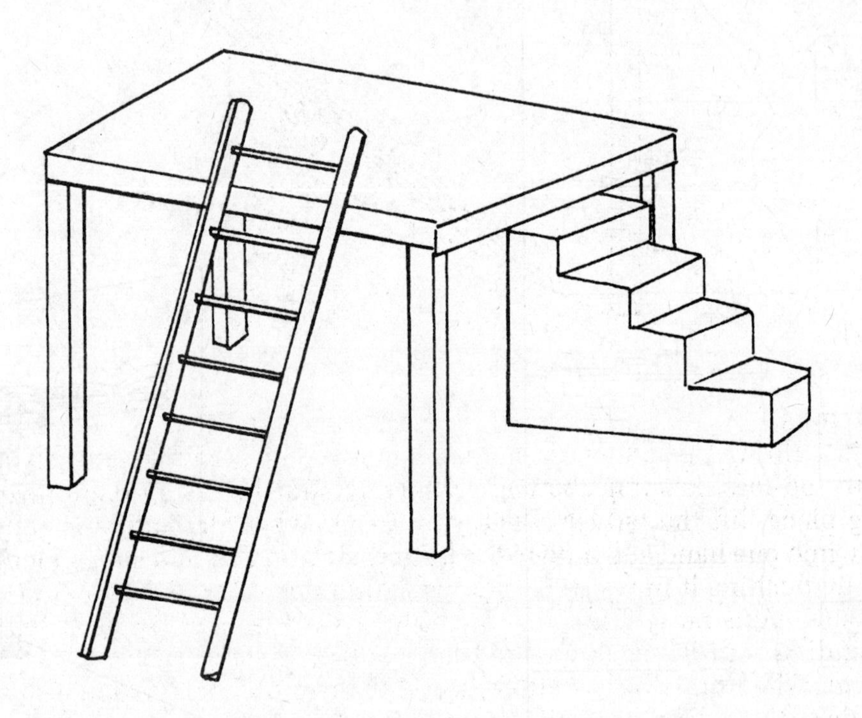

Tips/Tricks

When ascending or descending a ladder, use an upright position and avoid leaning into the ladder. DO NOT skip rungs as you ascend or descend the ladder. DO NOT try to slide down the beams of the ladder. Maintain a firm grip with at least one hand at all times.

Note: When practicing this event, there should be a safety person securing the ladder at the top and bottom of the ladder. The safety person should only step in and assist if you get into trouble.

LADDER RAISE

The ladder raise test is used to measure your ability to lift a ladder from a horizontal position into a vertical position.

From a standing position, you will pick up one end of a 20- to 24-foot ladder. Using a wall or another fixed point as a brace, you will lift the ladder from the horizontal to the vertical position.

Tips/Tricks

When lifting the ladder, do not bend at the waist. Squat with the knees and lift with the legs, keeping the back relatively straight. Using the top rung as a starting place, lift the ladder above your head. When the ladder is about head high, switch one hand at a time to the beams of the ladder and step under the ladder while pushing it upward. As you walk into the ladder, keep your hands and arms above your head. Lean forward into the ladder and push on the beams, while sliding your hands down the beams of the ladder as it raises.

Note: When practiced or attempted at the test site, a safety guide shall be at each side of the ladder, or a safety rope will be tied to the top of the ladder, to assist if you get into trouble.

LADDER CARRY

The ladder carry is used to measure your ability to carry a ladder from one point to another. The ladder carry simulates lifting a portable ladder from the side of a fire apparatus and transporting it to the place it will be used.

From a standing position, you will lift a 10- to 20-foot ladder and carry it a distance to a predesignated end point. The ladder will then be placed either on a rack or on the ground.

Questions

Before beginning this event, determine what restrictions apply to the way the ladder can be carried and placed down.

- Must you carry the ladder on your shoulder?
- Can you drop the ladder when you get to the completion point?
- Is there a penalty if the ladder hits the ground while in transit?
- What constitutes the completion of the event?

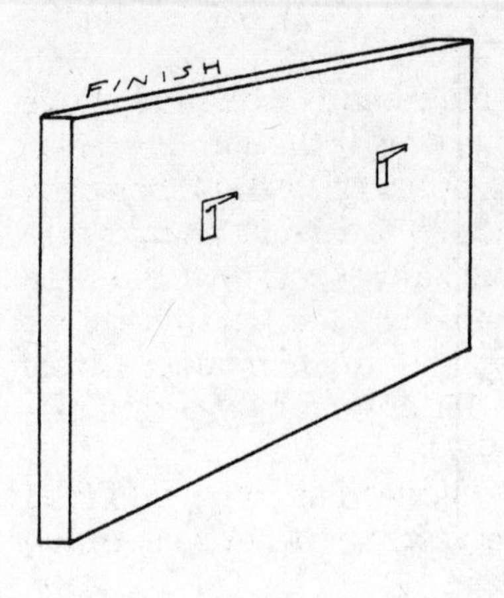

Tips/Tricks

When carrying the ladder, put one arm through a rung, just forward of center, and hold the upper beam with the hand that went through the rung. With your other hand, grip the lower beam as far forward as is comfortable for you. When you lift the ladder off the rack, it will rest on your shoulder with its weight slightly to the rear. Lean your body slightly forward to balance the weight of the ladder and use your arms to control the sway and direction of the ladder.

LADDER EXTENSION

The ladder extension activity is used to measure your ability to apply sufficient pulling force to raise the fly section of an extension ladder. This test simulates the fire fighting activity of raising the fly section of a portable ladder.

You will pull a haul rope downward until the fly ladder is extended three to six rungs. You may be required to wear an air pack while performing this test event.

Tips/Tricks

When raising the fly section, grip the rope with both hands extended above the head. Position the body so that you are in the center of the ladder rungs, with one foot just in front of the lowest rung and the other 12 to 16 inches back from the ladder. Use a hand over hand pulling motion (your palms should face away from your body and your thumbs should point down) to move the rope approximately 12 to 16 inches with each pull. Keep the rope parallel to the ladder as you pull. It is much harder to raise the fly section when you pull the rope away from the ladder. Steady, smooth pulls are better than fast jerking motions.

Note: The number of rungs or the height the fly ladder has to be extended may vary. The ladder should be anchored or held by a safety person on each side during this event.

DUMMY DRAG (DUMMY CARRY)

The dummy drag measures your ability to pull a dummy weighing approximately 110 to 180 pounds approximately 75 to 150 feet.

The dummy drag simulates a firefighter dragging an unconscious person from a fire area. This test may be conducted in a confined space simulation or on an open surface. You may be required to wear an air pack while performing this test event.

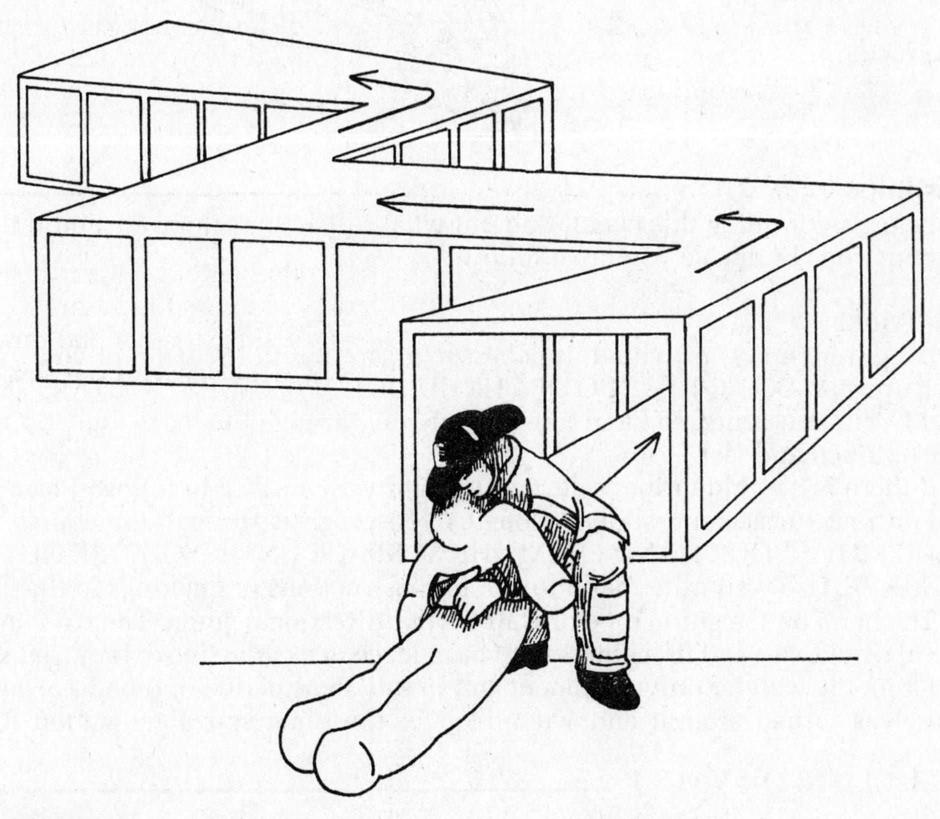

Tips/Tricks

When dragging a dummy, an initial good grip is very important. Stand behind the dummy towards the head. Pick up the dummy and prop it up on your knees. Grasp the dummy firmly under the arms and lean slightly backward. Use your body and legs to drag the dummy.

OBSTACLE COURSE—MAZE OR CONFINED SPACE

The obstacle course, or maze test, is used to measure your ability to navigate a 50- to 100-foot confined space that changes in direction and height at several points. Obstructions may be encountered throughout the course.

The obstacle course simulates the firefighter's activity of searching a fire area to find a trapped person or locate the seat of the fire. The obstacle course may be dark or you may be required to wear an eye shield that obscures your vision. You may also be required to wear an air pack while performing this test event. (See page 30.)

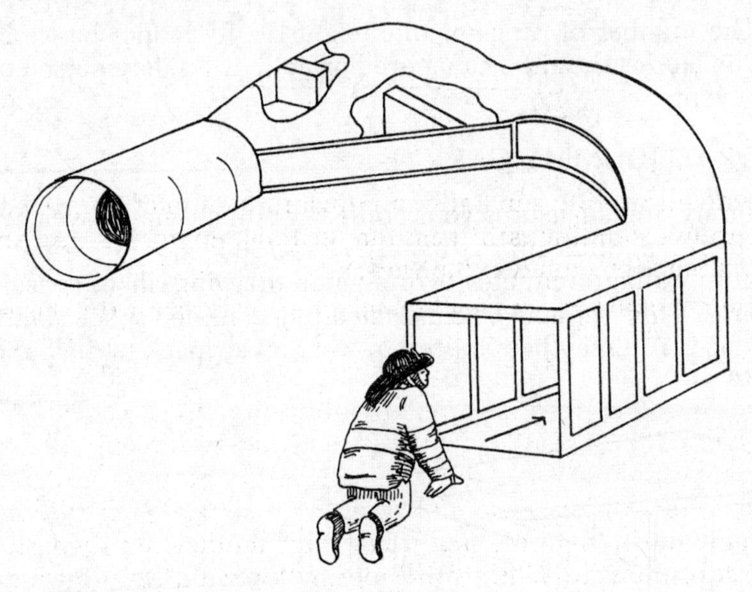

Questions

Before performing this event, find out what restrictions there are and if there is a guide rope or device you must follow.

Tips/Tricks

If a guide rope is provided, it is best to grasp it with the palm of your hand facing upward. DO NOT LET GO OF THE GUIDE ROPE ONCE YOU START INTO THE MAZE. Use your free hand to feel your way through the maze. Rely on your sense of touch and feel.

If there is no guide rope, select an upright wall surface to follow. Place one hand on that surface and slide it along as you progress through the course. DO NOT REMOVE YOUR HAND FROM THE SURFACE UNTIL YOU ARE OUT OF THE MAZE. Use your other hand to feel for obstructions or openings in the floor.

The hand on the guide rope or wall is your directional guide. There are many cases of candidates and firefighters who have let go of a guide line or removed their hand from the wall for only a moment and in this short period of time have gotten themselves turned around and wound up in the same spot they started from.

WALL CLIMB OR VAULT

The wall climb or vault is used to test your ability to get over a 4- to 6-foot fence or other obstacle.

The wall climb simulates the firefighter's activity of climbing over a wall or up on to a higher roof. You may be required to wear an air pack while performing this test event.

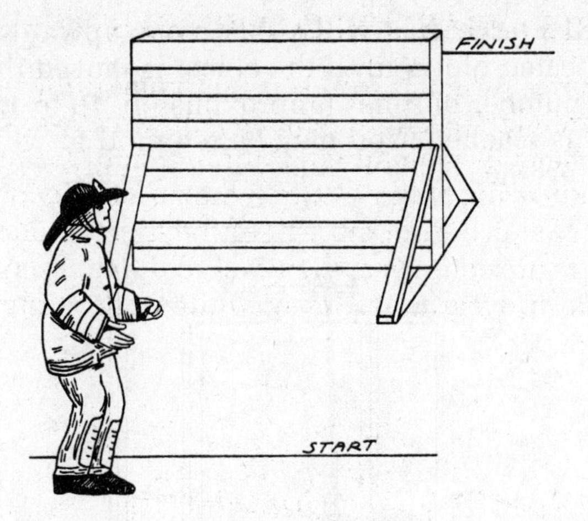

Questions

- Are you allowed to run and jump toward the top of the wall, or must you start from a standing still position, or lying down, and use your arm and body strength to get over the wall?
- Do you have to carry a weight pack or an air pack during the event?

Tips/Tricks

The key to success in this event is getting the upper portion of your body over the threshold of the wall. If you are permitted to run and jump, you should use your momentum to its full advantage. At a safe running speed, jump up toward the top of the wall, while simultaneously placing both hands on the crest of the wall. Using the strength of your arms and shoulders, pull your body up and over the crest of the wall. While still holding the wall, lean down and forward, swinging the rest of your body over the wall. Do not let go of the top of the wall until your body returns to a vertical position on the other side of the wall. Proceed to the event completion point.

If you are not permitted to run at the wall, then make sure your feet are firmly planted and make a strong initial jump at the wall. Get both hands on the crest of the wall, and pull your upper body up and over, using the strength of your arms and shoulders. While still holding the wall, lean down and forward, swinging the rest of your body over the wall. Do not let go of the top of the wall until your body returns to a vertical position on the other side of the wall. Proceed to the event completion point.

If you are unable to achieve enough momentum, or if you are having difficulty pulling yourself up and over the wall, use your legs to walk up the wall while you hold on to the top of the wall with your hands. Try to swing one leg over the top, and then pull hard with your arms while pushing down with the leg that is on top of the wall.

TOOL USE—HOOK (PIKE POLE) PUSH

The hook push activity is used to measure your ability to use a firefighter's hook to exert a pushing force of approximately 70 to 100 pounds on a metal plate in the ceiling.

The hook push simulates the fire fighting activity of pushing a firefighter's hook through a ceiling to prepare for pulling down the ceiling to search for fire burning in the spaces behind the ceiling.

You will hold a firefighter's hook (pike pole) with both hands and force the hook upward toward a target that will either move upward or register the force. You will deliver repeated blows until the object is moved the required distance, struck the required number of times (approximately 10 to 15 repetitions), or the required total force is reached. You may be required to wear an air pack while performing this test event.

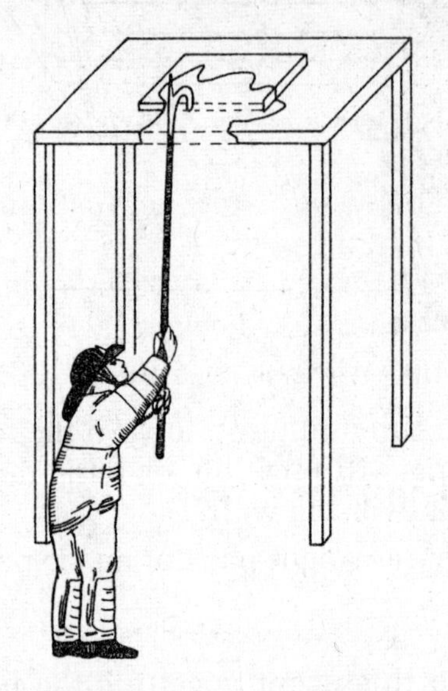

Tips/Tricks

A firefighter's hook (pike pole) is approximately 6 feet long and one end is significantly heavier than the other. Holding the weighted end up, grasp the pike pole with one hand about 3 feet from the unweighted end. Place your other hand one foot from the pike pole's unweighted end. Using a firm motion, force the hook upward and hit the designated target with the tip of the pole slightly forward. Use your body and legs to first go down into a slightly crouched position, and then using the legs, body and arms in a coordinated upward motion, force the tool in a narrow up-and-inward arc toward the target.

TOOL USE—HOOK (PIKE POLE) PULL

The hook pull test measures your ability to exert a pulling force of about 70 to 100 pounds on a firefighter's hook.

The hook pull event simulates a firefighter pulling open a ceiling to expose the fire burning behind the ceiling.

You will be required to grip the firefighter's hook with both hands and exert a pulling force that will move a portion of the ceiling downward or will lift a weight. Repeated downward pulls will move the ceiling or the weight the desired distance.

Tips/Tricks

Grip the firefighter's hook with one hand approximately 2 feet from the weighted end and the other hand a comfortable distance (12 to 18 inches) below the top hand. Exert a downward pulling force using your arms, shoulders, legs, and body weight in a coordinated smooth motion.

TOOL USE—FORCIBLE ENTRY

The forcible entry activity measures your ability to deliver sufficient force to open a locked door.

The forcible entry activity simulates the fire fighting activity of forcing entry into an occupancy through a locked door.

You will be required to swing a flat head ax, maul, or sledge hammer, weighing approximately 6 to 9 pounds, in a horizontal side-to-side motion and hit a target that will either move backward or register the force. You will deliver repeated blows until the object is moved the required distance, the required total force is reached, or for a specific number of repetitions. You may be required to wear an air pack while performing this test event.

Tips/Tricks

Hold the handle of the tool (ax, maul, or sledge hammer) horizontal to the ground. Place one hand with its palm facing up several inches behind the head of the tool. Your other hand should be comfortably placed approximately 15 to 20 inches behind the first hand with the palm facing down. As you swing the tool to strike the object, rotate your body, first away from and then toward the object. Use the power of your body, transmitted through the arms, hands, legs, and tool to strike the target. Avoid using only your arms.

TOOL USE—CHOP ROOF OR FLOOR

The chopping activity measures your ability to use a weighted tool with a downward force for a series of approximately 50 strokes.

The chopping activity simulates the firefighter's activity of cutting a hole in a roof or floor to vent the heat and smoke or to expose the fire burning in the floor.

You will be required to swing a flat head ax, maul, or sledge hammer weighing approximately 6 to 9 pounds and hit the end of a standard railroad tie or a mechanical device that will either move backward or register the force of the impact. You will deliver repeated blows until the object is moved the required distance (approximately 5 to 10 feet), the required force is reached, or for a specific number of repetitions (approximately 35 to 90 strokes within 90 seconds).

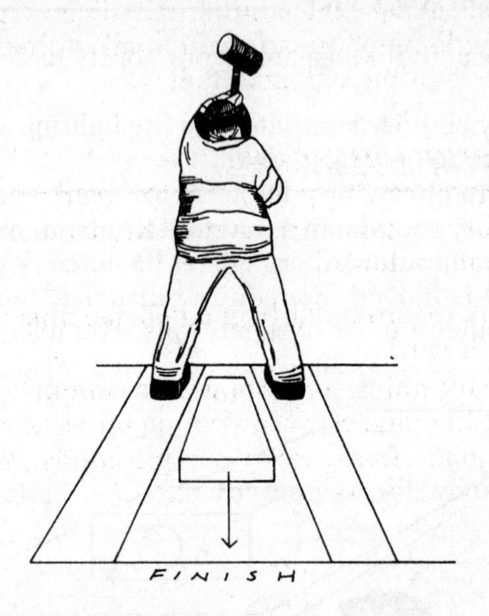

Tips/Tricks

Whenever you are required to use a tool in a chopping manner, straddle the object if possible and use a series of forceful strikes instead of a few really hard blows. Each strike with the tool should move the object or register the necessary force. You should not try to complete the activity in one or two strokes.

ADMINISTER CPR

The CPR activity measures your ability to perform effective cardiopulmonary resuscitation.

This test simulates the firefighter's activity of administering cardiopulmonary resuscitation to a nonbreathing person.

You will perform cardiopulmonary resuscitation on a rescue mannequin for a specified duration (approximately four minutes, delivering 60 compressions per minute).

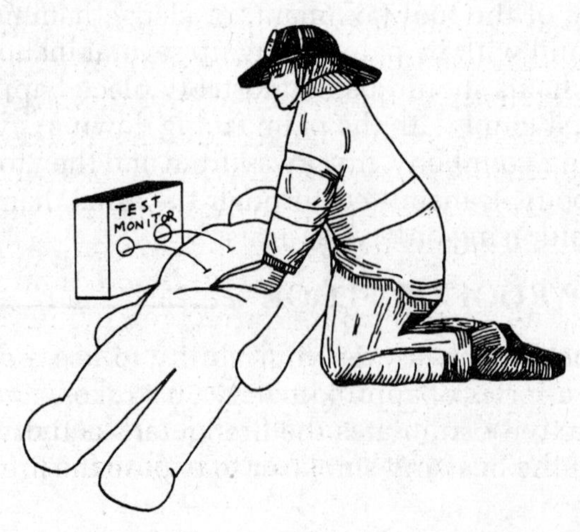

Tips/Tricks

This test will usually be required only if some form of certification is a prerequisite.

The preceding list is only a representative example of what can be asked of a prospective firefighter. The physical ability examination will consist of five to eight of these types of tasks. You will be made aware of the examination tasks before the test so that you can become familiar with them. The choice of the particular tasks will be varied enough to ensure that all of the muscles and physical abilities needed for fire fighting will be tested.

HOW TO PREPARE

There are four extremely important measures that you must follow to prepare for the physical ability examination.

1. Realize that even if you are in fairly good shape, this test will require you to practice and to work out.
2. Begin your physical training program with a complete medical examination by your doctor. Tell the doctor what you plan to do and how you plan to accomplish it. If the doctor foresees any complications, you should know about them and be guided by the doctor's advice.
3. In your training program include various types of exercises. A quick look at the tasks outlined in the preceding section should tell you that many different parts of the body will be involved in this examination.
4. Practice tasks that simulate those on the actual test. Don't be fooled into thinking that you can do everything that may be required; a trial run will boost your score.
5. Proper hydration (drinking water) and acclimatization (getting your body accustomed to the effects of higher body temperature) are important. Drink plenty of water in small amounts throughout your workout. Try to work out in an environment that simulates the conditions (temperature and humidity) that you expect on the day the test will be administered. Use caution when you're trying to simulate warm weather conditions and progress slowly to allow your body to adjust to the conditions. Do your exercises or practice in the clothing you expect to wear on the day of the test. If the equipment you wore during your practice workout sessions is worn out, unsafe, or unreliable, you should get replacements several weeks before the test to break them in.

There is no one best method for all people to use to get into shape. There are, however, some basics that, when understood and put into operation, will allow you to develop a comprehensive program for your individual needs.

Contrary to popular belief, you cannot achieve the fitness required to pass the firefighter physical examination by relying only on exercises such as push-ups, chin-ups, and weight lifting. The exercise program must prepare your body for jumping, running, twisting, and bending while pulling, lifting, and carrying heavy weights.

Before you begin your exercise routine, you should do stretching exercises to loosen your muscles and prevent injury. Stretching helps you prevent muscle strain, increases your range of motion, improves circulation, and prepares you for strenuous activities. Stretch for short periods with concentration on specific groups of muscles, but work all the groups. Work all the muscle groups for 20 to 30 seconds with three to five repetitions for each group.

A good training program consists of two major components: (1) cardiovascular fitness and (2) strength development.

CARDIOVASCULAR FITNESS

An effective cardiovascular program should be based on a heartbeat rate of 140 to 160 beats per minute. You will be doing well when you are able to maintain this heart rate for a period of 20 to 30 minutes.

Jogging and Interval Training

Stage One: A sound cardiovascular system is based on a systematic jogging program, which starts by finding out what your heart rate is when you begin and then bringing it up to the ideal range of 140 to 160 beats per minute. To avoid injuring your muscles, it will be necessary to warm up before you start your run. A series of stretching exercises for approximately 10 minutes to increase the flexibility of your tendons and muscles is required. Start slowly and stretch the muscle until it hurts, then back off slightly. Hold this new position for a short while, then relax for a moment and repeat the stretching exercise.

When you are ready to begin jogging, you must get your heartbeat up to the ideal rate. Start jogging slowly. When you begin to feel slightly breathless, stop and count your heartbeats for a 10-second period; multiply the number of beats by 6. This figure will be your heart rate per minute. For example, if your heart rate is 24 beats for 10 seconds, and you multiply 24 by 6, then your heart rate per minute is 144 beats per minute (BPM). Continue your workout, trying to maintain this ideal heart rate through your entire period. If your heart rate exceeds 160 BPM, slow down and walk leisurely until your heart rate returns to about 120 BPM. When you have reached the point where you can jog for 20 to 30 minutes without your heart rate exceeding 160 BPM, you are ready for stage two.

Stage Two: This is a form of interval training. Interval training is the simplest method to achieve a rapid improvement in performance. It is a system of intense exercise followed by an interval of rest. It entails (1) running fast on the straightaway of a track to get the heart beat way up and then jogging around the turn to recover, or (2) running up several flights of stairs and then walking back down to recover. It must be noted that this type of training can create a very rapid increase in your heart rate, and care must be exercised to avoid overexertion and damage.

Jogging should be done daily if possible. Interval training, however, should be limited to three times a week.

When you are approaching the end of your workout, it is necessary to cool down. This is done by means of a 5- to 10-minute period of limited exercise that allows your body to return to its normal condition.

Aerobic Dancing

Another excellent form of cardiovascular exercise, which also uses the interval training concept, is aerobic dancing. Aerobic dancing is energetic, continuous exercise that develops muscle tone and flexibility while increasing the cardiovascular capacities. It has the advantage that you can do it in the privacy of your home, in all types of weather and at almost any time of the day.

An aerobic dancing program uses music as an aide to develop rhythm in the exercise routine. The overall plan is similar to the jogging plan—stretch and warm-up, followed by getting the heart rate up to the ideal range, keeping it there for about ½ hour, and then cooling down over a 5- to 10-minute period.

There are a number of good aerobic dancing programs on nationwide television, and many excellent aerobic tapes are available at local libraries and better music stores. Give aerobic dancing a chance; it is a type of exercise that can produce significant improvement in as little as 6 weeks, even if you work out only three times a week.

STRENGTH DEVELOPMENT: CALISTHENICS AND WEIGHT TRAINING

Calisthenics

A routine set of calisthenic exercises lasting 10 to 20 minutes should be done daily. The number of repetitions should be based on your ability to complete the full program each day. Start with 10 repetitions, and try to work up to 25. If you can do 25, then add some weight to get you back down to 10. The exercises that follow are designed to develop strength.

- Sit-ups. Lie on your back, and either have someone hold your feet or anchor them under a couch. Grasp your hands behind your head and then sit up, trying to touch your elbows to your knees. Don't worry about keeping your legs straight until you can do 20 repetitions. As you get stronger, straighten your legs.
- Push-ups. Lie on your stomach with your hands on the floor at a point even with your chest. Push up while keeping your back and legs straight. Try to lift the body smoothly, without jerking. If this is not possible, keep your knees on the floor until you develop the necessary strength.
- Chins. Grasp an overhead bar with your hands in a comfortable position, palms facing the bar for the triceps, palms facing you for the biceps. Now pull yourself up slowly. Do not hook your chin on the bar; just get it up to the bar, and then lower yourself slowly.
- Leg raises. While lying on your back, grasp the end of a couch or other substantial object. Slowly raise both legs until they are pointing toward the ceiling. Now slowly lower your legs to a point just off the floor, and repeat the lift.
- Dips (this exercise requires parallel bars or dip bars). While standing between the two bars, grip them so that your palms are facing your sides. Now jump up so that your arms are straight, holding your weight with your feet off the floor. Lower your body until your chest is level with your hands, and then push yourself back up.

Weight Training

Weight training is based on a theory known as "progressive overload." The theory states that, if we repeatedly work our muscles, our body will respond by building strength. To be prepared for the strength aspects of the Firefighter Examination, you will have to put demands on your body that will enable it to react properly.

Basic Rules for Weight Training

- This type of training should be done under proper supervision. If this is not available, extreme caution must be taken. Safety measures include having someone act as a safety person during the lifts, not attempting to lift an excessive amount of weight, and not overworking your muscles.
- Weight training calls for increasing the amount of weight as the muscles

grow in strength. You must identify how much weight you should start with. The suggested starting weight should be that which you can lift and exercise with for 10 repetitions. As you work out, you will be able to increase the number of repetitions. When you reach 15 per session, you should add more weight. Add only enough weight to bring you down to 10 repetitions.

- Weight training should be done only three times a week. Progress will appear to be slow, but it will come if you continue faithfully. Keeping a chart of what you could lift when you started and how you have progressed will help to overcome your feeling of not getting anywhere.

The muscles you will need to develop are in the chest, back, shoulder, and arm areas. Therefore you should concentrate on exercises that apply to these areas.

- Chest—Bench press. Lie on your back, and have your safety person place the weight in your hands at a point over your chest. Press the weight up, and then lower it back to your chest. Repeat for three sets of 10 repetitions each.
- Back—Bent-over rowing. While in a standing position, bend over the weight so that your chest is over the weight bar, grasp the bar, lift it to your chest, and then lower. Repeat for three sets of 10 repetitions each.
- Shoulders—Standing press. While standing and holding the weight at chest level, push the weight overhead and then lower to your chest. Repeat for three sets of 10 repetitions each.
- Arms—Tricep and bicep curl. *Tricep curl:* while seated, and using both hands, hold a single dumbbell behind your head. Keeping your arms as close to your head as possible, raise and lower the weight for three sets of 10 repetitions each. *Bicep curl:* while seated, and using either a barbell or two dumbbells, slowly raise the weight to your chest and then lower. Do not jerk or bounce the weight; if you must, use less weight. Repeat for three sets of 10 repetitions each.

WHAT TO DO ON THE DAY OF THE TEST

Get a good night's sleep the night before the physical ability test. On the morning of the test eat a light breakfast, hopefully the same kind of breakfast that you have been eating for several days. It is best if, on the day of the test, nothing is different in your diet. Do your stretching routine and loosen up before you go to the testing site.

Arrive at the test site early enough to ensure that you can be checked in and still have plenty of time for your warm-up. Since the test is of short duration and high intensity, you must warm up before you start.

Wear proper footwear (avoid new shoes or new sneakers). Wear the clothing that you wore when you were working out; it will be familiar, comfortable, and dependable. While waiting at the test site, you should do stretching exercises to keep yourself relaxed and ready for the test.

If something goes wrong during the test, and it's not serious, do not stop, because the clock won't. In other words, if you get a bruise or a scrape that is not bleeding profusely, or if something embarrassing happens (say, your pants fall down), just recover as quickly as possible—DON'T STOP. Note that this advice does not apply if you are seriously injured and need medical help. In that case you should yell for assistance. It will be right there.

Medical and Psychological Testing

Because of the hard work and the danger involved in the occupation of fire fighting, many cities offer substantial health care, medical leave, and retirement programs for their firefighters. To ensure that the citizens of the community get the best life-saving services without unreasonably high tax burdens, cities require that all firefighters be medically and mentally sound when hired. Once a firefighter is employed, job-related injury or illness that leads to total disability can be assumed to be a result of the nonavoidable occupational hazards. To avoid or at least reduce the chance of a person entering the fire service with a precondition that would be likely to disable him/her, or a strong susceptibility to such a condition, a thorough physical examination is required of all firefighter candidates.

Also, because firefighters must work at great heights, under confined conditions, and under the most adverse weather conditions—in short, in extremely stress-provoking situations, it is necessary to ensure that the firefighter candidate is mentally capable of handling these types of situations. The psychological examination is used for this purpose.

MEDICAL EXAMINATION

The candidate is required to complete a medical questionnaire before the physical ability examination and must pass the medical examination before being employed. A typical examination includes but may not be limited to the following:

- Cardiovascular system
- Eye and ear conditions
- Blood condition
- Gland and gastrointestinal conditions
- Hernia and genitourinary conditions
- Upper and lower extremities
- Back, skull, teeth, and skin
- Muscular and skeletal system

If you feel that you may not qualify medically, you should consult your personal physician to get a qualified authoritative opinion. Do not disqualify yourself only on your belief that you may not be able to pass. If you do have a condition that could disqualify you, a visit to the doctor's office is even more appropriate. The doctor may be able to correct your condition through proper diet, special exercises, medication, or surgery.

PSYCHOLOGICAL EXAMINATIONS

A psychological test (or psychiatric evaluation, as it is also known) can be required as part of the employment conditions. As stated before, the occupation of fire fighting can and often does produce high levels of stress, which can lead to medical problems, alcoholism, hard drug use, and depression. Conditions such as these lead to problems on the job because they affect the person's judgment and performance. Poor judgment and performance are unacceptable in an occupation such as fire fighting, where the individual is a member of a highly interactive team, and death or serious injury can result if one member cannot do his/her part.

Usually the psychological test is the last part of the examination process. The candidate will have passed the written, physical ability, and medical tests. The number of eligible candidates will have become much smaller.

The psychological examination is given in two parts. The first part is a "personality questionnaire," requiring candidates to answer a relatively large number of short questions about themselves and their personalities. This is followed by a structured interview in which the candidate has a chance to explain his/her responses to a qualified psychologist.

There is very little a candidate can do to prepare for the psychological examination. Knowing what to expect, however, will help to reduce fear and anxiety. The best technique is to be yourself and answer the questions truthfully.

CHAPTER 3

Test Strategies

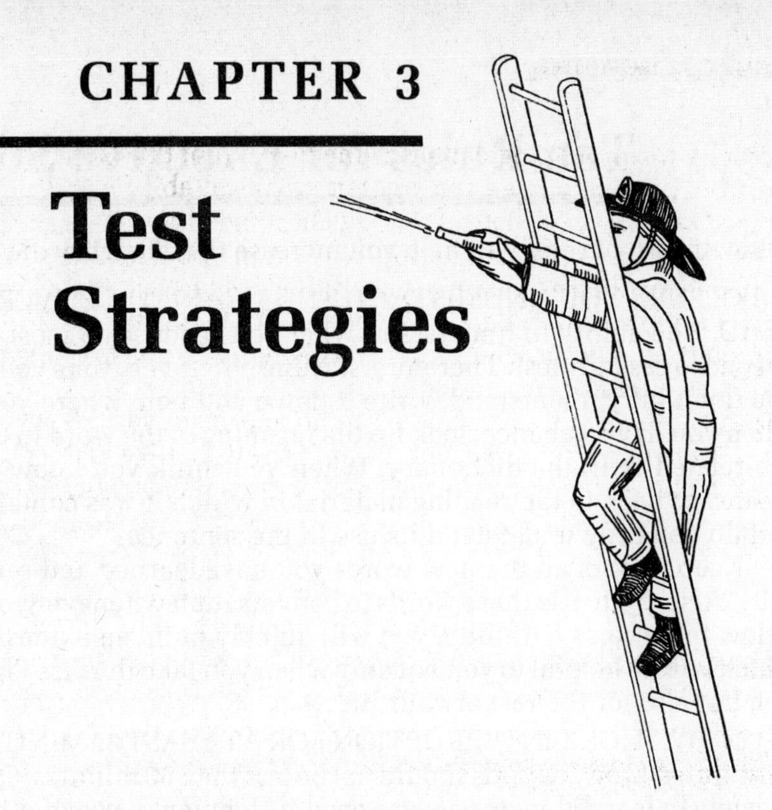

Achieving Your Highest Score

This chapter covers two related topics critical to obtaining a high score on the Firefighter Examination. The first part helps you develop good study habits, and the second provides you with a specific strategy for dealing with multiple-choice questions.

Because both of these subjects are so important, it will be best to review this chapter periodically as you study for the examination. At the very least, you should review the test strategies before taking each of the full-length examinations included in this book.

Developing Good Study Habits

Many people believe, incorrectly, that the amount of time spent studying is the most important factor in test preparation. Efficient study habits, not necessarily time, however, are the key to successful test preparation. Of course, all else being equal, the amount of time you devote to your studies is a critical factor; yet spending time reading is not necessarily studying. If you want to retain what you read, you must develop a system.

You must set aside time, preferably at a point in the day when you are most receptive to learning—mornings for some, early afternoons for others, perhaps late at night for a third group. A single, well-planned reading and question-practicing session of 1 hour will be more productive than several 15- to 20-minute sessions.

Once you have made a commitment to yourself, stick to it.

Ten Rules for Effective Studying

The following list of rules will help you increase the efficiency of your study time.

1. MAKE SURE YOU UNDERSTAND THE MEANING OF EVERY WORD YOU READ. The ability to understand what you read is the most important skill needed to pass any test. Therefore, starting now, every time you see a word that you don't fully understand, write it down and note where you saw it. Then, when you have a chance, look up the meaning of the word in Chapter 5 if it is fire-related or in the dictionary. When you think you know what the word means, go back to the reading material in which it was contained, and make certain you fully understand its use in the sentence.

   Keep a list of all the new words you have learned and review it periodically. Also, try to use these words in conversation whenever you can. If you do follow these steps faithfully, you will quickly build an extensive vocabulary, which will be helpful to you not only when you take the Fire Fighter Examination but also for the rest of your life.
2. STUDY WITHOUT INTERRUPTION FOR AT LEAST 30 MINUTES, preferably 60 minutes. Study periods should not be less than 30 minutes. Quick reviews of previously learned materials are good if they make use of otherwise wasted time, such as standing in line, or the last part of your lunch hour. These, however, should not be considered study periods.

   It is essential that you learn to concentrate for extended periods of time. The actual examination takes anywhere from 2 to 3½ hours to complete, and you must concentrate just as hard in the third hour of the test as you did in the first hour. Therefore, as the examination approaches, study without interruption for extended periods of time. Also, when you take the practice examinations in Chapters 12–14, complete each examination in one sitting, just as you must at the actual examination.
3. SIMULATE EXAMINATION CONDITIONS WHEN STUDYING. As far as possible, study under the same conditions as those of the actual examination. Eliminate as many outside interferences as you can. If you are a smoker, refrain from smoking while studying, since you will not be allowed to smoke in the classroom on the day of your examination. Failing to practice this abstinence from smoking may result in your being annoyed and frustrated during the examination—a distraction you do not need.
4. READ, STUDY, AND TAKE PRACTICE EXAMINATIONS ALONE. This is the time to find out what you don't know or understand, and the time to correct these weaknesses. Studying alone is the most effective way of learning; however, some people may need support. This support should take the form of a "study group," which meets periodically to review and clarify. If possible, the group should have from three to five serious students and should meet for 2 to 3 hours on a periodic basis, perhaps every other week. Everyone in the group should keep a list of items that he/she is having difficulty with or is confused about; these should be discussed at the study session. Items that no one understands should be referred to an outside source, such as a fire fighter, teacher, parent, or librarian. Arguing in a study group defeats the purpose of the group and must be avoided at all costs.
5. MAKE SURE YOU UNDERSTAND THE ANSWERS TO EVERY TEST QUESTION IN THIS BOOK. Every answer is accompanied by an explanation. When-

ever you answer a question incorrectly, be sure you understand why, so you won't make the same mistake again. It is equally important to make certain that you have answered a question correctly for the right reason; therefore study the answer explanation for every question in this book as carefully as you study the question itself.

6. ALWAYS FOLLOW THE RECOMMENDED TECHNIQUE FOR ANSWERING MULTIPLE-CHOICE QUESTIONS. On pages 43–46 in this chapter, there is a 17-point technique for answering multiple-choice questions.

7. ALWAYS TIME YOURSELF WHEN DOING PRACTICE QUESTIONS. Running out of time on a multiple-choice examination is a needless error that is easily avoided. Learn, through practice, to move to the next question after you have spent a reasonable period of time on any one question. When you are answering practice questions, always time yourself, and always try to stay within the recommended time limit. The correct use of time during the actual examination is an integral part of the technique that will be explained later in this chapter.

8. CONCENTRATE YOUR STUDY TIME IN THE AREAS OF YOUR GREATEST WEAKNESS. The diagnostic examination in Chapter 4 will give you an idea of the types of questions that are most difficult for you, and the greater part of your study time should be spent on these areas. Do not, however, ignore other types of questions completely.

9. EXERCISE REGULARLY, AND STAY IN GOOD PHYSICAL CONDITION. People who are in good physical condition have an advantage over those who are not. It is a well-established principle that good physical health improves the ability of the mind to function smoothly and efficiently, especially when taking examinations of extended duration. You must also keep in mind that the Firefighter Examination includes a physical examination as part of the overall test.

10. ESTABLISH A SCHEDULE FOR STUDYING, AND STICK TO IT. Do not relegate studying to times when you have nothing else to do. Schedule your study time, and try not to let anything else interfere with that schedule. If you feel yourself weakening, review Chapter 1 and remind yourself of why you would like to become a firefighter.

Strategies for Handling Multiple-Choice Questions

The rest of this chapter outlines a very specific test-taking strategy valuable for a multiple-choice examination. Study the technique, practice it, then study it again until you have mastered it.

1. COVER THE QUESTIONS ON WHICH YOU ARE NOT WORKING. This first procedure is a technique to help you concentrate only on the question at hand. There is a tendency to go back to quickly check what was done on the question before, or to look ahead to see what's coming. These distractions detract from the ability to concentrate on the question at hand.

Therefore, before you begin the examination, get a piece of scrap paper from the examiner, fold it in half horizontally, and tear it into two equal pieces. Use these two half sheets to cover the questions and the directions to the examination. When you begin the examination, remove the sheet that covers the instructions and read them carefully. After you understand fully what you are to do, cover the directions again. Then move the second sheet down to expose the stem (the part before the answer choices) of the first question, but do not expose the choices at this time. After you have read and understood the question, move the second sheet down to expose one answer choice at a time. (You will be shown how to make and mark your choice shortly.) After you have completed a question, slide the top sheet down to cover it and the bottom sheet down to expose the stem of the next question. Continue this process throughout the examination.

2. READ THE DIRECTIONS. Do not assume that you know what the directions are; make sure you read and understand them. Note particularly whether the directions differ from one section of the examination to another.
3. MAKE SURE YOU HAVE THE COMPLETE EXAMINATION. Check the examination page by page. Since examination booklets have numbered pages, make certain that you have all the pages.
4. TAKE A CLOSE LOOK AT THE ANSWER SHEET. Some answer sheets are numbered vertically, and some horizontally. The answer sheets on your practice examinations are typical of the one you will see on the actual exam. However, do not take anything for granted. Review the directions on the answer sheet carefully, and familiarize yourself with its format.
5. BE CAREFUL WHEN MARKING YOUR ANSWERS. Be sure to mark your answers in accordance with the directions on the answer sheet. Be extremely careful that:
   a. You mark only one answer for each question.
   b. You do not make extraneous markings on your answer sheet.
   c. You darken completely the allotted space for the answer you choose.
   d. You erase completely any answer that you wish to change.
6. MAKE ABSOLUTELY CERTAIN YOU ARE MARKING THE ANSWER TO THE RIGHT QUESTION. Many multiple-choice tests have been failed because of carelessness in this area. All that is needed is one mistake: if you put one answer in the wrong space, you will probably continue in the same way for a number of questions until you realize your error. When marking your answer sheet, the following procedure is recommended:
   a. Select your answer, circle that choice in the test booklet, ask yourself what question number you are working on, and write it next to the answer.
   b. If, for example, you select choice (C) as the answer for question 11, circle choice (C) in the test booklet, and say to yourself, "(C) is the answer to question 11," as you mark the answer on the answer sheet. Although this procedure may seem rather elementary and repetitive, after a while it becomes automatic. If followed properly, it guarantees that you will not fail the examination because of a careless mistake.
7. MAKE CERTAIN THAT YOU UNDERSTAND WHAT THE QUESTION IS ASKING. Read the stem of the question—the part before the answer choices—very carefully to make certain that you know what the examiner is asking. In fact, it is wise to read it twice. Underline or circle the key words, which tell you to look for a correct or incorrect, accurate or inaccurate, statement.

8. ALWAYS READ ALL THE CHOICES BEFORE YOU SELECT AN ANSWER. Don't make the mistake of falling into a trap when the most appealing wrong answer comes before the correct choice. Read *all* choices!

9. BE AWARE OF KEY WORDS THAT OFTEN REVEAL INCORRECT ANSWERS. *Absolute Words* are usually the wrong choice, because they are generally too broad and difficult to defend. *Examples* are: *never, none, nothing, nobody, always, all, everyone, everybody, only, must.*

Transitional words such as "and," "also," and "besides" indicate there may be more than one part to the answer and both must be true to be correct. Words such as "or," "but," and "however" indicate contrast and offer a choice of situations where only one part of the selection will be true and thus correct. Words such as "similarly," "sometimes," "generally," and "possibly" are comparison words and require you to make a judgment, thus leaving more latitude for the item to be correct.

10. NEVER MAKE A CHOICE BASED ON FREQUENCY OF PREVIOUS ANSWERS. Some students try to pay attention to the pattern of correct answers when taking an exam; this should be avoided. Answer each question without regard to what your previous choices have been.

11. CROSS OUT CHOICES YOU KNOW ARE WRONG. As you read through the answer choices, put an "X" through the letter designation of any choice you know is wrong. Then, after reading through all the choices, you have to reread only the ones you did not cross out the first time. If you cross out all but one of the choices, the remaining choice should be the answer. Read the choice one more time to satisfy yourself, put a circle around its letter designation (if you still feel it is the best answer), and mark the answer sheet to correspond. (See the procedure given under Rule 6.)

If you cross out all but two choices when you read through the question the first time, you have to reread only the two remaining choices, and decide between them. Frequently, the second time you read the remaining choices, the correct answer is clear. If that happens, cross out the wrong choice, circle the correct one, write the answer, and transpose it to the answer sheet.

If more than two choices still are not crossed out, reread the stem of the question to make certain you understand it. Then go through the choices again. (Keep in mind the key words mentioned in Rule 9, which may indicate the incorrect answers.)

When the instructions tell you to select more than one choice and to prioritize your choices, write at the beginning of each question "S" for strong choice, "O" for OK choice, and "W" for a weak choice. When you have designated each choice as STRONG, OK, or WEAK, go back and mark the strongest choice "1," the next choice "2," and the third choice "3." Then record your answer on the answer sheet.

12. SKIP OVER QUESTIONS THAT GIVE YOU TROUBLE. The first time through the examination, don't dwell too long on any one question. However, put a circle around the number on the answer sheet (to keep your answers from getting out of sequence), and write the question number on a piece of scrap paper (this will help you to locate skipped questions rapidly when you go back), and go to the next question. DO NOT GUESS AT THIS POINT.

13. WHEN YOU HAVE FINISHED THE REST OF THE EXAMINATION, RETURN TO THE QUESTIONS YOU SKIPPED. Once you have answered all the questions you are sure of on the entire examination, check the time remaining. If time permits (and it should if you follow the recommendations), return to each question you did not answer and reread the stem and the choices that are not crossed out. It should be easy to find these questions if you followed the instructions in Rule 12, as all of them will have their number designations circled on the answer sheet, and will be written down on your scrap paper list. If the answer is still not clear and you are running out of time, make an "educated guess" (see the guidelines in Rule 16) from the choices that you have not already eliminated.

14. DON'T LEAVE QUESTIONS UNANSWERED UNLESS THE INSTRUCTIONS INDICATE A PENALTY FOR WRONG ANSWERS. In almost all Firefighter Examinations, you do not lose credit for incorrect answers. Therefore, guess at any questions you are not sure of.

In rare instances, however, a penalty is assessed for wrong answers on multiple-choice examinations. Since this must be explained in the instructions, be sure to read them carefully. If there is a wrong-answer penalty on your examination, decide how strongly you feel about each individual question before answering it. (*Note:* This rarely happens on entry-level Firefighter Examinations.)

15. CHECK YOUR TIME PERIODICALLY. The typical Firefighter Examination consists of 50 to 100 questions, and has a time limit of 2 to 3½ hours. You will have no trouble with time if you average about 1½ minutes per question. In other words, take about 15 minutes for every 10 questions. Check yourself after every 10 questions to make sure you are not taking too long. *Never spend more than 2 minutes on any question.* In that way, you will have plenty of time at the end of the examination to go back to the questions you skipped the first time. (Always schedule your time to leave at least ½ hour for the questions you skipped.) Be careful that the time limit and the number of questions on the test are the same as the practice tests. If not, you have to make adjustments in your schedule.

16. RULES FOR MAKING AN EDUCATED GUESS—to be used only as a last resort and only if you are not penalized for guessing. Some questions may ask you to select from a list only those items that are correct. Incorrect items selected will be deducted from your total score. If the instructions tell you there will be a deduction for wrong answers, DO NOT GUESS. Your chances of choosing the correct answers to questions you are not sure of will be significantly increased if you obey the following rules:

a. Don't reconsider answer choices that you have eliminated. (See Rule 11.)
b. Be aware of key words that give you clues to the correct answer. (See Rule 9.)
c. If two choices have conflicting meanings, one of them is probably the correct answer. If two choices are very close in meaning, probably neither is correct.
d. The choice that contains significantly more or significantly fewer words than the other choices is very often correct.
e. If all else fails, and you have to make an outright guess at more than one question, guess the same lettered choice for each such question. The odds are that you will pick up some valuable points.

17. BE VERY RELUCTANT TO CHANGE ANSWERS. Unless you have a very good reason, do not change an answer. Studies have shown that all too often people change an answer from the right one to the wrong one.

General Instructions About the Test's Administrative Procedures

On the day of the test you will be required to complete several administrative forms. You can expect to be fingerprinted and you will need to have with you some form of valid proof of your identity.

When you first arrive you will be asked for your admission card. You will be assigned to a room and assigned a numbered seat in that room. You should go directly to that location and get prepared for the examination. When the examiners are ready to begin the examination, they will distribute a package of materials. Follow the examiner's instructions very carefully. You do not want to disqualifty yourself by inadvertently doing something wrong, such as opening the wrong envelope or prematurely looking at papers that you are not permitted to look at until sometime later in the examination process.

Generally the examiner will direct you to provide some preliminary information about yourself, i.e.: your name, address, social security number, the official test number (this will be provided by the examiner), seat number, room number and location of the test site. You must check to see if the information on your admission card is correct. If it is not, advise the examiner and follow the examiner's instructions for correcting it. You may be told to use the information on the card and to add the correct information below or on the back of the card.

When filling out the identification information on the ANSWER SHEET print carefully and provide only the information asked for. Many examiners do not want you to put your name on the answer sheet, and direct you to use your social security number or some other identification number. This provides security and assures that each candidate is treated fairly and equally.

Test Instructions

When the test begins you may be given several signals. The first is to open the packet and read the instructions, the second, to open the memory booklet. You will be told how long you have to memorize as much as possible of the floor plan or illustration. Generally you will not be able to write or make any notes while studying the memory diagram.

The next signal you receive will direct you to close the memory booklet and put it aside. It may be collected by the examiner or room monitor. When this is completed, the examiner will give the test question booklet and will direct you to the test. Answer the memory questions immediately. Do not look through the rest of the test at this time. You will want to concentrate your efforts on the memory portion of the test while the information is still fresh in your mind. After you have answered the memory questions, look through the test and make sure you have a complete test booklet. All the pages should be numbered and in sequential order, and each page should be printed clearly so that you can easily read it. If you find any problems, raise your hand and notify the examiner immediately.

The final signal will indicate that your test time is up and the examination has ended. Stop all work and put your pencil down.

How to Work With the Answer Sheet

Answer all the questions and record your answers on the answer sheet before the final signal. ONLY YOUR ANSWER SHEET WILL BE MARKED. You may be permitted to make a scrap copy of the answers for your records; do this after you have completed and checked the answer sheet.

If you want to change an answer, erase it and then mark your answer sheet with the correct answer. Avoid putting stray pencil marks, dots, or dashes any place on the answer sheet. If you accidentally put a stray mark on the paper, erase it.

You have read the question, selected an answer, and are now ready to mark the answer sheet. Examinations requiring only one selection for each question have rows and columns of answer boxes. It is important for you to closely review the format of the answer sheet. The order of the answers in the row and columns could run laterally across the page from column to column, or they could run vertically down the page, one under another until one column is complete. Usually there will be four (A, B, C, D) or five (A, B, C, D, E) spaces for each question. Use an index card, a piece of scrap paper, or one of the test sheets to cover the answer boxes below the question you will be working on and then mark your selection, being careful not to select a wrong choice or to go significantly outside the box you have selected.

Some examiners have developed tests that direct the candidate to select more than one choice for a question. When you are directed to do this, you must carefully look at the answer sheet and, using a technique similar to the one explained above, first record your number one choice in the column or row labeled "1st choice," then your number two choice in the column or row marked "2nd choice" and then your third choice in the column or row marked "3rd choice."

WARNING—You are not allowed to copy answers from anyone. You may only use books or reference materials the examiner provides you with or materials the examiner has given specific written approval to use.

Instructions: Use this answer sheet to record your answers to the practice exam. For each question, record your first, second, and third choice.

Follow the test instructions for candidates. Darken your answers in the ovals below. Do not write anything else on this page. USE NO. 2 SOFT PENCIL ONLY. See instructions for sample question.
WARNING: Be sure that the oval you darken under each choice corresponds to the question you are answering. BE SURE YOUR PENCIL MARKS ARE HEAVY AND BLACK. ERASE COMPLETELY ANY ANSWER YOU WISH TO CHANGE. DO NOT make stray pencil dots, dashes or marks ANYPLACE on this page.

START HERE AND WORK DOWNWARD

	1st Choice		2nd Choice		3rd Choice
1-1	Ⓐ Ⓑ Ⓒ Ⓓ	1-2	Ⓐ Ⓑ Ⓒ Ⓓ	1-3	Ⓐ Ⓑ Ⓒ Ⓓ
2-1	Ⓐ Ⓑ Ⓒ Ⓓ	2-2	Ⓐ Ⓑ Ⓒ Ⓓ	2-3	Ⓐ Ⓑ Ⓒ Ⓓ
3-1	Ⓐ Ⓑ Ⓒ Ⓓ	3-2	Ⓐ Ⓑ Ⓒ Ⓓ	3-3	Ⓐ Ⓑ Ⓒ Ⓓ
4-1	Ⓐ Ⓑ Ⓒ Ⓓ	4-2	Ⓐ Ⓑ Ⓒ Ⓓ	4-3	Ⓐ Ⓑ Ⓒ Ⓓ
5-1	Ⓐ Ⓑ Ⓒ Ⓓ	5-2	Ⓐ Ⓑ Ⓒ Ⓓ	5-3	Ⓐ Ⓑ Ⓒ Ⓓ
6-1	Ⓐ Ⓑ Ⓒ Ⓓ	6-2	Ⓐ Ⓑ Ⓒ Ⓓ	6-3	Ⓐ Ⓑ Ⓒ Ⓓ
7-1	Ⓐ Ⓑ Ⓒ Ⓓ	7-2	Ⓐ Ⓑ Ⓒ Ⓓ	7-3	Ⓐ Ⓑ Ⓒ Ⓓ
8-1	Ⓐ Ⓑ Ⓒ Ⓓ	8-2	Ⓐ Ⓑ Ⓒ Ⓓ	8-3	Ⓐ Ⓑ Ⓒ Ⓓ
9-1	Ⓐ Ⓑ Ⓒ Ⓓ	9-2	Ⓐ Ⓑ Ⓒ Ⓓ	9-3	Ⓐ Ⓑ Ⓒ Ⓓ
10-1	Ⓐ Ⓑ Ⓒ Ⓓ	10-2	Ⓐ Ⓑ Ⓒ Ⓓ	10-3	Ⓐ Ⓑ Ⓒ Ⓓ
11-1	Ⓐ Ⓑ Ⓒ Ⓓ	11-2	Ⓐ Ⓑ Ⓒ Ⓓ	11-3	Ⓐ Ⓑ Ⓒ Ⓓ
12-1	Ⓐ Ⓑ Ⓒ Ⓓ	12-2	Ⓐ Ⓑ Ⓒ Ⓓ	12-3	Ⓐ Ⓑ Ⓒ Ⓓ
13-1	Ⓐ Ⓑ Ⓒ Ⓓ	13-2	Ⓐ Ⓑ Ⓒ Ⓓ	13-3	Ⓐ Ⓑ Ⓒ Ⓓ
14-1	Ⓐ Ⓑ Ⓒ Ⓓ	14-2	Ⓐ Ⓑ		
15-1	Ⓐ Ⓑ				
16-1					

WORK DOWNWARD

	1st Choice		2nd Choice		3rd Choice
51-1	Ⓐ Ⓑ Ⓒ Ⓓ	51-2	Ⓐ Ⓑ Ⓒ Ⓓ	51-3	Ⓐ Ⓑ Ⓒ Ⓓ
52-1	Ⓐ Ⓑ Ⓒ Ⓓ	52-2	Ⓐ Ⓑ Ⓒ Ⓓ	52-3	Ⓐ Ⓑ Ⓒ Ⓓ
53-1	Ⓐ Ⓑ Ⓒ Ⓓ	53-2	Ⓐ Ⓑ Ⓒ Ⓓ	53-3	Ⓐ Ⓑ Ⓒ Ⓓ
54-1	Ⓐ Ⓑ Ⓒ Ⓓ	54-2	Ⓐ Ⓑ Ⓒ Ⓓ	54-3	Ⓐ Ⓑ Ⓒ Ⓓ
55-1	Ⓐ Ⓑ Ⓒ Ⓓ	55-2	Ⓐ Ⓑ Ⓒ Ⓓ	55-3	Ⓐ Ⓑ Ⓒ Ⓓ
56-1	Ⓐ Ⓑ Ⓒ Ⓓ	56-2	Ⓐ Ⓑ Ⓒ Ⓓ	56-3	Ⓐ Ⓑ Ⓒ Ⓓ
57-1	Ⓐ Ⓑ Ⓒ Ⓓ	57-2	Ⓐ Ⓑ Ⓒ Ⓓ	57-3	Ⓐ Ⓑ Ⓒ Ⓓ
58-1	Ⓐ Ⓑ Ⓒ Ⓓ	58-2	Ⓐ Ⓑ Ⓒ Ⓓ	58-3	Ⓐ Ⓑ Ⓒ Ⓓ
59-1	Ⓐ Ⓑ Ⓒ Ⓓ	59-2	Ⓐ Ⓑ Ⓒ Ⓓ	59-3	Ⓐ Ⓑ Ⓒ Ⓓ
60-1	Ⓐ Ⓑ Ⓒ Ⓓ	60-2	Ⓐ Ⓑ Ⓒ Ⓓ	60-3	Ⓐ Ⓑ Ⓒ Ⓓ
61-1	Ⓐ Ⓑ Ⓒ Ⓓ	61-2			
62-1	Ⓐ				

How to Make the Most of This Guide

To obtain maximum benefit from the use of this book, you should utilize the following approach:

1. Learn the "Strategies for Handling Multiple-Choice Questions" on pages 43–46.
2. Read through the glossary of common fire service terms in Chapter 5. Familiarize yourself with terms that are new to you, but don't spend an excessive amount of time on memorization. Rather, review the terms you do not understand from time to time.
3. Take the Diagnostic Examination. After completing this examination and checking your answers, fill out the diagnostic chart on page 82. This will indicate your strengths and weaknesses. You can then devote most of your study time to the areas in which you are weak.
4. Study Chapters 5 through 11. As mentioned above, concentrate your study efforts on your weak areas, but make certain you cover each chapter. Follow the "Ten Rules for Effective Studying" on pages 42–43. Also, make sure you employ the test-taking strategies (pages 43–46) when answering the practice multiple-choice questions at the end of each chapter.
5. Take Practice Examination One. After you have finished this examination and have reviewed the answer and their explanations, complete the diagnostic chart on page 84. Then restudy the appropriate chapters in accordance with the directions below the diagnostic chart.
6. Take Practice Examination Two. Then follow the same procedure you followed after completing Practice Examination One.
7. Take Practice Examination Three. Again, follow the same procedure you followed after finishing Practice Examination One.
8. When the actual examination is 2 weeks away, read Chapter 15. Follow the strategy recommended in this chapter for the 7 days immediately preceding the examination.

PART TWO

A Diagnostic Examination

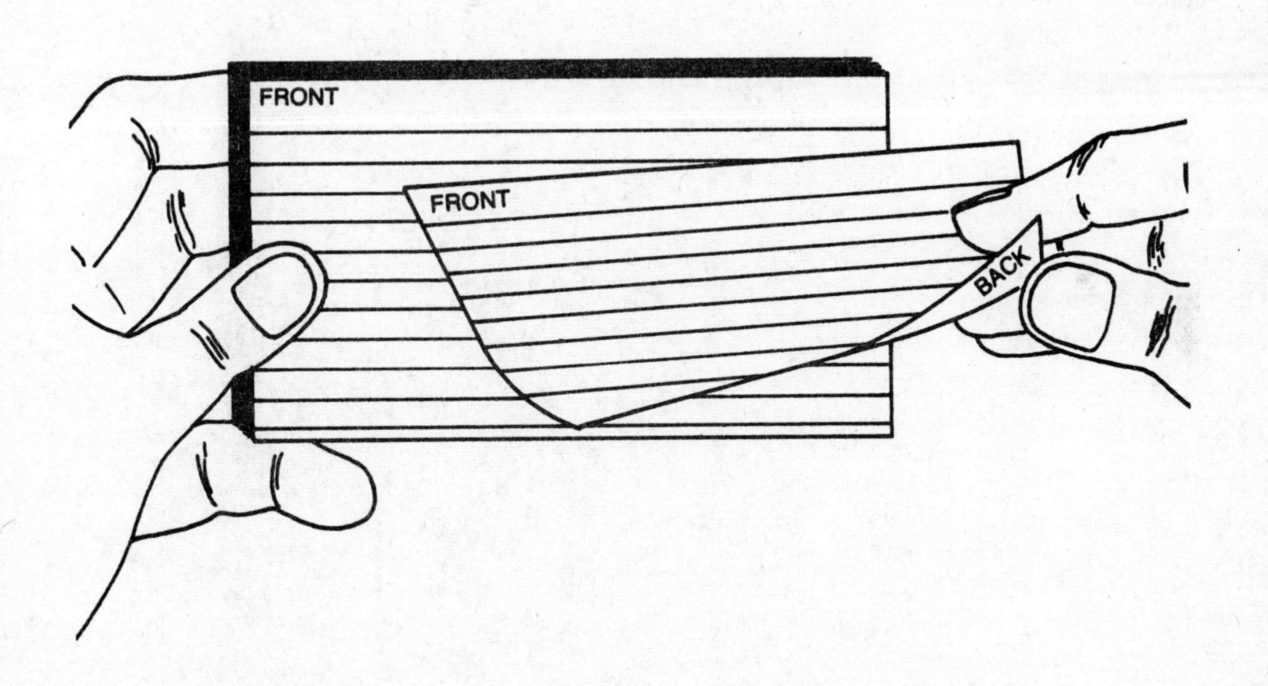

CHAPTER 4

Diagnose Your Problem

Diagnostic Examination

There are 100 questions on this Diagnostic Examination. You should finish the entire examination in 3½hours. For maximum benefit, it is strongly recommended that you take the examination in one sitting, as if it were an actual test.

The answers to this examination and their explanations begin on page 82. By completing the chart on page 84 in the section entitled "Diagnostic Procedure," you can get an idea of the types of questions that give you the most difficulty. You can then devote most of your time to these areas.

BEFORE YOU TAKE THE EXAMINATION

Before taking this examination, you should have read Chapters 1, 2, and 3. Make sure that you employ the test-taking strategy recommended in Chapter 3.

Remember to read each question and related material carefully before choosing your answers. Select the choice you believe to be correct, and mark your answer on the answer sheet provided at the beginning of this chapter. This answer sheet is probably similar to the one used on the actual examination you will take. The Answer Key, Diagnostic Procedure, and Answer Explanations appear at the end of this chapter.

Note: This Diagnostic Examination is divided into seven sections to facilitate your familiarization with the various question types and to make it easier for you to analyze the results.

ANSWER SHEET
DIAGNOSTIC EXAMINATION

Follow the instructions given in the test. Mark only your answers in the ovals below.
WARNING: Be sure that the oval you fill is in the same row as the question you are answering. Use a No. 2 pencil (soft pencil).
BE SURE YOUR PENCIL MARKS ARE HEAVY AND BLACK. ERASE COMPLETELY ANY ANSWER YOU WISH TO CHANGE.

START HERE DO NOT make stray pencil dots, dashes or marks.

1 Ⓐ Ⓑ Ⓒ Ⓓ	2 Ⓐ Ⓑ Ⓒ Ⓓ	3 Ⓐ Ⓑ Ⓒ Ⓓ
4 Ⓐ Ⓑ Ⓒ Ⓓ	5 Ⓐ Ⓑ Ⓒ Ⓓ	6 Ⓐ Ⓑ Ⓒ Ⓓ
7 Ⓐ Ⓑ Ⓒ Ⓓ	8 Ⓐ Ⓑ Ⓒ Ⓓ	9 Ⓐ Ⓑ Ⓒ Ⓓ
10 Ⓐ Ⓑ Ⓒ Ⓓ	11 Ⓐ Ⓑ Ⓒ Ⓓ	12 Ⓐ Ⓑ Ⓒ Ⓓ
13 Ⓐ Ⓑ Ⓒ Ⓓ	14 Ⓐ Ⓑ Ⓒ Ⓓ	15 Ⓐ Ⓑ Ⓒ Ⓓ
16 Ⓐ Ⓑ Ⓒ Ⓓ	17 Ⓐ Ⓑ Ⓒ Ⓓ	18 Ⓐ Ⓑ Ⓒ Ⓓ
19 Ⓐ Ⓑ Ⓒ Ⓓ	20 Ⓐ Ⓑ Ⓒ Ⓓ	21 Ⓐ Ⓑ Ⓒ Ⓓ
22 Ⓐ Ⓑ Ⓒ Ⓓ	23 Ⓐ Ⓑ Ⓒ Ⓓ	24 Ⓐ Ⓑ Ⓒ Ⓓ
25 Ⓐ Ⓑ Ⓒ Ⓓ	26 Ⓐ Ⓑ Ⓒ Ⓓ	27 Ⓐ Ⓑ Ⓒ Ⓓ
28 Ⓐ Ⓑ Ⓒ Ⓓ	29 Ⓐ Ⓑ Ⓒ Ⓓ	30 Ⓐ Ⓑ Ⓒ Ⓓ
31 Ⓐ Ⓑ Ⓒ Ⓓ	32 Ⓐ Ⓑ Ⓒ Ⓓ	33 Ⓐ Ⓑ Ⓒ Ⓓ
34 Ⓐ Ⓑ Ⓒ Ⓓ	35 Ⓐ Ⓑ Ⓒ Ⓓ	36 Ⓐ Ⓑ Ⓒ Ⓓ
37 Ⓐ Ⓑ Ⓒ Ⓓ	38 Ⓐ Ⓑ Ⓒ Ⓓ	39 Ⓐ Ⓑ Ⓒ Ⓓ
40 Ⓐ Ⓑ Ⓒ Ⓓ	41 Ⓐ Ⓑ Ⓒ Ⓓ	42 Ⓐ Ⓑ Ⓒ Ⓓ
43 Ⓐ Ⓑ Ⓒ Ⓓ	44 Ⓐ Ⓑ Ⓒ Ⓓ	45 Ⓐ Ⓑ Ⓒ Ⓓ
46 Ⓐ Ⓑ Ⓒ Ⓓ	47 Ⓐ Ⓑ Ⓒ Ⓓ	48 Ⓐ Ⓑ Ⓒ Ⓓ
49 Ⓐ Ⓑ Ⓒ Ⓓ	50 Ⓐ Ⓑ Ⓒ Ⓓ	51 Ⓐ Ⓑ Ⓒ Ⓓ
52 Ⓐ Ⓑ Ⓒ Ⓓ	53 Ⓐ Ⓑ Ⓒ Ⓓ	54 Ⓐ Ⓑ Ⓒ Ⓓ
55 Ⓐ Ⓑ Ⓒ Ⓓ	56 Ⓐ Ⓑ Ⓒ Ⓓ	57 Ⓐ Ⓑ Ⓒ Ⓓ
58 Ⓐ Ⓑ Ⓒ Ⓓ	59 Ⓐ Ⓑ Ⓒ Ⓓ	60 Ⓐ Ⓑ Ⓒ Ⓓ
61 Ⓐ Ⓑ Ⓒ Ⓓ	62 Ⓐ Ⓑ Ⓒ Ⓓ	63 Ⓐ Ⓑ Ⓒ Ⓓ
64 Ⓐ Ⓑ Ⓒ Ⓓ	65 Ⓐ Ⓑ Ⓒ Ⓓ	66 Ⓐ Ⓑ Ⓒ Ⓓ
67 Ⓐ Ⓑ Ⓒ Ⓓ	68 Ⓐ Ⓑ Ⓒ Ⓓ	69 Ⓐ Ⓑ Ⓒ Ⓓ
70 Ⓐ Ⓑ Ⓒ Ⓓ	71 Ⓐ Ⓑ Ⓒ Ⓓ	72 Ⓐ Ⓑ Ⓒ Ⓓ
73 Ⓐ Ⓑ Ⓒ Ⓓ	74 Ⓐ Ⓑ Ⓒ Ⓓ	75 Ⓐ Ⓑ Ⓒ Ⓓ
76 Ⓐ Ⓑ Ⓒ Ⓓ	77 Ⓐ Ⓑ Ⓒ Ⓓ	78 Ⓐ Ⓑ Ⓒ Ⓓ
79 Ⓐ Ⓑ Ⓒ Ⓓ	80 Ⓐ Ⓑ Ⓒ Ⓓ	81 Ⓐ Ⓑ Ⓒ Ⓓ
82 Ⓐ Ⓑ Ⓒ Ⓓ	83 Ⓐ Ⓑ Ⓒ Ⓓ	84 Ⓐ Ⓑ Ⓒ Ⓓ
85 Ⓐ Ⓑ Ⓒ Ⓓ	86 Ⓐ Ⓑ Ⓒ Ⓓ	87 Ⓐ Ⓑ Ⓒ Ⓓ
88 Ⓐ Ⓑ Ⓒ Ⓓ	89 Ⓐ Ⓑ Ⓒ Ⓓ	90 Ⓐ Ⓑ Ⓒ Ⓓ
91 Ⓐ Ⓑ Ⓒ Ⓓ	92 Ⓐ Ⓑ Ⓒ Ⓓ	93 Ⓐ Ⓑ Ⓒ Ⓓ
94 Ⓐ Ⓑ Ⓒ Ⓓ	95 Ⓐ Ⓑ Ⓒ Ⓓ	96 Ⓐ Ⓑ Ⓒ Ⓓ
97 Ⓐ Ⓑ Ⓒ Ⓓ	98 Ⓐ Ⓑ Ⓒ Ⓓ	99 Ⓐ Ⓑ Ⓒ Ⓓ
100 Ⓐ Ⓑ Ⓒ Ⓓ		

The Test

SECTION ONE—PART A

RECALL OF VISUAL MATERIAL

Directions: You are given 5 minutes to study the floor plan on the following page and to commit to memory as many details as you can. You are *not* permitted to make any written notes during the 5 minutes you are studying the illustration.

After 5 minutes stop studying the illustration, turn the page, and answer the questions without referring to the illustration. The next time you are permitted to look at the illustration is when you have completed the test and are verifying your answers.

Start your 5 minutes on a clock. Now turn the page and begin.

Illustration 1

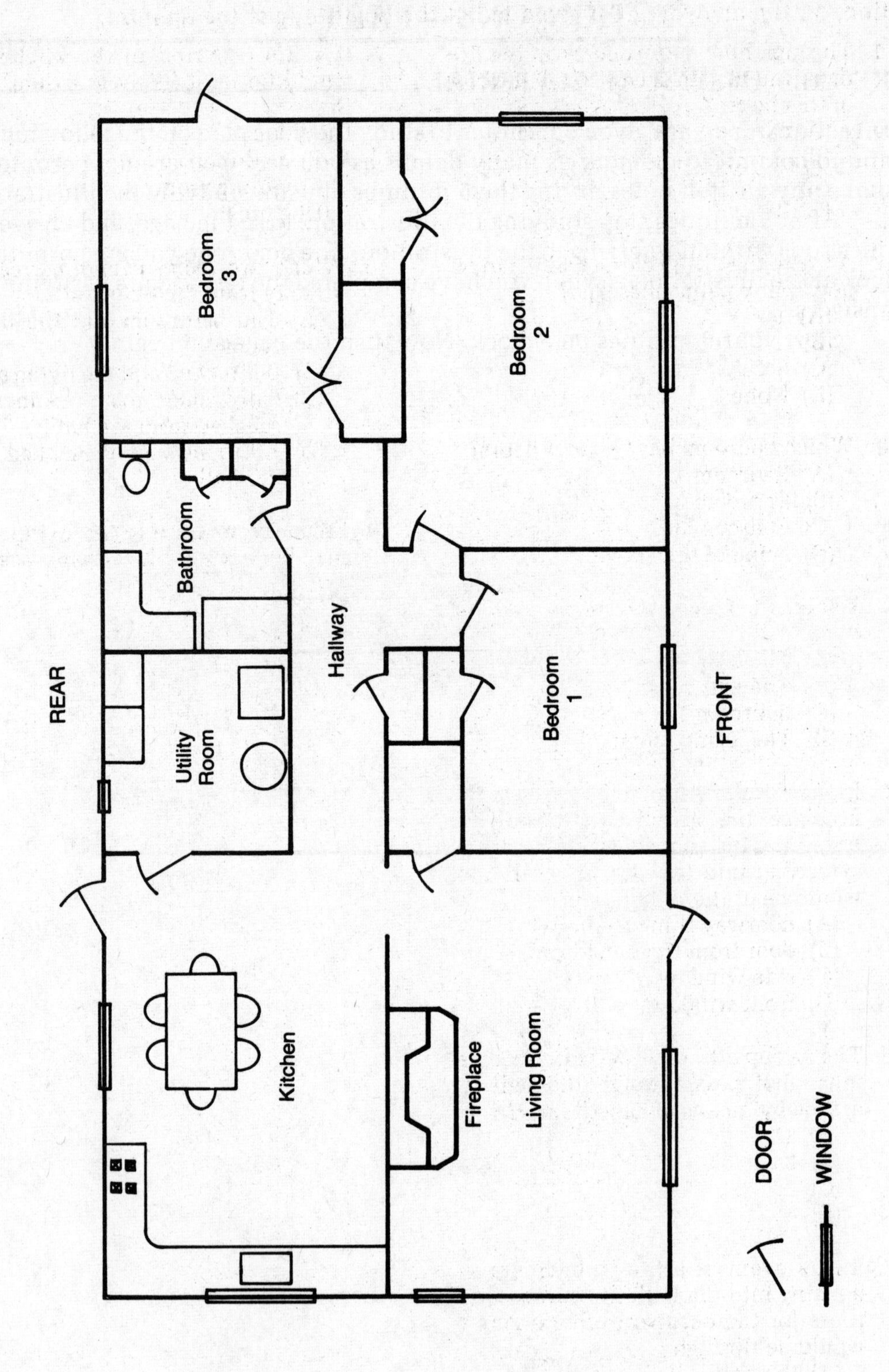

Answer questions 1 through 10 on the basis of the floor plan you just studied. For each question select the one *best* answer—(A), (B), (C), or (D). Then indicate your choice by blackening the appropriate letter next to the number of the question, on the answer sheet provided at the beginning of the chapter.

1. Through how many doors can the fire department gain access to the interior of this house?
   (A) 1
   (B) 2
   (C) 3
   (D) 4

2. A fire in the utility closet would cut off how many paths of escape?
   (A) 1
   (B) 2
   (C) 3
   (D) None

3. Which bedroom has two exit doors?
   (A) Bedroom 1.
   (B) Bedroom 2.
   (C) Bedroom 3.
   (D) None of the above.

4. Which room has only one means of escape?
   (A) The kitchen.
   (B) The bathroom.
   (C) Bedroom 1.
   (D) The living room.

5. In the event of a fire in the living room fireplace, the best place to position a hose line to prevent the fire from spreading into the rest of the house would be at the living room
   (A) doorway from the hall
   (B) door from the outside
   (C) side window
   (D) front window

6. The occupants of how many rooms must first pass through the hall to escape by means of an exit door?
   (A) 1
   (B) 2
   (C) 3
   (D) 4

7. In the event of a fire in bedroom 1, burning into the hall, the best escape route for the occupant of bedroom 2 would be through
   (A) the hall
   (B) the bathroom window
   (C) the bedroom 1 window
   (D) a bedroom 2 window

8. If a fire occurred in the kitchen, it would be most likely to extend into the
   (A) hall
   (B) bathroom
   (C) dining room
   (D) porch

9. Which two rooms cannot be reached directly from the hallway?
   (A) The bathroom and the dining room.
   (B) Bedroom 2 and the living room.
   (C) Only one room cannot be reached from the hall.
   (D) All rooms can be reached from the hall.

10. Escape from a fire in the living room may be made via how many possible exits?
    (A) 2
    (B) 3
    (C) 4
    (D) 5

SECTION ONE—PART B

RECALL OF WRITTEN MATERIAL

Directions: You are allowed 5 minutes to read the material and study the illustration on the following pages and to commit to memory as much about them as you can. You are *not* permitted to make any written notes during the 5 minutes you are reading the material.

At the end of the 5 minutes, stop reading the material, turn the page, and answer the questions without referring to the written material or the illustration.

Start your 5 minutes on a clock. Now turn the page and begin.

Computerized Dispatch

Computerized dispatch systems are rapidly replacing outdated fire alarm bells. The illustration shows the selector panel that can be found in many fire stations. The selector panel allows the fire fighter on house watch at the station to communicate with the central computer, to report the unit's availability and status, and to acknowledge response to a fire.

Key to Selector Panel

Unit Identification Buttons

1 = Engine Company
2 = Ladder Company
3 = Special (SPL) Unit—Rescue, Squad, etc.
4 = Chief

Status Buttons

10-4 = responding to incident as directed by alarm teleprinter
A–Q = available in quarters
10-8 = leaving quarters for non-fire-fighting reason
RCP = rest and recuperation
Verbal = responding to verbal alarm concerning a fire
Unlabeled = reserved for future use
10-14 = unit out of service

Function Buttons

Send button = release information to computer
10-5 = repeat message
10-11 = test of system
CLR/TEST = tests all lamps and clears errors if wrong button is pushed

The system requires that several buttons be pushed to transmit a message. For example, to let the computer know that the chief is responding to the incident requires pushing the Chief button, then the 10-4 button, and finally the send button.

Illustration 11

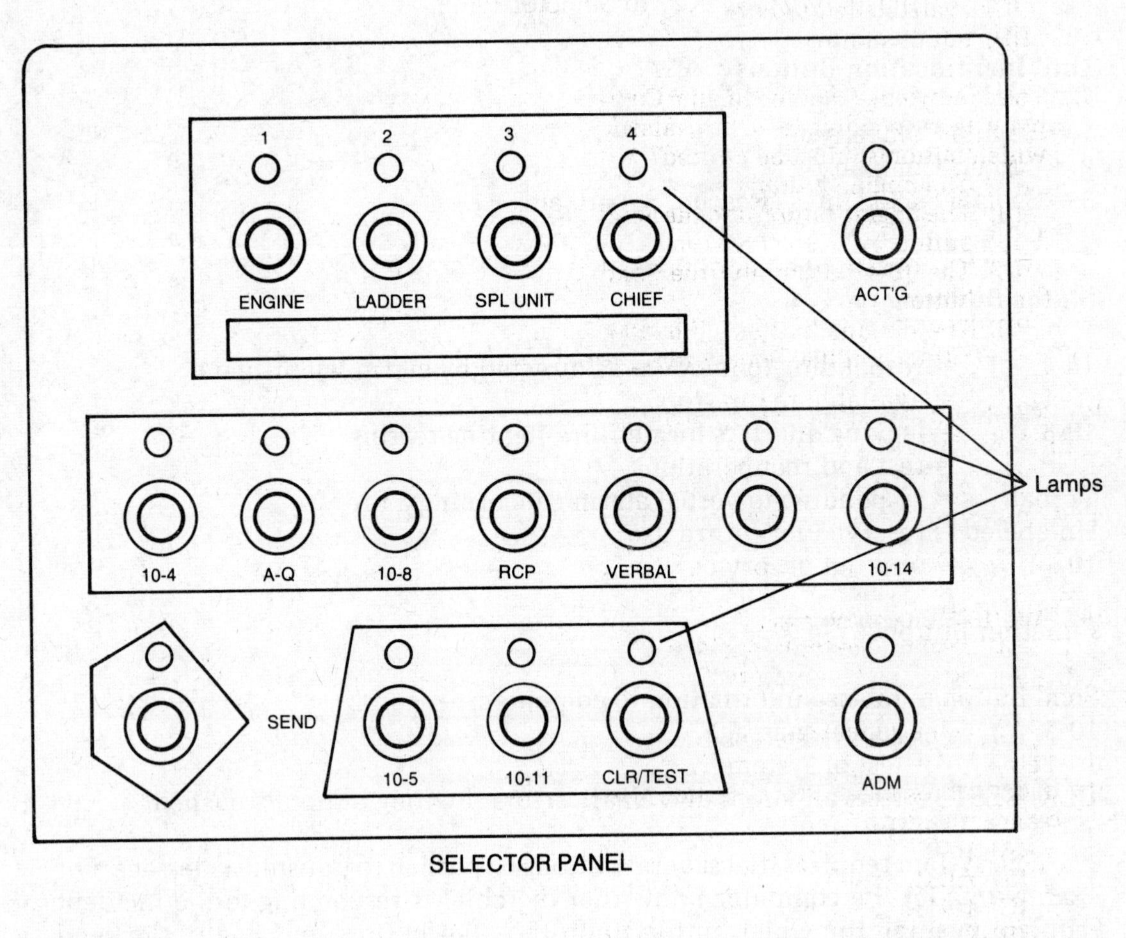

SELECTOR PANEL

Answer questions 11 through 15 using only the information in the preceding passage and illustration. For each question select the one *best* answer—(A), (B), (C), or (D). Then indicate your choice by blackening the appropriate letter, next to the number of the question, on the answer sheet provided at the beginning of the chapter.

11. The button marked 10-14 is used when the unit is
    (A) out of service
    (B) on rest and recuperation
    (C) available in quarters
    (D) out of quarters

12. To acknowledge that the Engine Company is responding to a fire alarm, which button(s) must be pushed?
    (A) The Engine button.
    (B) The Engine button and the Send button.
    (C) The 10-4 button and the Send button.
    (D) The Engine button, the 10-4 button, and the Send button.

13. The button used for a Rescue Unit is the one labeled
    (A) Verbal
    (B) SPL Unit
    (C) ACT'G
    (D) ADM

14. The 10-5 button means
    (A) "responding to the incident"
    (B) "available at quarters"
    (C) "out of service"
    (D) "repeat message"

15. If an error is made, which button will cancel the error?
    (A) 10-8.
    (B) ADM.
    (C) 10-11.
    (D) CLR/TEST.

SECTION TWO

READING COMPREHENSION

Directions: The questions in this section test your ability to comprehend what you read. After reading each passage carefully, answer the questions that follow it, using only the information in the passage. For each question choose the one *best* answer—(A), (B), (C), or (D). Then indicate your choice by blackening the appropriate letter, next to the number of the question, on the answer sheet provided at the beginning of the chapter.

Answer questions 16 through 20 using only the information in the following passage.

Ladder Safety

The safe use of ladders is an everyday problem, and with good weather and the need for work on and about the home, the use of ladders by off-duty department members will increase. However, there seems to be a certain amount of Missouri-mule stubbornness in every individual when it comes to using any kind of ladder; a large percentage of us will go to great lengths to misuse them. As a result, we end up with broken bones, cracked heads, mashed noses, and a general assortment of cuts and bruises. All of the above result in the loss of a member's services to the Fire Department.

Unfortunately, most ladder accidents occur as a result of overconfidence on the part of the member. A few helpful hints on correct ladder usage follow:

Stepladders

1. Always open the ladder wide enough so that the spreader locks itself in the open position.
2. Never stand on top of a stepladder; use a ladder tall enough to let you stand at least two steps from the top.
3. Always make sure that the feet of the ladder are on a firm, level foundation.
4. Never lean a stepladder against a wall and use it as a straight ladder.

Straight and Extension Portable Ladders

1. When setting up a ladder, place it so that the horizontal distance from its base to the vertical plane of the support is approximately one-fourth the ladder length. (For example, place a 12-foot ladder so that the bottom is approximately 3 feet away from the object against which the top is leaning.)
2. Set the ladder on a firm foundation. Make sure that it does not wobble before you climb it.
3. Never place a ladder against a window pane, a window sash, or a loose box, barrel, etc.
4. When using a ladder for access to a high place, securely lash or otherwise fasten it to prevent slipping.
5. Be aware that it is not advisable to use a metal ladder in close proximity to live electric wiring.
6. Do not climb higher than the third rung from the top on a straight or an extension ladder.

All Portable Ladders

1. Make sure that the ladder is equipped with a nonslip base.
2. Resist the temptation to overreach. It is better to get down and move the ladder.
3. Always face the ladder when ascending or descending.
4. Hold on with both hands when climbing up or down. Use a rope to raise or lower tools or equipment.
5. Never use defective ladders. When a defective ladder is found, it should be destroyed immediately. Many people have been injured by using a ladder retrieved from a junk pile. Resist the urge to patch a ladder with wire or make-shift steps or rungs "until another one can be obtained."

16. Which of the following statements about ladder safety is *incorrect?*
    (A) Never lean a stepladder against a wall and use it as a straight ladder.
    (B) The base of the ladder should be a distance equal to one-fourth the length of the ladder from the wall when in position.
    (C) The top two steps of a stepladder may be used only if the ladder is less than 5 feet high.
    (D) The stepladder should always be opened sufficiently to engage the spreader locks.

17. The improper use of ladders around the home leads to injuries and loss of ability to perform as a fire fighter. Injuries commonly result from all of the following *except*
    (A) overreaching
    (B) not holding on properly
    (C) using a rope to raise and lower tools
    (D) using a metal ladder in proximity to live electrical wiring

18. For safety, the ladder should be
    (A) placed against the window sash
    (B) brand-new
    (C) made of heavy metal
    (D) set on a firm foundation

19. When using a straight or extension ladder, the highest rung that should be climbed to is the
    (A) next to the top
    (B) second from the top
    (C) third from the top
    (D) fourth from the top

20. There seems to be a certain amount of stubbornness in every individual when it comes to ladder safety. It would be correct to state that:
    (A) Few accidents occur because of overconfidence.
    (B) Most injuries caused by improper use of ladders are minor.
    (C) Most fire fighters hold on with one hand while climbing ladders.
    (D) Many people have been injured by using ladders they found in a junk pile.

Answer questions 21 through 25 using only the information in the following passage.

Smoking

Smoking in bed is a dangerous habit. Cigarette, cigar, or pipe ashes often fall undetected and smolder in the upholstery or bedding while the smoker sleeps. Hours later, the hidden fire results in deadly smoke which overcomes the smoker and often other people as well. Falling asleep in a comfortable overstuffed chair or on a couch while smoking can have the same deadly results.

It is important to keep a sufficient number of ashtrays in areas of the home where smoking is permitted. The ashtrays should be large, deep, and designed to prevent cigarettes or cigars from accidentally falling out. After you light your cigarette, the burning match should be extinguished and placed in the ashtray. Paper, empty match-books, and cellophane wrappers should not be put in the ashtray. Ashtrays should be emptied frequently, but never into wastebaskets. To properly dispose of the contents of ashtrays, you should douse them with water and put them into a metal container.

21. Falling asleep in an overstuffed chair while smoking a pipe is
    (A) all right if there is a large ashtray nearby
    (B) not as dangerous as smoking in bed
    (C) as deadly as smoking in bed
    (D) much more dangerous than smoking in bed

22. Ashtrays should be
    (A) kept in every room
    (B) made of paper
    (C) emptied into wastebaskets
    (D) large and deep

23. Smoking in bed is deadly because
    (A) smoke replaces the air and kills the smoker

(B) smoke damages the drapes and ruins the bed
(C) the cigarette burns the person in the bed
(D) the ash may start a fire in the smoker's clothes or bedding

24. The contents of a properly used ashtray may include all of the following *except*
(A) cigarette butts
(B) matches
(C) cigarette wrappers
(D) cigar ashes

25. The preferred method for disposing of the materials in an ashtray is to
(A) douse them with water and put them into a metal container
(B) empty them into a wastepaper basket
(C) throw them out the window
(D) place them in a plastic garbage bag

Answer questions 26 through 28 using only the information in the following passage.

On the fire ground, who is responsible for making the decisions? It is the officer in command, regardless of rank. The first officer on the scene is in command until relieved by a superior; the officer may be a captain, lieutenant, or acting officer. He has the responsibility for making and carrying out initial decisions. Therefore all officers should be trained in the use of an effective action plan that is practical for all ranks at all types of fires.

26. The writer of this passage believes that:
(A) All fire fighters should be trained as officers.
(B) Captains and lieutenants need to be reminded of their fire fighting responsibilities.
(C) Only senior fire officers have fire-ground command responsibility.
(D) All officers should be trained to implement an effective fire-ground action plan.

27. The officer in command, until relieved by a superior, is the first
(A) officer to arrive
(B) lieutenant to arrive
(C) captain to arrive
(D) chief officer to arrive

28. An effective action plan should be drawn up to handle
(A) unusual types of fires
(B) dangerous types of fires
(C) all types of fires
(D) some types of fires

Answer questions 29 and 30 using only the information given.

29. Without proper breathing equipment, you should never attempt to fight a fire in a cellar or basement that has become filled with smoke. The products of combustion may be poisonous, and the lack of oxygen may quickly overcome you.
According to this passage, you
(A) should never extinguish a fire in a cellar
(B) should not fight a fire in a smoke-filled cellar without proper equipment
(C) can overcome the fumes produced by a cellar fire
(D) could attempt to fight a fire in a smoke-filled basement

30. School fire drills should be held early in the term, and should be well planned and frequently rehearsed. They should be conducted as surprise drills, without prior warning to the teaching staff.
The theme of this passage is:
(A) Surprise fire drills should be held in schools.
(B) School fire drills should be held early in the day.
(C) Frequent fire drills surprise teachers.
(D) Teachers won't drill without prior warning.

Answer questions 31 through 35 using only the information in the following passage.

To understand the role of a fire fighter, it is first necessary to understand the mission of the fire service. The objectives of the fire service fall into four broad categories. In order of priority, they are as follows:

1. Prevent unwanted fires.
2. Prevent loss of life and property when a fire does occur.
3. Confine the fire to its place of origin.
4. Control and extinguish the fire.

We prevent fires by doing something about the factors that cause them. These causes are usually hazardous and unstable materials, improper use of materials, and human carelessness. The hazards are removed by enforcing laws and educating the public.

The prevention of loss of life and property is accomplished by enforcing building and fire prevention codes. These codes require a sufficient number of properly lighted and marked exits, limits on the number of occupants, and control over the kinds of materials used in construction, as well as those that may be brought into the building after it is occupied. The fire department reduces the loss of life by making rescues, confining the fire to allow people to get out, and extinguishing the fire.

When a fire does start, damage can be limited by rapid extinguishment. However, if the fire can be prevented, there will be no property damage and no loss of life.

31. What title best fits the theme of this passage?
    (A) Preventing Fires.
    (B) Rescue and Fire Fighting.
    (C) The Objectives of the Fire Service.
    (D) The Extinguishment of Fire.

32. What is the most important objective of the fire service?
    (A) Fire prevention.
    (B) Fire fighting.
    (C) Fire education.
    (D) Fire control and confinement.

33. The prevention of fire is accomplished by
    (A) understanding the role of the fire fighter
    (B) doing something about the factors that cause fires
    (C) making rescues and confining and extinguishing the fire
    (D) rapid extinguishment

34. To understand the role of a fire fighter you must
    (A) study fire fighting and fire prevention guides
    (B) know the responsibilities of the fire service
    (C) experience the daily work of a fire fighter
    (D) have extinguished a fire or saved a life

35. After a fire starts, damage can be limited by
    (A) restricting the number of people who enter the area
    (B) using less water
    (C) rapid extinguishment
    (D) none of the above; extensive damage cannot be avoided

Answer questions 36 through 38 solely on the basis of the following table.

FLAMMABLE LIQUIDS

Material	Flash Point (degrees F)	Ignition Temperature (degrees F)	Upper Flammable Limit (percent)
Gasoline (100 octane)	−36	853	7.4
Acetone	0	869	12.8
Benzene	12	1040	7.1
Ethyl alcohol	55	689	19.0

36. The flammable liquid with the lowest ignition temperature is
    (A) gasoline
    (B) acetone
    (C) benzene
    (D) ethyl alcohol

37. The flash point of benzene is
    (A) 0 degree
    (B) 12 degrees
    (C) 35 degrees
    (D) 55 degrees

38. If the flash point indicates how dangerous a substance may be by telling us at what temperature the liquid will change into an ignitable gas, the most dangerous substance on the list is
    (A) ethyl alcohol
    (B) benzene
    (C) acetone
    (D) gasoline

Answer questions 39 and 40 using only the information given.

39. Most fire departments require fire fighters to be fully dressed and equipped before boarding the apparatus for response. What is the main reason for this rule?
    (A) Dressing while responding has proved to be dangerous.
    (B) Unprepared fire fighters arriving at a fire scene make a poor impression.
    (C) The probability of leaving important safety equipment in quarters is reduced.
    (D) The possibility of dropping or damaging equipment while responding is reduced.

40. "Fire fighters assigned to housewatch duty shall remain at the housewatch desk at all times, except when it is necessary to observe conditions in front of quarters. In the event a relief is required for any purpose, the officer on duty shall be notified." According to the passage, it would be correct to state
    (A) The fire fighter on housewatch can never leave quarters.
    (B) The lieutenant on duty relieves the housewatch.
    (C) Fire fighters and officers shall do housewatch.
    (D) If a relief is needed, the officer on duty must be notified.

Answer questions 41 through 45 using only the information in the following passage.

Fire fighters shall execute the salute in the following manner:

1. It shall be executed at attention in accordance with infantry drill regulations of the U.S. Army.
2. It shall be executed within six paces of the person saluted, or at the nearest point of approach. The salute shall be held until the person saluted has passed, or until the salute has been properly acknowledged.

3. In passing in review at parades or in other formations, the hand salute shall be executed only by the commanding officer, members of the staff, or officers in command of integral units. All others in formation shall, on command, execute "EYES RIGHT" or "EYES LEFT." The hand salute is executed within six paces of the reviewing party, and is held as the march is continued in rigid formation past the reviewing party.

4. Six paces beyond the reviewing party, arms and hands are snapped to the position of attention. Officers in command of integral units shall issue the command "FRONT," at which time heads and eyes are returned to the normal position, and the march is continued in formation at attention. At formal reviews or parades, the boundaries of the reviewing stand are usually marked by guidons.

41. Fire fighters should salute
(A) all officers
(B) at twelve paces
(C) at the nearest point of approach
(D) after another member has saluted them

42. While passing in review at parades, the hand salute should be executed by
(A) all members
(B) commanding officers only
(C) commanding officers and staff officers
(D) commanding officers, staff officers, and officers in command of integral units

43. The command "FRONT" is used to have the members turn their
(A) heads and eyes to the normal position
(B) heads toward the reviewing stand
(C) march formation toward the reviewing stand
(D) march formation toward the guidons

44. While passing in review, it is correct to execute the salute
(A) three paces from the reviewing stand
(B) six paces from the reviewing stand
(C) at the reviewing stand
(D) at the guidons

45. Arms and hands are snapped to the position of attention
(A) after passing in review
(B) six paces beyond the guidons
(C) after passing a superior officer
(D) at formal parades

SECTION THREE

JUDGMENT AND DECISION

Directions: Questions 46 through 60 test your ability to make sound judgments and decisions. For each question select the one *best* answer—(A), (B), (C), or (D). Then indicate your choice by blackening the appropriate letter, next to the number of the question, on the answer sheet provided at the beginning of the chapter.

46. While searching burning homes or apartments, firefighters make a practice of checking inside closets, under beds, and behind furniture. What is the most likely reason for this practice?
    (A) Usually the cause of the fire can be found there.
    (B) Valuable possessions may be found there.
    (C) Children often hide from danger in these places.
    (D) The family pet usually seeks out these places.

47. The fire officer in command of operations has the responsibility of analyzing and evaluating the fire problem. Which of the following would not enter into this process?
    (A) The occupancy of the burning building.
    (B) The fire insurance coverage.
    (C) The time of day.
    (D) The height and area of the building.

48. A firefighter who had just arrived at the scene of a visible fire grabbed his tools from the apparatus, ran to the door of the building, and broke it down. The action of the firefighter was
    (A) correct—this allowed the firefighters with the hose to gain entry and extinguish the fire.
    (B) incorrect—the firefighter should have broken a window first.
    (C) correct—quick, decisive action leads to the rescue of fire victims.
    (D) incorrect—the firefighter should have tried the door knob first to see whether the door was locked.

49. While riding on a subway train, you smell and see heavy smoke coming into the car through the windows. The best action you are to take would be to
    (A) announce that you are a firefighter and ask the other passengers to close their windows.
    (B) announce that you are a firefighter and tell everyone to vacate the car and go to the last car in the train.
    (C) get up quietly, leave the car, and find the conductor.
    (D) go quietly to the front of the car, pull the emergency stop cord, announce that you are a firefighter, and direct the people to leave the train.

50. In illustration 50, the firefighter is carrying the axe
    (A) incorrectly—the firefighter should be using two hands.
    (B) correctly—this position is safest.
    (C) incorrectly—the blade should be nearer the ground.
    (D) correctly—the axe is lighter this way.

Illustration 50

51. Illustration 51 shows a firefighter operating a hose line from a portable ladder. A correct statement to make about this firefighter is that:
    (A) At least one hand should be on the beam of the ladder for safety.
    (B) Both feet should be kept on the same rung.
    (C) The leg wrapped around the rung will hold the firefighter safely in place.
    (D) The leg wrapped around the rung is used to hold the hose against the side of the ladder.

Illustration 51

52. While waiting for a bus on a Sunday afternoon, you are told by an excited neighborhood youth, who seems to be always in trouble, that there is a fire in a store six blocks away. The best action for you to take at this time is to
    (A) call the fire department from the nearest telephone.
    (B) disregard this statement as another prank and get on the bus.
    (C) call the police and take a ride with the youth to the store.
    (D) run to the store to investigate this report.

53. A fire company advanced a dry hose line to the fourth floor of an apartment house in preparation for fighting a fire on the fifth floor. What is the most likely reason for stretching this hose line dry?
    (A) The dry hose line was less likely to damage the firefighters' protective clothing.
    (B) The dry hose line was lighter and therefore could be put into place more quickly.
    (C) The dry hose line was less likely to be damaged.
    (D) The chance of water damage was removed.

54. While inspecting a building, a firefighter was told, "This building is fireproof and doesn't need to be inspected." What is the most appropriate reason for completing the inspection?
    (A) The building may not be fireproof.
    (B) The contents of the building may not be fireproof.
    (C) The statement was a deliberate attempt to prevent a fire prevention inspection.
    (D) The fire department will not give credit for the inspection if the firefighter does not finish.

55. You are a firefighter vacationing at a hotel. You discover a fire that has been burning for some time and is rapidly spreading. The *first* action you should take is to
    (A) leave the hotel and go to a safer place.
    (B) run through the hotel yelling "Fire!"
    (C) call the hotel management, identify yourself, and report the fire.
    (D) use available fire appliances in an attempt to extinguish the fire.

56. Fire departments investigate all fires to determine the cause. The *least* im-

portant reason for doing this is to
(A) determine the amount of insurance loss
(B) determine whether the fire was accidental or arsonous
(C) determine whether the fire-safety standards need improvement
(D) determine whether there was a violation of the fire prevention laws

57. To pull the ladder in illustration 57 up to the roof of the building,
(A) the knot of the rope should be between the building and the ladder
(B) the knot should be on the outside and the ladder should be between the building and the knot
(C) the position of the knot is not important
(D) the knot should be positioned where the rope loops the beams of the ladder

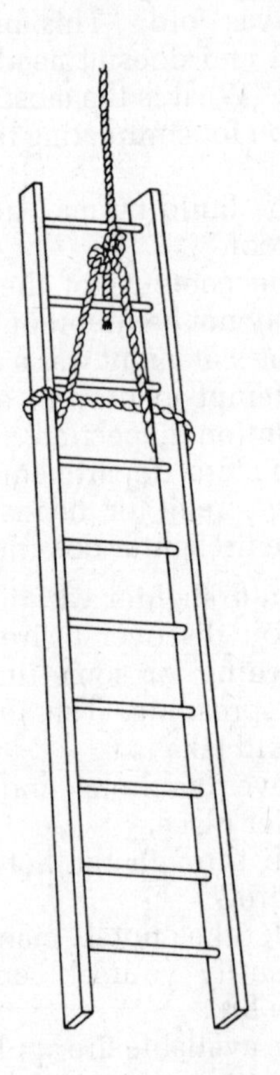

Illustration 57

58. The fire department's portable fan, shown in illustration 58, is most likely to be positioned on the
(A) floor above the fire, bringing air in
(B) fire floor, blowing smoke out
(C) floor below the fire, blowing smoke out
(D) fire floor, bringing air in

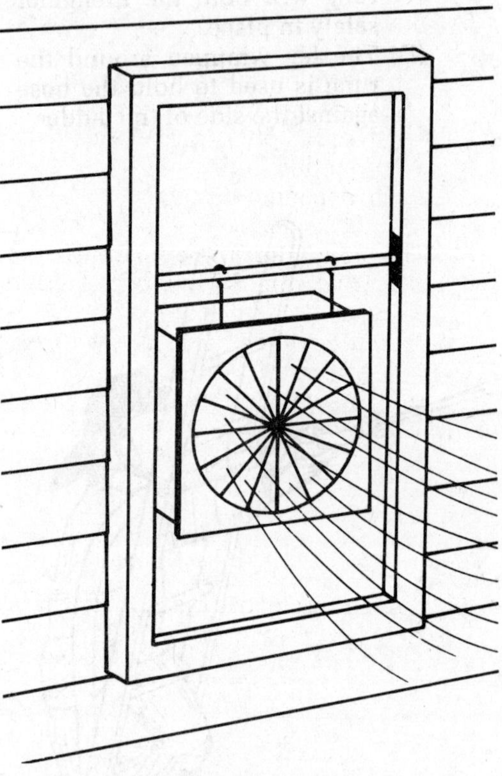

Illustration 58

59. While fighting a visible fire, you find that the water running away from the fire is cold. What conclusion should you draw from this discovery?
(A) The water main has broken.
(B) The fire streams are not reaching the fire.
(C) The fire streams are ineffective in fighting this type of fire.
(D) The water should be cold.

60. You are fighting a fire on the second floor of a four-story building. What area would be most exposed to the heat from this fire?
(A) The walls.
(B) The ceiling.
(C) The floor.
(D) All areas would be equally exposed.

SECTION FOUR

MATHEMATICS AND SCIENCE

Directions: Base your answers to questions 61 through 70 on information that may be given in the question, as well as on your knowledge of math and science. For each question select the one *best* answer—(A), (B), (C), or (D). Then indicate your choice by blackening the appropriate letter, next to the number of the question, on the answer sheet provided at the beginning of the chapter.

61. Flammable liquids are generally heavier than air and tend to
    (A) accumulate at the ceiling
    (B) spread out very slowly
    (C) seek out low points when spilled
    (D) decompose rapidly

62. The prevention of combustion in areas where volatile, flammable liquids are stored is accomplished by
    (A) keeping oxygen in the air away from the liquid
    (B) keeping nitrogen in the air away from the liquid
    (C) keeping hydrogen in the air away from the liquid
    (D) lowering the temperature of the liquid

63. Regarding electric fuses, it is correct to state that they
    (A) often blow and can be replaced by putting a penny behind them
    (B) often blow and should be replaced by a larger size
    (C) are a safety device and should be handled only by a licensed electrician
    (D) are a safety device and blow when there is an overload of electricity

64. Hydrogen produced by charging a storage battery has been found to diffuse rapidly in the air above the battery. What does this indicate?
    (A) There is a need for increased ventilation in areas where batteries are charged.
    (B) The battery is rapidly approaching a fully charged state.
    (C) There is a need to supply additional pure oxygen to the area.
    (D) The battery is rapidly deteriorating.

65. The best procedure for extinguishing a fire in an electric broiler is to
    (A) pull out the plug and throw a large bucket of water on the fire
    (B) pull out the plug and use an ABC-rated fire extinguisher
    (C) use any fire extinguisher
    (D) pull out the plug and use baking soda

66. Shocks from touching a high-voltage electrical appliance
    (A) are caused by carelessness
    (B) will not occur if rubber gloves are worn
    (C) may cause death
    (D) will be caused by a short circuit

67. In illustration 67, water flows down the outlet pipe of the water tower through the faucet. The pressure in the system is greatest at the point marked
    (A) H
    (B) F
    (C) D
    (D) J

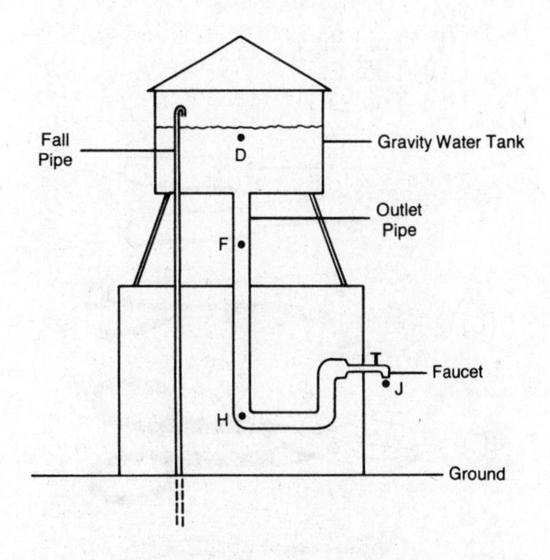

Illustration 67

68. If a length of fire hose does not exceed 50 feet and is no less than 40 feet, to reach a fire in a building that is 350 feet away will require a minimum of
    (A) 3 lengths of hose
    (B) 7 lengths of hose
    (C) 9 lengths of hose
    (D) 15 lengths of hose

69. If a fire apparatus has a tank with 1000 gallons of water and pumps water onto a fire at a rate of 150 gallons per minute, about how long will the water supply last?
    (A) 10 minutes.
    (B) 8½ minutes
    (C) 6½ minutes.
    (D) 5 minutes.

70. If it takes approximately ½ pound of pressure to pump water 1 foot, how much pressure will be required to pump water to the top floor of a 12-story hotel, if each story is 10 feet high?
    (A) 50 pounds.
    (B) 60 pounds.
    (C) 120 pounds
    (D) 240 pounds.

SECTION FIVE

TOOLS AND EQUIPMENT

Directions: Base your answers to questions 71 through 80 on information given and on your own knowledge of tools and equipment. For each question select the one *best* answer—(A), (B), (C), or (D). Then indicate your choice by blackening the appropriate letter, next to the number of the question, on the answer sheet provided at the beginning of the chapter.

71. The tool that you should use to test your automobile battery is a
    (A) barometer
    (B) hydrometer
    (C) thermometer
    (D) dynamometer

72. In illustration 72, what is the best tool to use to remove insulation from a wire?
    (A) Tool A.
    (B) Tool B.
    (C) Tool C.
    (D) Tool D.

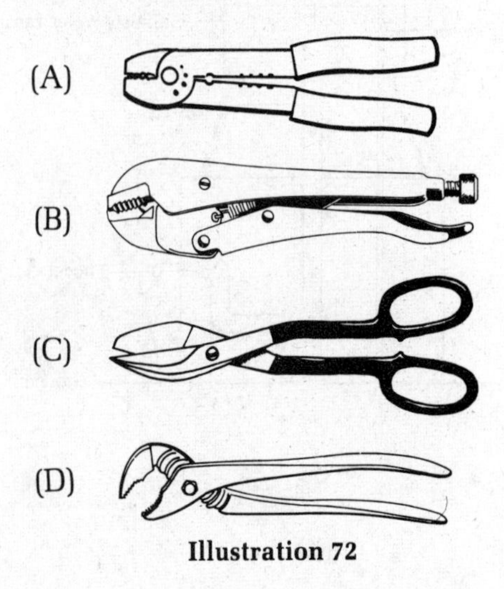

Illustration 72

73. In illustration 73, which tool should be used to cut the carriage bolts on the plate?
    (A) Wrench
    (B) Knife
    (C) Chisel
    (D) Pipe cutter

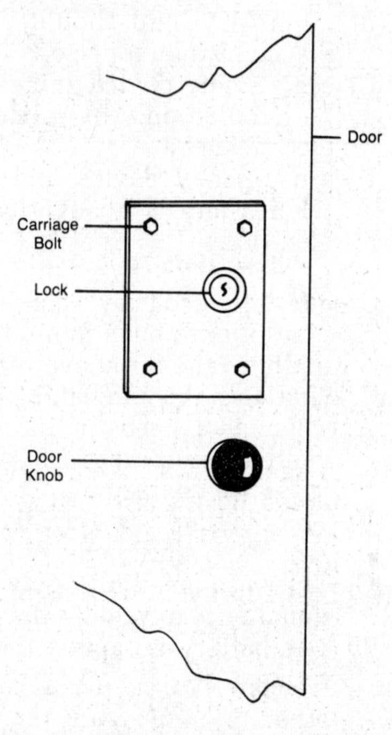

Illustration 73

74. The tool shown in illustration 74 would be used to
    (A) tighten a screw
    (B) fix a loose pipe
    (C) drive a nail
    (D) repair a car

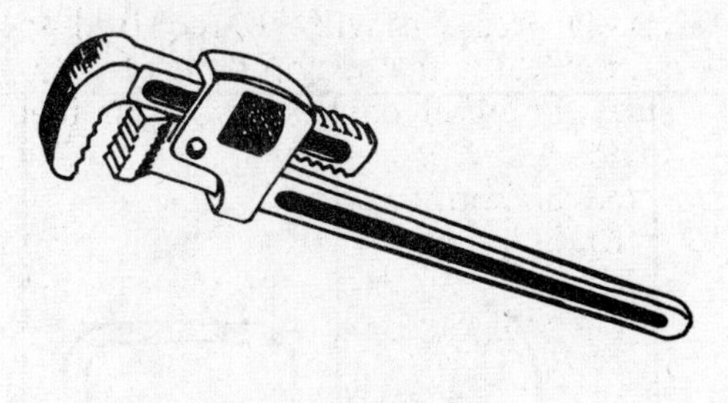

Illustration 74

75. The term "Phillips head" would be used most appropriately in reference to
    (A) a wrench
    (B) a screwdriver
    (C) a clamp
    (D) a saw

76. The tool in illustration 76 is best known as a
    (A) hacksaw
    (B) crosscut saw
    (C) backsaw
    (D) coping saw

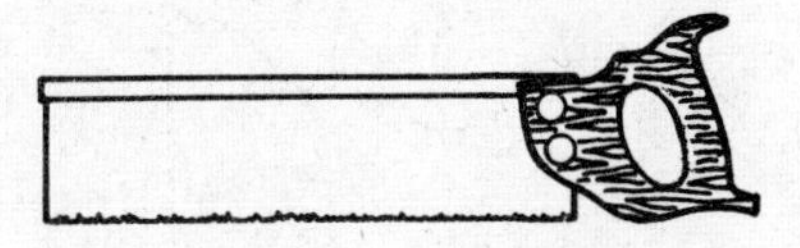

Illustration 76

77. Regarding illustration 77, it would be correct to state that:
    (A) A greater force would be obtained by the fire fighter if he moved his hand close to the lock.
    (B) Less effort would be required if the fire fighter moved his hand farther from the lock.
    (C) A greater force could be obtained by squeezing the vice grips tighter.
    (D) The amount of effort required to remove this lock would remain the same regardless of hand position or pressure.

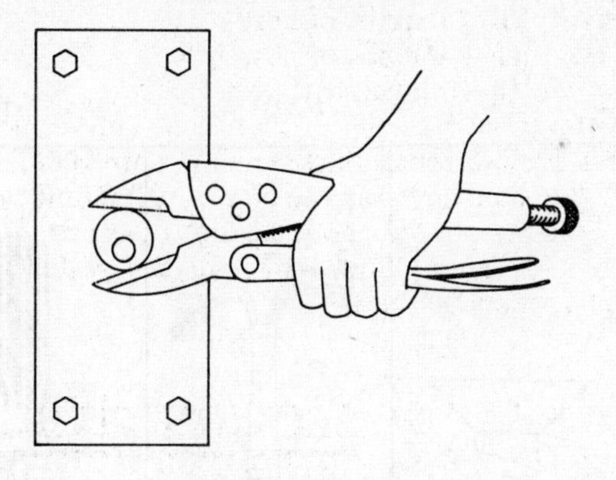

Illustration 77

78. The fire fighter in illustration 77 would remove the lock cylinder to
    (A) gain access to the door-locking mechanism
    (B) make sure the door cannot be relocked while he is operating at a fire
    (C) teach the owner a lesson about leaving the key with the building manager
    (D) make sure that the door cannot be opened

Answer questions 79 and 80 by indicating which item—(A), (B), (C), or (D)—on the right in the illustrations is most closely associated with the item on the left.

79.

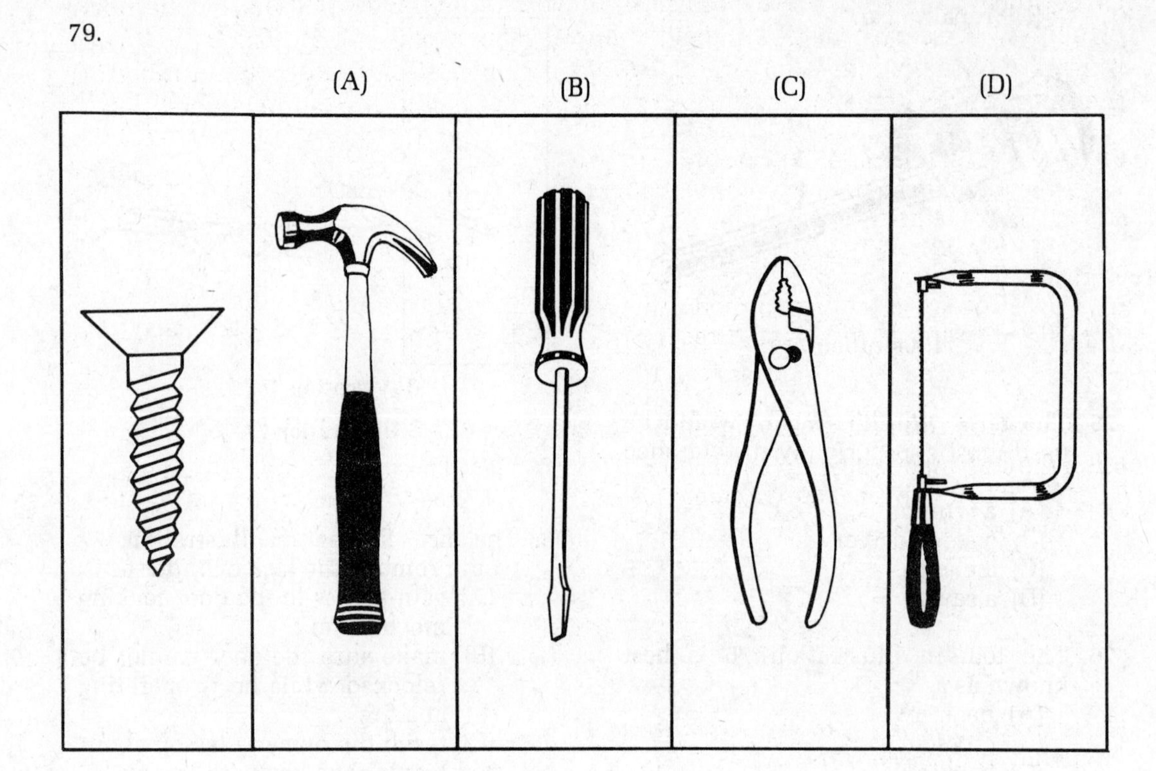

80.

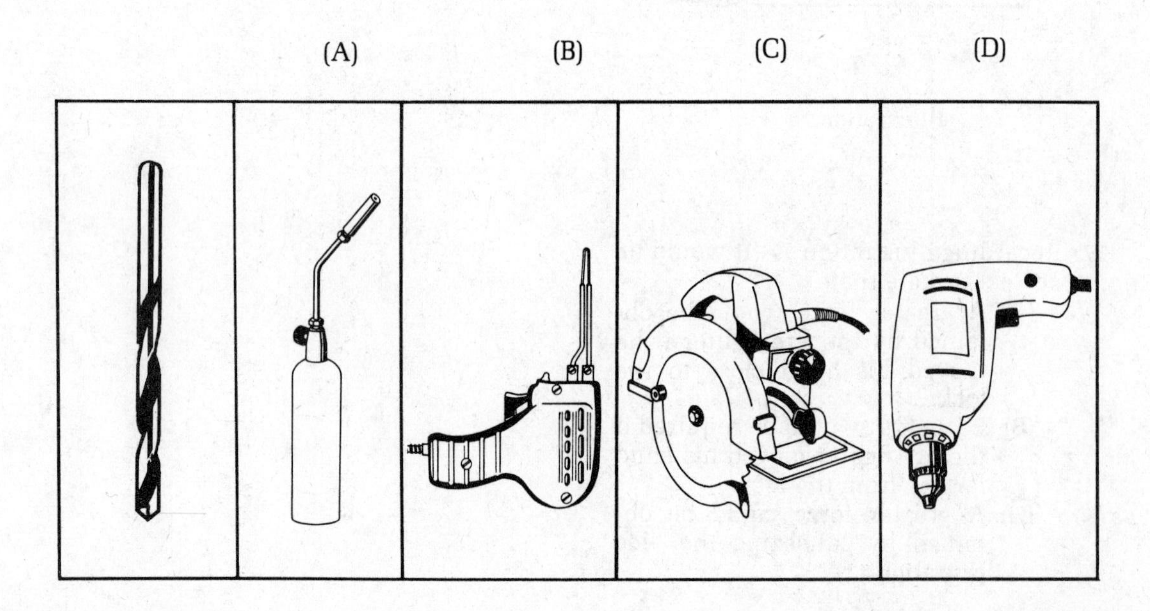

SECTION SIX

MECHANICAL DEVICES

Directions: Questions 81 through 90 test your knowledge of basic mechanical principles. For each question select the one *best* answer—(A), (B), (C), or (D). Then indicate your choice by blackening the appropriate letter, next to the number of the question, on the answer sheet provided at the beginning of the chapter.

81. If gear A in illustration 81 turns in the direction indicated, in what direction will gear C turn?
    (A) The same direction as gear A.
    (B) The same direction as gear B.
    (C) Toward gear C from direction X.
    (D) Toward gear A from direction X.

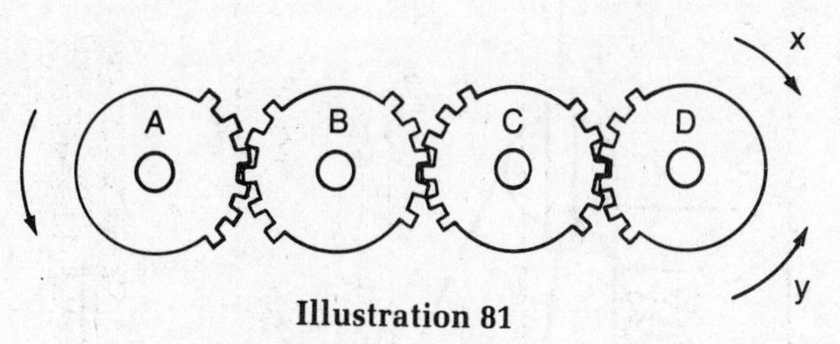

Illustration 81

Questions 82 and 83 are based on illustration 82–83.

82. If the amount of weight on each pulley system is equal, the pulley system that requires the greatest effort to lift the weight is
    (A) 1
    (B) 2
    (C) 3
    (D) 4

83. It would be correct to say about the pulley systems shown that:
    (A) They all have the same mechanical advantage.
    (B) The mechanical advantages of pulley systems 2 and 3 are equivalent.
    (C) The mechanical advantage is determined by counting the number of pulleys.
    (D) The mechanical advantage is determined by the length of the rope.

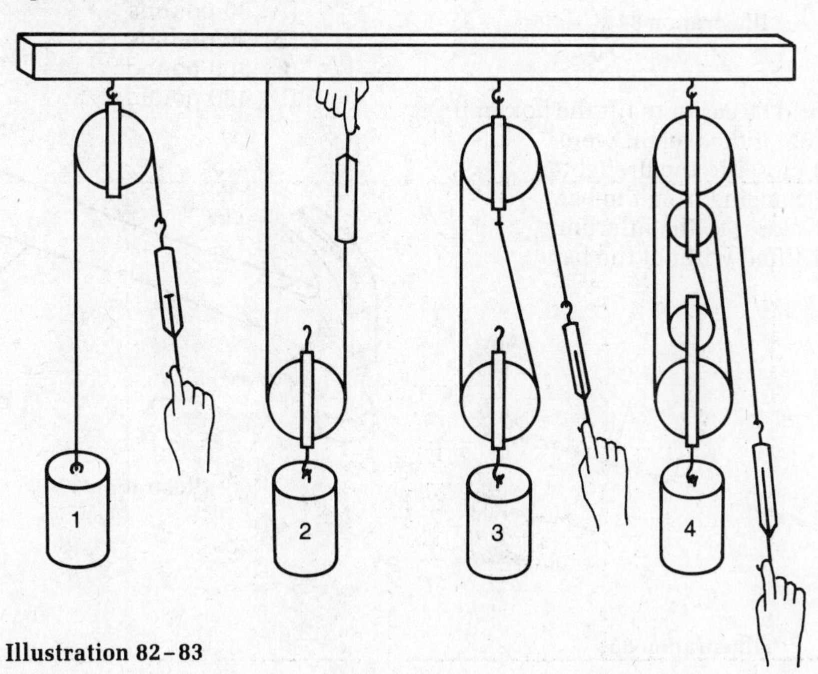

Illustration 82–83

84. In illustration 84, if a firefighter pulls 4 feet of rope down, how high will the weight rise?
 (A) 1 foot
 (B) 2 feet
 (C) 3 feet
 (D) 4 feet

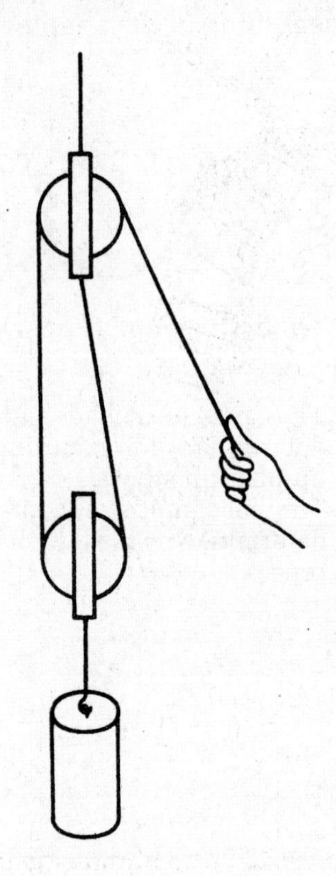

Illustration 84

85. It would be easier to lift the box in illustration 85, if the weight were
 (A) closer to the firefighter.
 (B) hanging from the bar.
 (C) closer to the fulcrum.
 (D) lifted without the bar.

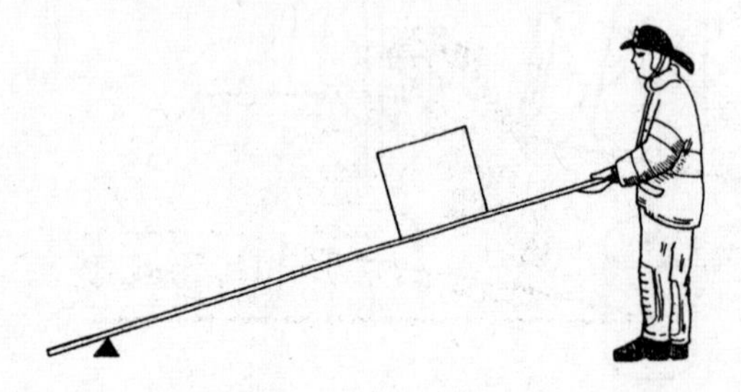

Illustration 85

86. Illustration 86 shows a simple lever system used as a safety device on many steam boilers. The pressure at which the steam escapes to the outer air is controlled by moving the weight. To reduce the pressure at which the valve works how should the weight be moved?
 (A) toward the valve
 (B) to the top of the lever
 (C) away from the valve
 (D) so that it swings

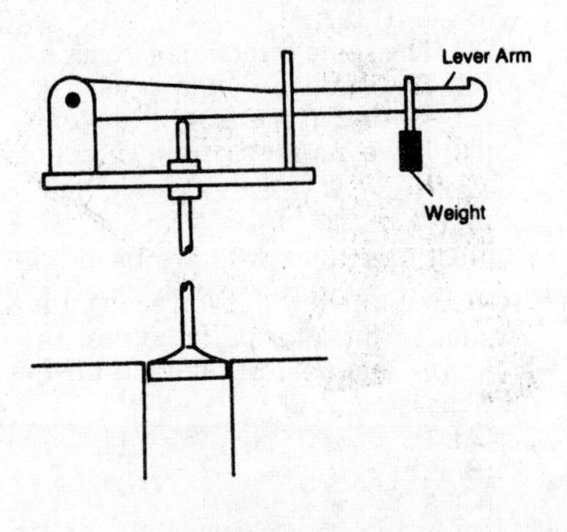

Illustration 86

87. If the distance from the claw to the fulcrum in illustration 87 is 3 inches, and the handle measures 12 inches, how much force will be needed to pull out a nail with a resistance equal to 120 pounds?
 (A) 30 pounds
 (B) 90 pounds
 (C) 360 pounds
 (D) 480 pounds

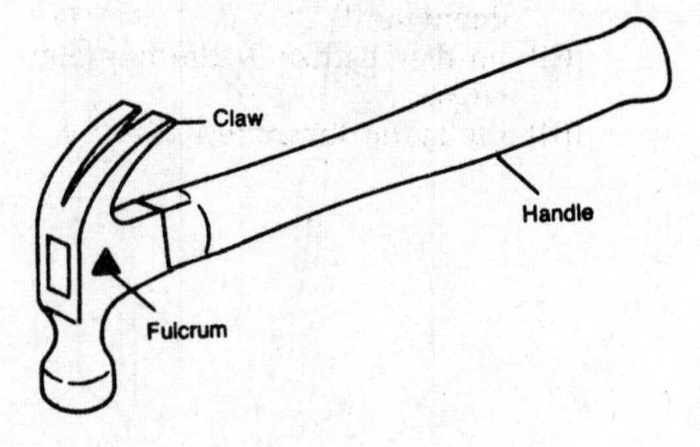

Illustration 87

Illustration 88

88. Illustration 88 shows two fire fighters, one under a ladder and one at the base of the ladder. As the fire fighter under the ladder approaches the ladder,
    (A) more force must be exerted
    (B) the tip of the ladder will bend downward
    (C) the fire fighter at the base can let go
    (D) the ladder becomes lighter

89. An axe gains its mechanical advantage by making use of the principle of the
    (A) lever
    (B) wedge
    (C) screw
    (D) axle

90. If the mechanical advantage gained in using the ramp in illustration 90 is 5, how much force must the fire fighter use to push the drum up the ramp?
    (A) 25 pounds.
    (B) 50 pounds.
    (C) 100 pounds.
    (D) 250 pounds.

Illustration 90

SECTION SEVEN

DEALING WITH PEOPLE

Directions: Questions 91 through 100 test your ability to deal with people. For each question select the one *best* answer—(A), (B), (C), or (D). Then indicate your choice by blackening the appropriate letter, next to the number of the question, on the answer sheet provided at the beginning of the chapter.

91. Assume that you are a firefighter out with friends. One of them asks you a technical question about fire protection—a question to which you do not know the answer. The most appropriate response for you to give would be one that
    (A) may not be accurate but sounds good.
    (B) admits that you don't know the answer.
    (C) points out that the subject is too technical and not important.
    (D) raises another question and thus changes the subject.

92. While conducting a routine fire prevention inspection of a department store, a firefighter is told by the store manager that the entire inspection procedure is a waste of time. What is the most appropriate action for the firefighter to take?
    (A) Tell the manager that he is wrong.
    (B) Explain the benefits of the inspection program.
    (C) Agree with the store manager and make only a cursory inspection.
    (D) Continue the inspection and ignore the manager.

93. Assume that you are a firefighter conducting a regular inspection of a factory, and you have uncovered several minor infractions of the state fire prevention laws. The building manager does not agree with you, tells you that his brother is on the city council, and also mentions the names of several high-ranking fire department officials who he claims are his friends. The best action for you to take at this time is to
    (A) disregard the manager's remarks and continue with your inspection.
    (B) go back the next day and make an even more thorough inspection.
    (C) call a police officer and have the man arrested.
    (D) forget about the infractions you saw, as long as they were only minor.

94. After a fire on the first floor of a three-story building has been extinguished, the woman who lives on the top floor complains that there was no need to force open her apartment and search it, and no reason to break the skylight over the stairs. What is the most appropriate response that you as a firefighter can make to this person?
    (A) Explain that smoke and hot poisonous gases rise and are stopped by the ceiling of the top floor.
    (B) State that firefighters use this method to attack the fire from above.
    (C) Admit that this is an example of excessive damage, and promise to speak to the chief about it.
    (D) Tell the occupant that she was lucky; the firefighters really should have cut the roof and pushed the ceiling down to let out the smoke.

95. While operating at a fire, a newly appointed firefighter receives an order from the lieutenant that appears to be in conflict with what the firefighter has been taught at the fire academy. In this instance, the best action for the firefighter to take is to
    (A) ask another firefighter what to do.
    (B) refuse to follow the order.
    (C) tell the lieutenant about the apparent conflict and ask the lieutenant what to do.
    (D) tell the chief about the apparent conflict and ask the chief what to do.

96. During a fire in a house where neither the owner nor any of the family were at home, a firefighter found $450 under a rug while he was looking for hidden fire. What should the firefighter do?
    (A) Notify the chief to call the police.
    (B) Turn the money over to the lieutenant and explain how and where it was found.
    (C) Put the money back where it was found and say nothing to anybody.
    (D) Give the money to a neighbor to hold until the owner returns.
97. After a difficult fire has been extinguished, while you are putting the hose back on the apparatus, a group of six youths begins to ask you a lot of questions. The best action for you to take under these conditions is to
    (A) answer all their questions in great detail.
    (B) answer one or two questions, then chase them away.
    (C) answer several questions and then tell them to visit you at the fire station.
    (D) pay no attention to them.
98. While you are on housewatch, a distraught woman enters quarters and tells you that her 6-year-old son is lost. The most appropriate action for you to take is to
    (A) turn out the company to conduct a search of the neighborhood.
    (B) send the woman to the social services department.
    (C) get some information about the woman and the child, and aid the woman in notifying the police department.
    (D) explain to the woman that the fire department does not search for children and send her to the police department.
99. While on duty, you are told by your superior that you must remain on duty beyond the end of the shift. You are supposed to go to your sister's wedding that night. What is the most appropriate action for you to take?
    (A) Tell the officer you refuse to work.
    (B) File a grievance.
    (C) Explain the particular circumstances of why you need the night off.
    (D) Just leave at the end of the tour.
100. After having worked with a group of firefighters for some time, you feel that you are disliked by the group. The *first* thing for you to do is to
    (A) request a transfer to another house.
    (B) tell the officer to have the group treat you better.
    (C) confront the group and tell them to treat you better.
    (D) examine your own thoughts and feelings to determine whether you are at fault.

END OF EXAMINATION

ANSWER KEY

1. **C**	21. **C**	41. **C**	61. **C**	81. **A**
2. **A**	22. **D**	42. **D**	62. **A**	82. **A**
3. **D**	23. **A**	43. **A**	63. **D**	83. **B**
4. **B**	24. **C**	44. **B**	64. **A**	84. **B**
5. **A**	25. **A**	45. **B**	65. **B**	85. **C**
6. **C**	26. **D**	46. **C**	66. **C**	86. **A**
7. **D**	27. **A**	47. **B**	67. **A**	87. **A**
8. **A**	28. **C**	48. **D**	68. **B**	88. **A**
9. **C**	29. **B**	49. **A**	69. **C**	89. **B**
10. **C**	30. **A**	50. **B**	70. **B**	90. **B**
11. **A**	31. **A**	51. **C**	71. **B**	91. **B**
12. **D**	32. **A**	52. **A**	72. **A**	92. **B**
13. **B**	33. **B**	53. **B**	73. **C**	93. **A**
14. **D**	34. **B**	54. **B**	74. **B**	94. **A**
15. **D**	35. **C**	55. **C**	75. **B**	95. **C**
16. **C**	36. **D**	56. **A**	76. **C**	96. **B**
17. **C**	37. **B**	57. **A**	77. **B**	97. **C**
18. **D**	38. **D**	58. **B**	78. **A**	98. **C**
19. **C**	39. **A**	59. **B**	79. **B**	99. **C**
20. **D**	40. **D**	60. **B**	80. **D**	100. **D**

DIAGNOSTIC PROCEDURE

Use the following diagnostic chart to determine how well you have done and to identify your areas of weakness.

Enter your number of correct answers for each section in the appropriate box in the column headed "Your Number Correct." The column immediately to the right will indicate how well you did on each section of the test.

Below the chart you will find directions as to the chapter(s) in the guide you should review to strengthen your weak area(s).

Section Number	Question Number	Area	Your Number Correct	Scale	
One	1–15	Recall		15 right	Excellent
				14 right	Good
				13 right	Fair
				Less than 13 right	Poor
Two	16–45	Reading Comprehension		28–30 right	Excellent
				27 right	Good
				26 right	Fair
				Less than 26 right	Poor
Three	46–60	Judgment and Decision Making		15 right	Excellent
				14 right	Good
				13 right	Fair
				Less than 13 right	Poor
Four	61–70	Mathematics and Science		10 right	Excellent
				9 right	Good
				8 right	Fair
				Less than 8 right	Poor
Five	71–79	Tools and Equipment		10 right	Excellent
				9 right	Good
				8 right	Fair
				Less than 8 right	Poor
Six	81–90	Mechanical Devices		10 right	Excellent
				9 right	Good
				8 right	Fair
				Less than 8 right	Poor
Seven	91–100	Dealing with People		10 right	Excellent
				9 right	Good
				8 right	Fair
				Less than 8 right	Poor

1. If you are weak in Section One, concentrate on Chapter 6.
2. If you are weak in Section Two, concentrate on Chapter 7.
3. If you are weak in Section Three, concentrate on Chapter 8.
4. If you are weak in Section Four, concentrate on Chapter 10.
5. If you are weak in Section Five, concentrate on Chapter 9.
6. If you are weak in Section Six, concentrate on Chapter 10.
7. If you are weak in Section Seven, concentrate on Chapter 11.

Note: Consider yourself weak in a section if you receive other than an "Excellent" rating.

ANSWER EXPLANATIONS

SECTION ONE—PART A

1. **C** Living room, kitchen, and bedroom 3.
2. **A** The kitchen.
3. **D** None of the above.
4. **B** Bathroom—it does not have a window.
5. **A** This protects the rest of the house; any of the other positions could push the fire into the hall.
6. **C** Bedrooms 1 and 2 and the bathroom. All other rooms contain exit doors or lead to exits without passing through a hallway.
7. **D** This would be the safest. Trying to get out of the door may expose the occupant to heat and smoke.
8. **A** The kitchen leads into the hall and the living room (not a choice). The only acceptable choice available to you is the hall.
9. **C** All rooms except the utility room can be reached from the hall.
10. **C** There are two doors and two windows in the living room.

SECTION ONE—PART B

11. **A** Note under *Status Buttons.*
12. **D** This question requires you to remember the "Key" and the instructions given in the last paragraph of the reading material.
13. **B** Note under *Unit Identification Buttons.*
14. **D** Note under *Function Buttons.*
15. **D** Note under *Function Buttons.*

SECTION TWO

16. **C** In item 2 under "Stepladders," a clear statement indicates that the top must never be used, and that a person should be at least two steps below the top to be safe. No mention is made of the height of the ladder.
17. **C** Under the heading "All Portable Ladders," item 4, an instruction tells us to use two hands to climb the ladder, and to raise or lower tools with a rope. Using a rope to raise and lower tools is a correct procedure, which does not result in injuries.
18. **D** In this case, item 2 of "Straight and Extension Portable Ladders" tells us to make sure the ladder is placed on a firm foundation. Choice A is therefore incorrect. Choice B may be a correct statement, but it is not mentioned or implied in the passage. It is therefore not an acceptable choice. Choice C is incorrect—wood and metal ladders can be safe, although metal ladders can be unsafe around electricity.
19. **C** This fact is found in item 6 of "Straight and Extension Ladders."
20. **D** Most ladder accidents occur because of overconfidence, ruling out choice A. Choice B is incorrect in view of the broken bones, cracked heads, and mashed noses mentioned in the passage. Choice C is not indicated in the passage and therefore is not acceptable.
21. **C** The last sentence in the first paragraph tells us that smoking in an overstuffed chair can have the same results as smoking in bed.
22. **D** While the passage tells us that the "ashtray should be large, deep, and designed to prevent cigarettes . . . from accidentally falling out," the examiner can, and often will,

give in the question only part of the statement.

23. **A** The question asks why smoking is deadly, and choice A is the only one that answers it. Choices B and C may be true, but they are not mentioned in the passage. Choice D may occur, but would not necessarily result in death.

24. **C** The second paragraph in the passage states that "paper, empty match books, and cellophane wrappers should not be put in the ashtray."

25. **A** This is explained in the last sentence of the second paragraph.

26. **D** "In command" and "in charge" mean the same thing. The passage clearly states that the first officer on the scene is in command.

27. **A** This is stated in the second sentence of the passage.

28. **C** The last sentence of the paragraph gives this information. Although you should be skeptical of the word *all*, the author clearly used it in the passage.

29. **B** This answer is given in the opening statement of the passage.

30. **A** This answer reflects the two main themes of the passage: hold drills and conduct surprise drills. Choice B—not early in the day, but early in the term. Choice C—the drill itself should surprise the teacher, not the frequency of the drills. Choice D—teachers comply with training drills; in many states this is law.

31. **A** The subject of prevention is stressed by the author in every paragraph. The opening paragraph indicates fire prevention as the number 1 priority, and the last passage ends with a strong statement on prevention.

32. **A** This is stated in the first paragraph.

33. **B** This is the opening statement of the second paragraph. If you chose choice D—you probably made this selection based upon your own knowledge about the practice of fire fighting, and not the information in the passage. You must avoid the temptation to do this.

34. **B** In the first paragraph, the author states, ". . . it is first necessary to understand the mission of the fire service." The other three choices may be helpful, but are not mentioned in the passage.

35. **C** This statement is found in the last paragraph. Choices A and B are not mentioned in the reading. If you chose D, you misread the question and answer choice. The stem refers to limiting the damage, not avoiding it.

36. **D** Locate the column titled "Ignition Temp.," then read down the column and find the lowest temperature—689. Now, using the straight edge of a piece of paper placed immediately below the lowest temperature, read across to the column marked "Material." This is your answer.

37. **B** This answer may be obtained in a manner similar to that used for answer 36.

38. **D** Gasoline emits a vapor at −36 degrees. This means that it will ignite under almost all normal conditions and climates.

39. **A** Trying to dress while riding on a speeding fire apparatus is unsafe.

40. **D** This answer is clearly indicated in the last sentence of the paragraph. Choice A is incorrect because the fire fighter may leave to check the front of quarters. Choice B is incorrect, as the lieutenant is notified, but he does not relieve. Choice C is incorrect because the first sentence indicates that only fire fighters stand housewatch.

41. **C** The answer is stated in item 2—only information that is clearly stated should be used, unless otherwise indicated.

42. **D** This is clearly spelled out in item 3. You must be careful with this type of question, which can be very confusing. To help reduce the confusion, first find the relevant section in the passage and number each occurrence. Then go through each choice, making absolutely sure that it agrees with what you have identified in the passage.

43. **A** The command "FRONT" is used to cancel a prior command of either "EYES RIGHT" or "EYES LEFT" (see item 4).

44. **B** This is explained in the last sentence of item 3.

45. **B** This question requires a somewhat closer reading of the passage. The beginning of item 4 indicates that at a point six paces beyond the reviewing party, the salute is dropped and one marches at attention. However, the last sentence describes the boundaries of the reviewing stand as being marked by guidons.

SECTION THREE

46. **C** Children commonly use these areas as places of safety to hide from something that frightens them. It is not uncommon, therefore, for them to retreat there during a fire.

47. **B** This is a factor to be considered after the fire has been determined to be of suspicious origin.

48. **D** No time is lost in trying the door knob. If it opens, entry will be faster and safer, and will result in less property damage. *Note:* Choice C is true, but does not relate directly to the stem of this question. The question concerns itself with the method of opening the door.

49. **A** This will establish the fire fighter as a qualified, knowledgeable leader, and will reduce the possibility of panic, as well as controlling the smoke in the car. The train can proceed to the station, which is the safest and best place to disembark.

50. **B** The illustration shows the safest and most appropriate method for carrying the axe. It allows the other hand to remain free to handle doors or obstructions, and maintains control of the cutting blade. If you chose D, you may have thought that it was easier to carry the axe this way because it was balanced. However, the statement is that the axe is lighter. No matter how the axe is carried, the weight will not change.

51. **C** This is known as a leg lock, and is the method used to take a firm position on a portable ladder so as to allow the hands to be free. Choice A is wrong because both hands will be required to operate the nozzle, as indicated in the illustration. Choice B is not practical and would be unsafe. Choice D is eliminated because the hose is on the side of the left leg (the weight of the water holds the hose in place).

52. **A** "Call the fire department first." This is an action that will solve the problem. Don't assume it is a false alarm. If it is, the necessary action can be taken afterward to discipline the youth.

53. **B** Water weighs 8.34 pounds per gallon, and even a hose of small diameter can hold many gallons of water. In addition, when a hose is full of water and under pressure, it becomes much less flexible and thus more difficult to move.

54. **B** This question tests the candidate's knowledge of fire hazards. The concept of a fireproof building is misleading. Even though the building will not contribute combustible materials to the fire, the contents in it will. Choice A is immaterial—whether the building is fireproof or not, it must be inspected.

55. **C** This results in prompt notification of the fire department, and allows the hotel fire brigade to make the necessary evacuations and take fire extinguishment action. Choice A is a form of running away from one's responsibilities. Choice B would cause panic, would delay the response of the fire department, and would be impractical in a large, multistory hotel. Choice D—this fire is beyond the firefighter's ability to extinguish alone, and the smoke and gases generated by it would also be a serious problem.

56. **A** The amount of insurance and the cost of the loss are private matters between the owner and the insurance company. Only if the fire is suspicious, does the role of insurance come into play.

57. **A** When raised, the ladder does not hang straight. It therefore will swing outward, and the bottom will ride alongside the wall. In this posi-

tion, it will not get caught or hung up while raising. If you choose B, the ladder tip could get caught on any projection on the building (window sill, brick, clapboard, etc.). Choices C and D are incorrect, as the knot should be centered, and the change of direction of the forces on the rope should be gradual to reduce undue stress.

58. **B** This tactic is used to rapidly improve the visibility and living conditions on the fire floor. Choice A would result in drawing in the smoke that would be coming out of the fire windows. Choice C would pull the smoke down from the fire floor into the floor below, an area that is normally clear. Choice D would not clear out the toxic smoke and gases, and could provide a fresh supply of oxygen, causing embers to flare up.

59. **B** After a fire stream has hit the fire, the water should be warm or hot, thereby indicating that it is absorbing the heat from the fire and extinguishing it. Choice A—a water main would break in the street, while run-off water would be tested at the building. Choice C—water is the universal extinguishing agent and is effective on most types of fires. However, when it is not the correct extinguishing agent, a violent reaction occurs.

60. **B** Heat rises until it reaches an obstruction, at which time it spreads out laterally. Choice D could be misleading if you give consideration only to the radiation component of fire transmission. However, conduction and convection must also be considered.

SECTION FOUR

61. **C** Most flammable liquids are in either a vapor or a liquid state and are heavier than air; therefore they stay close to the ground. Flammable liquids seek the lowest areas, just as water does when it is spilled.

62. **A** Without sufficient oxygen, a fire cannot occur. Oxygen is kept away from liquids by confining the liquid to a tightly sealed container. Choice B—nitrogen is not flammable and is used as an extinguishing agent. Choice C—hydrogen is a highly flammable gas and is not found abundantly in air, which is made up of one part of oxygen and four parts of nitrogen. Choice D—lowering the temperature may temporarily reduce the chance of vapor emission, but maintaining this low temperature is very difficult. Gasoline gives off a flammable vapor at −36 degrees F.

63. **D** The electric fuse is a short length of material that conducts electric current. It will melt when a certain temperature is reached at which the carrying capacity of the electric wire is exceeded. Choices A and B are dangerous techniques for bypassing the breaking of the fuse—dangerous because they allow the high current to flow through the wire. This causes the wire to heat to a point at which it will start a fire. Choice C—any competent adult should be able to change a fuse. An electrician usually is not needed.

64. **A** Hydrogen is a highly flammable gas, which diffuses (or mixes with) air and can cause an explosion. Adequate ventilation allows the gas to escape harmlessly to the outer air before sufficient accumulation of the gas is reached.

65. **B** Pulling out the plug removes the electrical hazard and makes the fire a class A or B fire. While the ABC-rated extinguisher is capable of handling electrical fires, pulling out the plug makes the condition safer. Choice A is inaccurate, because the fire may be in grease or fat that could splatter. Choice C would include a water-based extinguisher or a class D extinguisher, neither of which would be appropriate. Choice D might work; however, choice B is a better answer because it provides an all-purpose extinguisher, thereby reducing the chance of error.

66. **C** High-voltage shocks are extremely

dangerous and have often caused death. Choices A, B, and D should not be absolute statements. They would be correct if the words *may be*, *may*, and *may be*, respectively, were used.

67. **A** A major factor in determining water pressure is the height of the water above the point being measured. The greater the vertical distance, the greater is the pressure. When measured vertically, the distance from point D to point H is greatest. Although the distance from point D to point J is greatest overall, this is not a vertical measurement. To find the vertical distance of point J from point D, draw a line through point J parallel with the ground and measure the vertical distance from point D to the line.

68. **B** To find the answer, divide 350, the distance from the building, by 50, the longest length of hose. If you chose choice C, you probably divided the distance, 350, by the shorter length of hose, 40 feet. If you did, you came up with an answer of 8.75, which you then rounded off to 9. This, as you know now, was incorrect. This answer would be correct if the question had asked for the maximum length of hose.

69. **C** To find the answer, divide the amount of water in the tank, 1000 gallons, by the discharge rate, 150 gallons per minute. The answer is 6.66, or *about* 6½ minutes.

70. **B** To solve this problem requires several steps. First, find the height of the building in feet: twelve stories multiplied by the number of feet for each story (10 feet) = 120 feet. This, however, is not the answer to the question; it only reveals how high the building is. We must now multiply the height of the building by ½:

$$\frac{1}{2} \times \frac{120}{1} = \frac{120}{2} = 60$$

Note: We put 120 over 1 to create a fraction. This is correct, because dividing any number by 1 yields the same number.

SECTION FIVE

71. **B** This device measures the specific gravity of the battery acid on a density scale inscribed on the tube. A high reading indicates a well-charged battery. Choice A—a barometer is used to measure atmospheric pressure. Choice C—a thermometer is used to measure heat. Choice D—a dynamometer is an electrical device capable of acting as a motor or a generator.

72. **A** This electrician's all-purpose tool is used for cutting and stripping all types of electric wire. Choice B is vice grips. Choice C is duck-billed snips. Choice D is adjustable pliers.

73. **C** A chisel is used in conjuction with a hammer or flat-headed axe to shear off the bolts. This allows the plate to be removed, thus giving access to the lock cylinder.

74. **B** This tool is known as a pipe wrench (sometimes called a Stillson wrench). It should not be mistaken for a monkey wrench. Note in the illustration that there are serrated teeth at the head of the wrench, which are used to grip the pipe. A monkey wrench does not have serrated teeth.

75. **B** There are two basic types of screwdrivers: one that has a flat, straight-bottomed head and is known as a standard screwdriver, and one that has a head resembling a four-pointed star (a Phillips screwdriver).

76. **C** The backsaw is easily recognized by the rigid plate across the top of the saw. This saw is used to make fine cuts.

77. **B** This would take advantage of the principle of the lever.

78. **A** Removing the lock cylinder permits the fire fighter to rapidly gain access to the locking device and to gain entry with the least amount of damage to the door and lock. To accomplish the objective of choice B would require only a piece of tape

over the locking device, a chock to wedge the door open, or a small rope to tie it open. Removing the lock would result in excessive damage. Choice C should be avoided—firefighters do not try to get even with people, and tenants should not be expected to leave a key with the building manager. Choice D is incorrect; at a fire, the doors should remain openable to permit a search for life and fire.

79. **B** The illustration shows a wood screw. The proper tool to be used with this screw is the screwdriver.

80. **D** The illustration shows a high-speed twist drill bit. The proper tool to be used with this bit is a power drill.

SECTION SIX

81. **A** In order to answer this question, you must follow the direction of each turning gear. As each gear turns, it will turn the adjoining gear in the opposite direction. Gear A is turning in direction Y, gear B is turning in direction X, gear C is turning in direction Y, and gear D is turning in direction X.

82. **A** This simple pulley serves only to change the direction of motion. It does not give any mechanical advantage.

83. **B** To determine the mechanical advantage, count all the ropes except the rope that is being pulled downward. Pulley system 2 has two ropes; system 3 has three ropes, but one rope is being pulled downward and therefore is not counted. Choices A, C, and D are all incorrect.

84. **B** Divide the length of rope pulled (4) by the mechanical advantage (2, because there are two other ropes) to find the height that the weight will be raised.

85. **C** By moving the weight closer to the fulcrum, it is possible to increase the length of the effort arm and reduce the force needed by the firefighter.

86. **A** By moving the weight closer to the valve, we are moving it closer to the fulcrum. Therefore, it will be easier for the valve to lift.

87. **A** 3 = Distance of claw from fulcrum
12 = Distance of handle from fulcrum
x = Force
120 = Resistance
(Cross multiply)

$$\frac{3}{12} \times \frac{x}{120}$$

$$12x = 360$$

$$x = 30$$

88. **A** Keep in mind the principle of the lever. The firefighter at the base is at the fulcrum. The firefighter raising the ladder will require more force as the distance from the fulcrum is diminished.

89. **B** The wedge is a double-incline plane that, when inserted into or between two objects, exerts great force, spreading them apart.

90. **B** Divide the mechanical advantage into the weight of the drum. This will tell you how much force the firefighter must exert to roll the drum up the ramp.

SECTION SEVEN

91. **B** As a firefighter, you should have knowledge of fire protection. You are not expected to know everything. You should not pretend that you do, nor should you disregard or make light of a question. Instead, you should state that you will find the answer and get back to the questioner.

92. **B** Many people do not know the true purpose of fire prevention inspection. They often feel it results in a summons or an interference with business. By explaining to the manager that the goal of fire prevention is to reduce or eliminate the hazards that can put him out of business and lead to serious injuries to his customers and employees, a firefighter often gains cooperation.

93. **A** Although the manager does not agree with your findings, the manager has not made any illegal attempt to have you conceal them. Mentioning people may be an attempt to build creditability as a responsible person. Choices B and C would be overreactions to this situation. Choice D would be a deliberate failure on your part to perform your duties properly.

94. **A** Failure to take these actions could have resulted in loss of life on an upper floor. Choice B—firefighters attempt to fight fires from below or on the same level. Fighting a fire from above is self-punishing and difficult. Choices C and D are incorrect and will further excite an already upset individual. All comments of this nature should be avoided.

95. **C** By discussing the matter with the company officer, the firefighter will find out whether there is a conflict and will learn from the officer's experience.

96. **B** The lieutenant will take the necessary and proper actions to safeguard the owner's property. Choice A violates the chain of command. Firefighters should go through their immediate supervisors. Choice C is incorrect because the money may get lost or accidentally destroyed. Choice D is obviously inappropriate; items of value should be turned over only to a supervisor, who will then return them to the owner.

97. **C** Because this was a difficult fire, and because you have not completed the work needed to ready your unit for another alarm, you must set priorities. Getting back into service is the first priority. Answering questions is important, but it can and should be limited and then postponed until a more appropriate time.

98. **C** You must obtain information about who the woman is and where she lives, as well as information about the lost boy. This process will help to calm the woman, and will ensure that you have some information for the police if she leaves. Second, you should help the woman to contact the police; this is a serious matter and should be handled immediately. However, the unit would not necessarily go out and begin to search. This decision would be made by the officer in command after having gathered all the facts.

99. **C** By explaining why you need the night off, you allow the officer to make other arrangements with another firefighter. Choice A is incorrect. The fire service is a 24-hour service and sometimes requires members to work beyond their tours. Refusing to do so is not permitted, and is considered a disciplinary offense. Choices B and D are obviously inappropriate.

100. **D** Before you can make any changes, you must identify the reasons underlying the problem. In this case, you may or may not be the cause, but before you take any of the actions outlined in choice A, B, or C, you must know what the cause is.

PART THREE

Correcting Your Weaknesses

CHAPTER 5

Understanding Fire Service Terminology

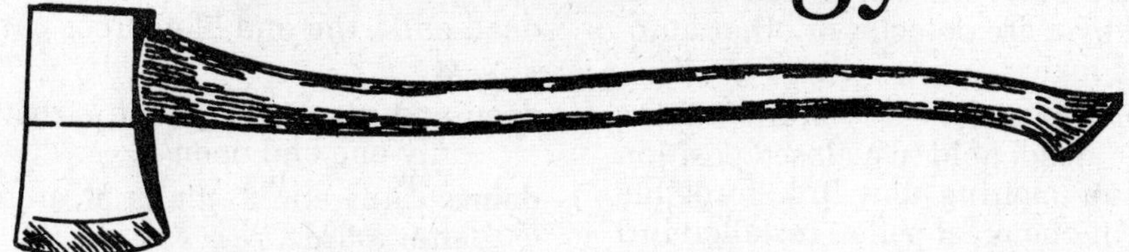

The Language of the Fire Service

The firefighter examination does not assume any prior fire fighting or fire science knowledge. It can, however, expect you to have a familiarity with the common fire safety principles and terminology.

The following terms will help you understand those words and phrases that might be found on a firefighter examination.

ability the capacity to perform a task

ablaze on fire; burning briskly

abrasion the act of wearing away by rubbing, as the rubbing of hose against the curb

abrasive an extremely hard material used for grinding, cutting, or polishing

academy the educational institution where firefighters receive their initial training and subsequent formal retraining

accelerant a substance used to start a fire and/or make it spread rapidly

adapter a device for connecting, joining, or fitting one object to another; *example:* the mechanism that changes a two-prong electrical outlet so that it will accept a three-prong plug

affidavit a written statement of facts that is supported by another; an affirmation

air pack a protective device that uses a face piece, compressed air tank, and an air flow regulator to provide clean air to the wearer

air regulating valve an adjustable valve used to regulate airflow to the facepiece; a helmet supplied with air through an air hose attached to a compressed air cylinder.

alarm an audible or visual signal that indicates the existence of a fire or emergency condition

alleged the word commonly used when the police are not certain of the facts, or when the matter is currently being decided in court; *example:* "an alleged crime," or "the alleged perpetrator"

alternating current electrical current which reverses direction with each cycle

ambient surrounding; especially pertaining to environment around a body, as in "ambient air" or "ambient temperature"

antifreeze extinguisher a fire extinguisher that contains a compound or solution that lowers the freezing point below that of water

apartment house a building or structure that contains three or more dwelling units

apparatus fire vehicle tank trucks, ladder trucks, pumpers, crash and rescue trucks, and others

arson willful burning of a building, vehicle, or structure, including one's own

arsonist a person who attempts or commits an act of arson

atmospheric pressure 14.7 lbs. per inch square; the pressure exerted by the weight of the air surrounding the earth

attic the area between the roof of a building and the ceiling of the top floor

automatic alarm an alarm that is activated by a fire detector or other automatic device

automatic sprinkler a water-spraying nozzle head, held in a closed position by a low-melting alloy link, frangible bulb, or chemical pellet, installed in a pipe grid; NOTE: when hot gases or flames weaken or destroy the holding element, the nozzle opens and sprays water or other extinguishing agents on the surrounding area

avoidable accident an accident that should have been prevented with proper equipment and behavior

axe a tool with a heavy blade-edged head designed to cut an object when forcibly struck

balcony a platform projecting from a wall

barge an unpowered, flat-bottom, shallow-draft vessel

base line a term used to identify an arbitrary line used for reference and control purposes

battalion a fire department district made up of several fire stations

battery a device consisting of cells for converting energy into electrical current

bonding interconnection of metal parts of a structure to form a continuous electrical circuit; used principally to prevent buildup of static charges and arcing

building grade the ground level of a building

business district an area of a city or town set apart for commercial and business purposes

chain of command the order of rank and authority in an organization

check valve a valve that permits flow in only one direction; also known as a clapper valve

civilian a person other than a uniformed fire fighter

combustible capable of burning, generally in air under normal conditions of ambient temperature and pressure, unless otherwise specified

constant a parameter that has a single numerical value

corrosion slow deterioration or destruction of metallic materials by chemical or electrochemical processes; contrasted with erosion, which is a mechanical process

damage material injury or loss resulting from fire or other disasters

dead end the end of a street without an exit

dead-end street a public street having only one end open

debris unwanted material; refuse left after a fire

decision making the process that occurs prior to a commitment being made to a course of action

decontamination removal of a polluting or harmful substance

deepseated fire a fire that has penetrated deep into bulk materials; a persistent fire that is difficult to extinguish

detector see fire detector

direct current electrical current that flows in one direction

drill practice of fire fighting techniques, including laying hose, raising ladders, and operating pumps

dry chemical extinguisher a fire extinguisher containing a dry chemical powder that is ejected by compressed gas and that extinguishes fire

dwelling any building used exclusively for residential purposes, with no more than two living units, or serving no more than 15 persons as a boarding or rooming house

dwelling unit one or more rooms used by one or more persons

egress a continuous path of travel from any point in a building to the outside at ground level; see exit; fire escape

electrical insulation a material possessing a high degree of resistance to the passage of an electric current

elevating platform a basket-like platform attached to mechanical or hydraulic power-operated boom mounted on an apparatus chassis, used to raise fire fighters and equipment to upper floors for fire fighting and rescue work

emergency a situation in which human life or property is in jeopardy and prompt action or assistance is essential

emergency exit any of several means of escaping from a building, including doors, windows, hatches, fire escapes, and the like, leading to the outside and not used under normal conditions; also known as secondary means of exit or egress

emergency procedure a plan for action or steps to be taken in case of an emergency

escape route the egress path that occupants must follow when evacuating a building in the event of fire

exit a way or means for leaving a building and reaching ground level outside; the portion of an evacuation route that is separated structurally from other parts of the building to provide a protected path of travel to the outside

exit marking a prominently displayed sign that identifies access to an exit, or one that identifies a door passage, or a stairway that is not an exit but could be mistaken for an exit

exposure the portion or whole of a structure that may be endangered by a fire because of many factors, including proximity to a hazard; the likelihood of property catching fire from neighboring fire hazards

extinguish to put out the flames; to quench a fire

eye bolt a closed-eye fitting with a threaded end, used to secure a wire, rope, or hook

facepiece the part of a breathing apparatus that fits over the face

false alarm an alarm for which there is no fire; NOTE: false alarms may be accidental, as in the case of a defective alarm system, or malicious, as in the case of a person calling the fire department to report a fictitious fire

felony a serious criminal offense punishable upon conviction by confinement in a state prison for periods of time in excess of one year; upon conviction of a felony, certain citizen's rights are forfeited

fire academy a comprehensive fire department training facility

fire alarm an audible or visible fire emergency signal

fire control fire protection suppression, including efforts to reduce fire losses

fire detector a mechanical or electrical device that senses the presence of a fire from one or more of its signatures (smoke, heat, gas, etc.) and generates a signal or trips another device

fire drill a practice exercise by a fire service unit in fire fighting procedures and the use of fire service equipment; the practice of evacuation of a building in the manner to be followed in case of a fire or other emergency

fire escape a means for leaving a building in case of a fire, such as an exterior or interior fire-protected stairway; an emergency fire exit

fire fighter an active participating member of a fire department, including volunteers and part-time firemen

fire hazard the relative danger of the start and spread of fire; the danger of smoke or gases being generated; the danger of explosion or other occurrence potentially endangering the lives and safety of the occupants of a building or structure

fire prevention precautionary fire protection activities carried out before fires occur, intended to prevent outbreak of fire, facilitate early detection, and minimize life and property loss if fire should occur; activities include public education, inspection, law enforcement, and reduction of hazards

fireproof the property of a material, such as concrete or iron, that does not burn or decompose when exposed to ordinary fire; NOTE: the term "fire resistive" is recommended by the NFPA in preference to fireproof because no material can withstand heat of sufficient intensity and duration

fire resistant having the capability of withstanding fire for a specified period of time, (usually in terms of hours) with respect to temperature; possessing the quality of resistance to damage or destruction by fire

fire scene the location of a fire, which must be carefully preserved until a search for evidence is completed

fire science the systematic body of knowledge and principles drawn from physics, chemistry, mathematics, engineering, administration, management, and related branches of arts and sciences, required in the practice of fire protection and prevention

fire service an organization that provides fire prevention and fire protection to a community; the members of such an organization; the fire fighting profession as a whole

fire station a building housing firefighters and fire equipment, serving as headquarters for a fire department company

fire triangle the three factors necessary for combustion: fuel, oxygen, and heat; NOTE: a fourth factor has been proposed to account for the chemical chain reaction in combustion processes

fire watch one or more firefighters detailed to be present at a public gathering as a fire safety precaution

first alarm the initial alarm of a fire, or the first signal calling for the first alarm response

first arrival the first unit or officer to report at the scene of a fire

first line referring to the apparatus and fire fighting units normally responding to fire alarms, as contrasted to reserve units

flammable subject to easy ignition and rapid flaming combustion

flammable gas any gas that can burn

flash point the minimum temperature at which a liquid vaporizes in order to form an ignitable mixture with air (flash points are determined in the laboratory by cup tests); NOTE: In the fire practice, flammable liquids are sometimes classified as high and low flash point liquids, depending on whether they flash above or below 100 degrees F. (37.8 degrees C.)

forcible entry tool any of a large assortment of axes, door breakers, pry bars, mauls, chisels, saws, and the like, used to gain entrance through walls and other obstructions to carry out fire fighting operations

forensic science the scientific activity related to the development of evidence which is suitable for introduction as evidence in a judicial proceeding

friction the resistance to relative motion between two bodies or surfaces in contact

general order a standing order that remains in effect until specifically amended or canceled; contrasted with a special order that is issued for a limited purpose or temporary situation

generalist a firefighter who performs all facets of general fire fighting work; contrasted with specialist

gravity the force of mutual attraction between masses; the weight of a body, with g representing the force of gravitational acceleration

gravity tank an elevated water storage tank provided for fire protection or used as a community water supply

gridirion the process of an aerial search for a small fire by systematically passing over an area along parallel grid lines

grievance a complaint related to conditions of work, such as unfair or unequal treatment or a situation affecting safety or health

ground a conductor that provides an electrical path for the flow of current into the earth or to a conductor that serves as the earth

handline a small hose line handled manually by a firefighter; a light rope used to hoist fire tools and to secure ladders and equipment

heat exhaustion physical distress characterized by fatigue, nausea, heavy perspiration, pale clammy skin, subnormal temperature, vacant eyes, and general weakness brought on by overexpose to high temperature and deficiency of salt

helmet protective headgear for a firefighter

hose a flexible conduit used to convey water under pressure; available in various diameters from ¾-inch to 5-inch or more and in standard lengths of 50 feet

hose line two or more lengths of hose connected and ready for use or in use

house watch a tour of duty at the fire house watch desk

hydrant a cast metal fitting attached to a water main below street or grade level; normally with a control valve and one or more gated or ungated threaded outlet connections for supplying water to fire department hoses and pumpers

ignition the initiation of combustion

ignition temperature the lowest temperature at which sustained combustion of a substance can be initiated

incendiary a fire that a trained arson investigator believes has been deliberately set; NOTE: see *suspicious*

incinerator a furnace-like apparatus used to burn waste substances

incipient referring to a small fire or a fire in its initial stages

incombustible an improper form of noncombustible; *see* noncombustible

inert deficient in active properties

inertia the tendency of a body at rest to remain at rest, or of a body in motion to resist a change in speed or direction

inflammable an improper form of flammable; *see* flammable

initial alarm the first notification received by a fire department that a fire or other emergency exists

injury physical harm or damage to a person resulting in disfigurement, pain or discomfort, infection, or physical impairment

in service a company or other fire fighting unit that is ready and able to respond to assignments

in-service training the continual training given to fire fighters to keep up their skill level and to keep them abreast of new laws, procedures, etc.

jackknife a folding pocket knife; a folding action of something upon itself

joist a structural member, often spanning beams or walls, used to support a floor or ceiling

kink to bend hose upon itself to temporarily block the flow of water

ladder company a fire company equipped with a ladder truck and trained in ladder work, ventilation, rescue, forcible entry, and salvage work

lanyard a rope suitable for supporting one person; one end is fastened to a safety belt or harness and the other end is secured to a substantial object or a safety line

latent heat the heat absorbed or liberated when a substance changes from one state to another at a fixed temperature

life gun a gun that fires a cord attached to a projectile to persons trapped in inaccessible places; the cord is used to pull in a lifeline for rescue

life safety the preservation and protection of life from the hazards of flame, heat, smoke, etc.

loss prevention a program designed to identify and correct potential accidents before they occur

mnemonic a cue for remembering, usually an association made with something familiar

molecule the smallest physical unit of a substance, consisting of atoms held together by chemical forces and possessing the properties of the substance

moonlighting work performed by fire fighters on their off-duty time for employers other than the fire department

multiple alarm a second or greater alarm

night shift a nighttime tour of duty

noncombustible resistant to burning under normal conditions

nozzle a tubular, metallic, constricting attachment fitted to a hose to increase fluid velocity and form a jet; nozzles are often adjustable to provide a solid stream, a spray, or fog

nozzleman a fire fighter assigned to handle the nozzle of a fire hose

occupancy the purpose for which a building, floor, or other part of a building is used or intended for use

officer a member of a fire department with supervisory responsibilities

officer in charge the highest ranking officer at a fire or commanding an on-duty fire department unit

off shift off duty

open circuit a discontinuity in an electrical circuit in which the path of the flow of electrical current between two points is broken

ordinary combustibles commodities, packaging, or storage aids that have a heat of combustion similar to wood, cloth, or paper and which produce fires that may be extinguished by the quenching and cooling effect of water

organization the association of people with specified functions and responsibilities for pursuing an agreed purpose

overhauling the terminal phase of fire fighting, in which the fire area is carefully examined for embers or for traces of fire; in addition, steps are taken to protect remaining property from further damage

owner the person who holds title to particular land or property

parapet a wall extending above the roof line that provides a fire barrier

pawl a mechanism or catch that prevents a wheel from turning in a backward direction by engaging a ratchet tooth or by utilizing a cam friction or a smooth wheel

pedestrian a person traveling on foot

permit an official document issued by a fire department or other agency authorizing the use of fire or a hazardous process

personal property a person's possessions that are movable, such as household goods, jewelry, pets, furniture, and other similar objects

pneumatic operated by compressed air

post indicator valve a valve used in a sprinkler system that indicates open and shut positions

prevention reduction or elimination of hazards, usually as part of a planned program to lessen the probability of harmful or dangerous events

protective clothing fire fighting clothing, including coats, boots, turnout pants, gloves, and helmet

pumper a fire department pumping engine equipped with 500-gallon-per-minute pump, hose, ladders, and other fire fighting equipment

pump operator a firefighter trained and assigned to operate a pumper

pump panel the console of instruments, gauges, and controls used by a pump operator

pump pressure the water pressure produced by a fire department pumper

resistance the opposition to the flow of electrical current

response the act of answering an alarm

roll call the taking of attendance before each tour of duty

roll call training daily training of firefighters prior to the start of their tour, given simultaneously with roll call

rules and regulations an official fire department book of rules specifying the responsibility of officers and subordinates and the proper conduct of fire department operations

safety can a container of not more than five gallons capacity having a flash-arresting screen, a spring-closing lid, and spout cover, so designed that it will safely relieve internal pressure when exposed to fire

safety shut off a device that shuts off the gas supply to a burner (and sometimes also to the pilot light) if the ignition source fails

salvage procedures or measures for reducing damage from smoke, water, and weather during and following a fire by the use of salvage covers, smoke ejectors, deodorants, etc.

salvage cover a waterproof tarpaulin of standard size made of cotton duck, plastic, or other material, used to protect furniture, goods, and other property from heat, smoke, and water damage during fire fighting operations

SCBA Self Contained Breathing Apparatus; sometimes called an *air pack*

search a thorough examination and exploration of a fire scene for life and fire

search warrant a legal writ that authorizes a law enforcement officer to search specified premises for evidence

siamese a hose fitting that joins two hose lines

smoke detector a device, usually of the ionization or photoelectric type, that triggers an alarm when a sufficient concentration of smoke upsets a balanced electronic circuit or interrupts or otherwise affects a light beam

smoke ejector a power-driven fan used to remove smoke from a burning building or to blow fresh air into a building to expel smoke and heat

specialist a firefighter whose efforts are concentrated in a specific field of fire fighting, such as rescue, emergency medical services, or marine fire fighting; contrasted with generalist

spontaneous combustion combustion resulting from spontaneous heating and spontaneous ignition

sprain an injury in which the ligaments are torn, accompanied by pain, discoloration, and swelling

spray nozzle a nozzle that is designed to break up a water stream into fine droplets

sprinkler system a grid system of water pipes and spaced discharge heads installed in a building to control and extinguish fires

strain an injury to a muscle caused by overexertion

streetbox a fire alarm signal box mounted on a post on a public street

subpoena an order signed by a judicial official commanding a person to appear in court under penalty of law

summons a written notice served to a person, legally obligating that person to answer a charge; a summons is in lieu of an arrest

suspicious a fire that is suspected of being arson but has not been declared as arson by a trained arson investigator; NOTE: see incendiary

thermal pertaining to heat

tour fire fighters' jargon used to refer to one work day

toxic poisonous; destructive to body tissues and organs or interfering with body functions

turn out the alert to a fire company to answer a fire alarm

under control a stage reached in fire fighting in which the fire has been contained and extinguished to the extent that fire authorities are confident of its complete extinguishment

uniform fire reports statistical summaries indicating the types and amount of fires occurring in the nation

unsafe act something that is not consistent with normal or correct practice; can lead to death and injury, as well as to waste and destruction

utility gas gas, manufactured gas, liquified petroleum gas, or a mixture of gases

variable a quantity that can assume any number of values

ventilation the act of opening windows and doors or of making holes in the roof of a burning building to allow smoke and heat to escape

visibility the relative clearness with which objects visually stand out from their surroundings under normal conditions

volatility the readiness with which a substance vaporizes

volt a unit of electrical potential; a potential difference of one volt causes a current of one ampere to flow through a resistance of one ohm

warehouse any building, structure, or area within a building used principally for storage purposes

wash down to flush away spilled combustible materials

watch the tour of duty at the watch desk, including the responsibility for keeping the company journal, receiving visitors, receiving alarms, and turning out the company when required

watch desk the area where the housewatch is assigned; normally near the entrance to quarters

watch line a hose line left at the scene with a detail of fire fighters to guard against rekindling of the fire

water curtain a screen of water thrown between a fire and an exposed surface to prevent the surface from igniting

water damage the damage done to goods and materials by the water from fire streams

water hammer a pressure wave traveling along a column of liquid when flow is suddenly halted or the velocity of the fluid is changed abruptly

water motor gong a gong driven by a water motor in a water motor alarm to indicate that water is flowing in a sprinkler system

welding a process of joining two metallic pieces together by heat

winch a powered hoisting machine

working fire a fire that requires fire fighting efforts from fire personnel assigned to the alarm

yard a distance of 3 feet; an open space usually in the rear of the building

yard hydrant an industrial type of hydrant with independent gates for each outlet

CHAPTER 6

Handling Recall, Visualization, and Spatial Orientation Questions

This chapter will help you develop your ability to handle questions involving recall. There are two basic types of recall questions: those that test your ability to recall what you have seen, and those that test your ability to recall what you have read. For questions of this type, each item of pictorial or written material is presented in a separate booklet (we will call this a "memory booklet") which you are permitted to study for a specified time. It is then collected, and you must answer the questions.

Since the purpose of these questions is to test your ability to concentrate and recall, the only notes you can make are mental notes. For this reason it is very important that you become familiar with this type of questions and develop a procedure for handling them.

Strategies for Recalling Pictorial Details

Let's look first at how you can handle visual materials. In this question type you will be given a sketch, picture, or other illustrative material to study for a period of time ranging from 5 to 20 minutes. Five minutes may not sound like much time, but it will be more than enough if you apply the strategies outlined below. You are *not* permitted to write anything down at this point; all details and observations must be kept in your mind. When the time period is up, the illustrative material will be collected and you will not be able to refer to it again.

The sketch, picture, or diagram you are given will not be extremely complicated, and your objective will be to form a clear image of the material in your mind. To do this requires a strategy, concentration, and some practice.

1. DEVELOP A STANDARDIZED METHOD OF STUDYING THE VISUAL MATERIAL. If you want to remember the details of pictures, sketches, diagrams, you must look at them in an organized fashion. It is a mistake to think that just staring at the material will record all the details in your mind. You must be methodical, and you must practice.

In the fire service this is called "size-up"; the term refers to mental evaluation of the fire situation while giving consideration to many different factors. To make this process routine, most new fire fighters are taught some mnemonic (memory-improving) devices to aid them in handling these many factors. One effective mnemonic is "COAL WAS WEALTH," which stands for the most important 13 points of the size-up

While you will not be sizing up fires just yet, this mnemonic device can be easily learned and will prove very helpful. Before we look at the 13 points, however, let's analyze the mechanics of studying pictorial material.

First, look at the sketch, picture, or diagram as a whole and try to get a feeling for what is shown and what is going on.

Now begin to scan for details. When the U.S. Army, Navy, or Coast Guard conducts an air search, it uses a number of specific search patterns. One of these, the "parallel track pattern," provides an excellent method of identifying all the details. This pattern is used when

(1) the search area is large and relatively level
(2) uniform coverage is desired
(3) information about the area is limited
(4) the area is rectangular or square

These criteria meet your situation: the visual material you will be observing will be level, you will want uniform coverage, you will not have any previous information about the material, and its shape will be rectangular or square.

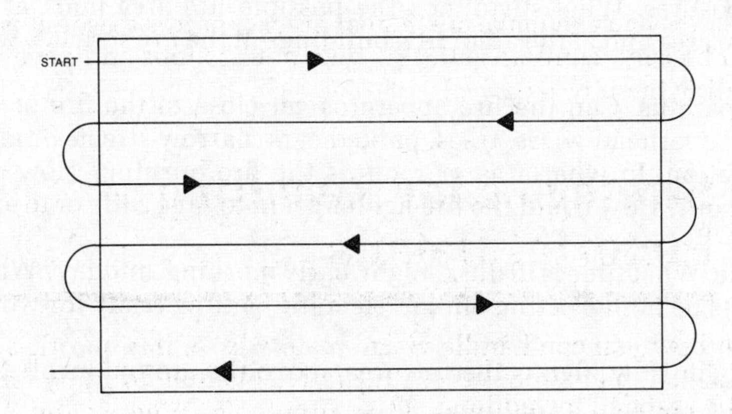

Parallel Track Pattern

There are three steps to applying the parallel track pattern:

a. Starting in an upper corner (left or right, depending on which way is most comfortable for you), identify, in your mind, each item as you scan across the picture.
b. Now repeat the process, but this time:

(1) Count repeated items such as windows, doors, and safety devices, and keep a mental tally for each such item.

(2) Observe carefully all readable material—street signs, addresses, room identifications, and exit signs.

c. Repeat the first step; that is, look with concentration at the material as a whole.

2. APPLY THE MEMORY AID: "COAL WAS WEALTH." As explained earlier, the mnemonic "COAL WAS WEALTH" refers to 13 specific factors critical to most fire operations and emergencies. They have been adapted here for use with recall questions.

The first letter of each of the 13 points is represented by a letter of "coal was wealth," as follows:

C = Construction. What is the building made of: wood, brick, concrete? What are the building features: roof, windows, doors, fire escapes, etc.?

O = Occupancy. What is the building or area used for: residence, factory, sales area, hospital, etc.?

A = Area. How large is the building or area: number of stories high, number of windows across?

L = Life. Is a life in danger? If so, where? Is escape possible? If so, how? If no life hazard is present, what are the most dangerous locations?

W = Weather. What are the weather conditions, and what effect might they have? Is it raining, snowing, cold, or warm? Would the windows be open or closed? Would the heat be on?

A = Auxiliary Appliances. Is there a sprinkler system or a standpipe hose system? Are there portable fire extinguishers?

S = Size. How many rooms are there? How do they differ? How many apartments are there?

W = Water. Is a water supply available to extinguish the fire: a fire hydrant, a water tank on the roof, or a swimming pool to draft from?

E = Exposures. What surrounds the possible fire area (look at the floor above the fire area and at the adjoining building)? If the fire spreads, where will it go?

A = Apparatus. Can the fire apparatus get close to the fire or are there obstructions: overhead wires, trees, parked cars, narrow streets or alleys?

L = Location. In what area or room is the fire burning? How could the occupant get out? How could the fire fighter get in to save a life or to extinguish the fire?

T = Time. What time is it: dusk, night, early morning, midday? What would the occupants be doing: eating dinner, sleeping, getting ready for work, working?

H = Height. How high is the fire area above the ground level? Could the occupant(s) be reached by ladders?

3. STUDY LISTS METHODICALLY. When you are given an illustration that contains a series of numbered or labeled items, you must take time to go through each item and identify what it relates to. There are two techniques for doing this. The first is to go through the numbers or items in sequential order. This method ensures that no item is missed but sometimes leads to jumping around the illustration in haphazard fashion. The second method is to identify items while following a geometrical pattern. A common pattern is to start at the outer border and go around in a spiral until you reach the center and have identified all the items.

If the number of items is limited, try to develop a simple key word or phrase, that is, a mnemonic device, to help you remember them.

Strategies for Recalling Written Material

Questions based on written material are not as difficult as you may think, since the material is never complicated or hard to understand. Performance on this test area, perhaps more than on any other, can be improved significantly by practice. If you follow the guidelines listed below, you should be able to do very well on this part of your examination.

1. DON'T JUST READ THE MATERIAL; BECOME PART OF IT! When you are reading, you must clear your mind of everything except the material at hand. You *must concentrate*. The kind of intense concentration that is needed is best achieved by "putting yourself into the story." Create a mental picture of what you are reading. For example, in reading about stepladder safety on the Diagnostic Test, you should have pictured yourself setting the latter on a firm foundation, opening it until the spreader locked, and then ascending no further than the second step from the top.
2. RELATE THE UNKNOWN TO THE KNOWN. You will find it easier to "put yourself into the story" if you create mental images involving persons, places, and things that you are familiar with.
3. DON'T TRY TO MEMORIZE THE ENTIRE PASSAGE. Some persons attempt to memorize the material verbatim. For most of us this is an impossible task. The trick is to identify the key facts and to remember them. All the examiner is interested in is how will you recall vital information.
4. DON'T STOP CONCENTRATING WHEN THE BOOKLETS ARE COLLECTED. The time between the closing of the memory booklet and the answering of the questions is the most critical. Be sure that you maintain your concentration during this time. Many inexperienced test takers forget what they read during these few minutes.
5. AS SOON AS YOU ARE PERMITTED, WRITE DOWN EVERYTHING YOU RECALL FROM THE WRITTEN MATERIAL. After the memory booklet is collected, there will be a delay before you are allowed to open the actual test booklet and begin taking the examination. This is the time during which you must continue to concentrate. However, once the signal is given to begin the examination, don't read the first question immediately. Instead, quickly write in a blank space in your test booklet all the details you can remember or simply make a quick sketch. Only after you have done this, should you start answering the memory questions, which usually come early in the examination.
6. USE ASSOCIATIONS TO HELP YOU REMEMBER. Rote memory will not suffice in most cases; you must make associations to help you remember. The

technique is to associate or relate what you are trying to remember *to something you already know* or that you find easy to remember. The type of association that is made will vary tremendously from individual to individual, depending on background, interests, and imagination. Some experts say "far out" associations are best; others recommend associations with familiar things. You must find what works best for you.

Association is an area where practice will help you greatly. For this purpose your daily newspaper will do just fine. Study a one-column news story for about 10 minutes, relate it to something you know, and then put the paper down. Now list the details you can remember.

7. CONSTANTLY ASK YOURSELF QUESTIONS ABOUT THE MATERIAL AS YOU READ. This will help you remember details, especially if you try to anticipate the questions you may be asked. Note that you can be fairly certain of being asked to compare items and to identify items in groups. If an illustration is included, make sure to identify all items in the illustration and to observe all references to them in the passage.

Developing Observation Skills

There is only one known way to improve your skills in observation: practice, practice, and more practice!

During the course of our lives we are continuously bombarded with different sights, sounds, and smells. Some of these become fixed in our minds, while others are disregarded. The stimuli associated with pleasurable experiences are easy to recall. Unconsciously we have committed them to memory; if necessary, we could recall these stimuli in great detail.

For some people it is very easy to remember large quantities of information in great detail, but for others it is very difficult. If you have difficulty in recalling information, do not despair. Through practice you too can become proficient in this area.

The first step is to be aware of your surroundings and what is happening in them. You can become skilled at observing by (1) making a deliberate effort to do so, and (2) asking yourself questions about your environment and the daily changes in it. The questions you must begin to consciously ask are *who, what, when, where, why*, and *how*.

Try this exercise: On a sheet of paper, using only the information in your mind, draw the layout of your home. How many windows are there? What is the safest way to exit? What other ways are there for you to escape? What does the floor or apartment layout immediately above or below you look like? Now answer these questions: What hazardous materials are stored in your home? When was the last time you held a fire drill in your home? When was the last time you checked the smoke detectors?

Asking yourself questions about your immediate surroundings is the first step. Now extend your range of observation. While on the way to work or other activities, stop and look at one or two buildings, using the parallel track pattern to scan the area quickly yet thoroughly. Your next step is to use the "13 points" system, and then ask yourself the questions *who, what, where, why,* and *how*. Ask yourself questions about the buildings and make mental notes. When you get to work take 5 minutes to draw a sketch of the buildings and to fill in details or write down as much as you can recall about the buildings. On your way home compare your sketch and data with the actual scene. What did you miss? Why did you miss it? The next time you try this exercise see whether you can look at the same buildings in a different perspective in order to create a new challenge. Repeat this exercise several times, using a different group of buildings each time.

Once you have honed your skills to some degree, try using pictures from books or magazines. They need not be related to firefighting but should be detailed in design. Observe closely; then try to recall.

Recall questions are one portion of the test on which you can definitely improve your performance. It requires nothing but dedication to improve a skill you already use in everyday life.

Visualizing Images

Firefighters must be able to look at an object and figure out how it would appear from the back, side, and top. The ability to do this is called visualization and is the process of recording a mental picture of an image and then transforming that image into other orientations. You may be called upon to demonstrate this talent by answering questions that show an object from one angle and then ask you to identify the way it would look from a different angle. This skill takes practice, but can be learned. Start by constructing a single-view image of the object in your mind. Pay particular attention to the parts of the object that can be see from other viewpoints. Focus your attention on the height and spatial relationship of objects extending over a roof or out from the front and sides of the building, as these are critical reference markers.

If you wanted to add depth to a two-dimensional image, you would project rays or lines backward from each major point in the outline of the image. Try adding depth to a simple two-dimensional object in your mind. Once you have done this, try to picture the object from the back. Remember, it will be reversed. To help you do this, first think of looking into a mirror and then project yourself inside the mirror looking out at where you were when you were looking in. Keep in mind, when you view the object from the back, that some of the items in the center of the image will not be visible from the front and others may appear to be new or different, but the outline that mirrors the image of the front should still be the same.

When you try to picture an image from the side, keep in mind the angles of the items that were projecting outward or above from the front. They may now be at your left or right, but they should be at the same height and at a complementary angle.

When you look at an object from above, as when you look down onto a roof, remember that flat surfaces will appear as large undivided squares or rectangles, while peaked roofs will appear as subdivided rectangles with lines at an angle indicating where the roof joins another section of the building. Cylindrical objects with high peaks will appear as circles.

If you are asked to describe the outside of a building as seen from the inside of the structure, keep in mind that it is a mirror image. You will not be able to see items on the face of a building unless they are on a glass panel, such as a sign on a window, and in this case, the writing will appear in reverse.

Spatial Orientation

In fire fighting it is important to know where you are, where you were, and how to get from one location to another by using a map or floor plan. The spatial orientation questions test your ability to make judgments about how to traverse areas in the most direct route without breaking the laws or rules. The laws could be traffic laws like one-way streets, traffic restrictions, or weight restrictions. The rules may be obstructions such as dead ends, narrow paths, unstable ground, bridges or walkways, or circular corridors or roads.

When working with maps and floor plans, find out if you are allowed to mark the map or floor plan. Orient the map so that North faces in the correct direction with respect to your current position. Locate the starting and end points of the route you must travel. Using your finger, trace a path toward the intended destination. When you come to a point that blocks your passage, either because it violates a rule or a law, put a mark (X) to indicate this is not a valid route. Now go back to the starting point and begin following a new route. You may be tempted to just backtrack over a small portion of the invalid route and then proceed once you have avoided the blockage, but this is not recommended. Go back and start from the beginning, and make sure the new path is correct from start to finish. When you have identified the correct path, use a pencil (if allowed) to trace the path you just negotiated. Keep in mind the rules about following traffic laws when you are drawing this line.

Practice Exercises

GROUP ONE

Directions: Answer questions 1 through 5 based on the illustration on the next page. You are permitted 5 minutes to study and commit to memory as much as you can of the illustration. You are *not* permitted to make any written notes during the 5 minutes you are studying the illustration.

After 5 minutes, stop studying the illustration and answer the questions without referring to the illustration. For each question, choose the one *best* answer—(A), (B), (C), or (D)—and write the corresponding letter on your answer page next to the number of the question.

Now start your 5 minutes on a clock and begin.

A. P. Smith Low-Pressure Hydrant

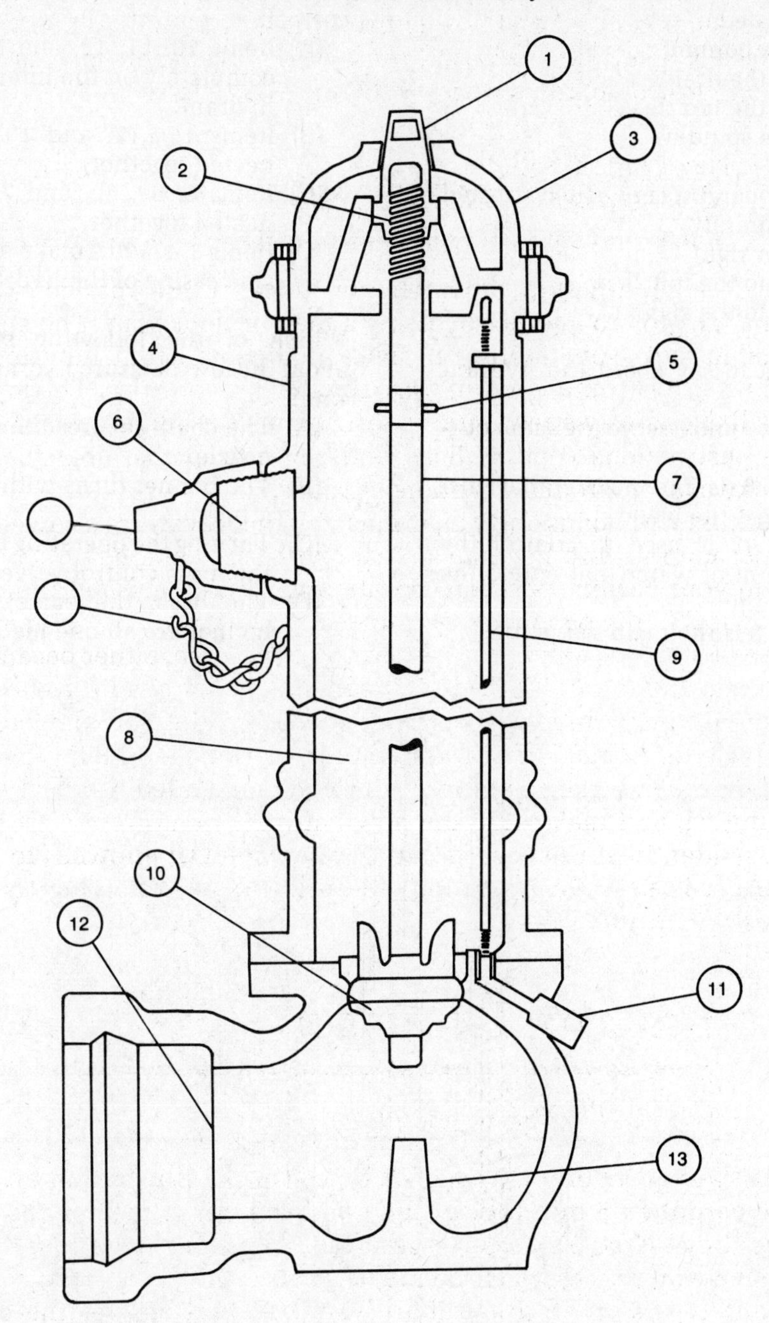

1. Operating nut
2. Spindle
3. Bonnet
4. Barrel
5. Shear pin
6. Nozzle

6a. Cap
6b. Chain

Turn:
right to open
left to close

7. Main valve rod
8. Standpipe
9. Drain rod
10. Main valve
11. Drain
12. Shoe and connecting hub
13. Main valve stop

Use a full 5 minutes (no less). Then turn the page and answer the questions. Do not refer to the illustration again.

1. The main valve in a low-pressure hydrant is located
   (A) at the bonnet
   (B) near the drain
   (C) near the nozzle
   (D) at the spindle

2. Item 1, the operating nut, can be turned
   (A) indefinitely
   (B) left or right
   (C) only to the left
   (D) only to the right

3. It is correct to say about the hydrant that:
   (A) Items 7 and 9 serve the same purpose.
   (B) Item 6 is the nozzle that discharges the water.
   (C) Item 13 is used to control the amount of water that enters the hydrant.
   (D) Item 8 is the main valve rod.

4. Which of the following statements is correct?
   (A) Items 10, 11, 12, and 13 are all completely on the interior of the hydrant.
   (B) Items 1, 5, 7, and 10 are connected together.
   (C) Items 6, 6a, 6b, and 7 are connected together.
   (D) Items 3, 4, and 9 make up the outside casing of the hydrant.

5. Which of the following statements about the low-pressure hydrant is accurate?
   (A) The chain secures the cap to the nozzle.
   (B) The bonnet turns with the operating nut.
   (C) Turning the operating nut causes the main control valve to open.
   (D) The base, the barrel, and the bonnet are all one piece.

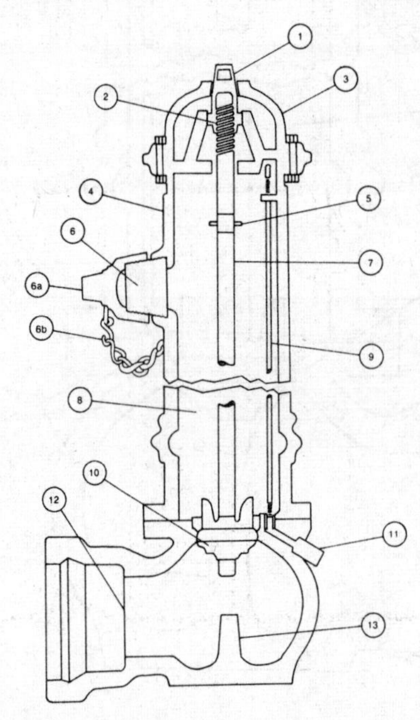

GROUP TWO

Directions: Answer questions 6 through 15 based upon the illustration on the next page. You are permitted 5 minutes to study and commit to memory as much as you can of the illustration. You are *not* permitted to make any written notes during the 5 minutes you are studying the illustration.

After 5 minutes, stop studying the illustration and answer the questions without referring to the illustration. For each question, choose the one *best* answer—(A), (B), (C), or (D)—and write the corresponding letter on your answer paper next to the number of the question.

Now start your 5 minutes on a clock and begin.

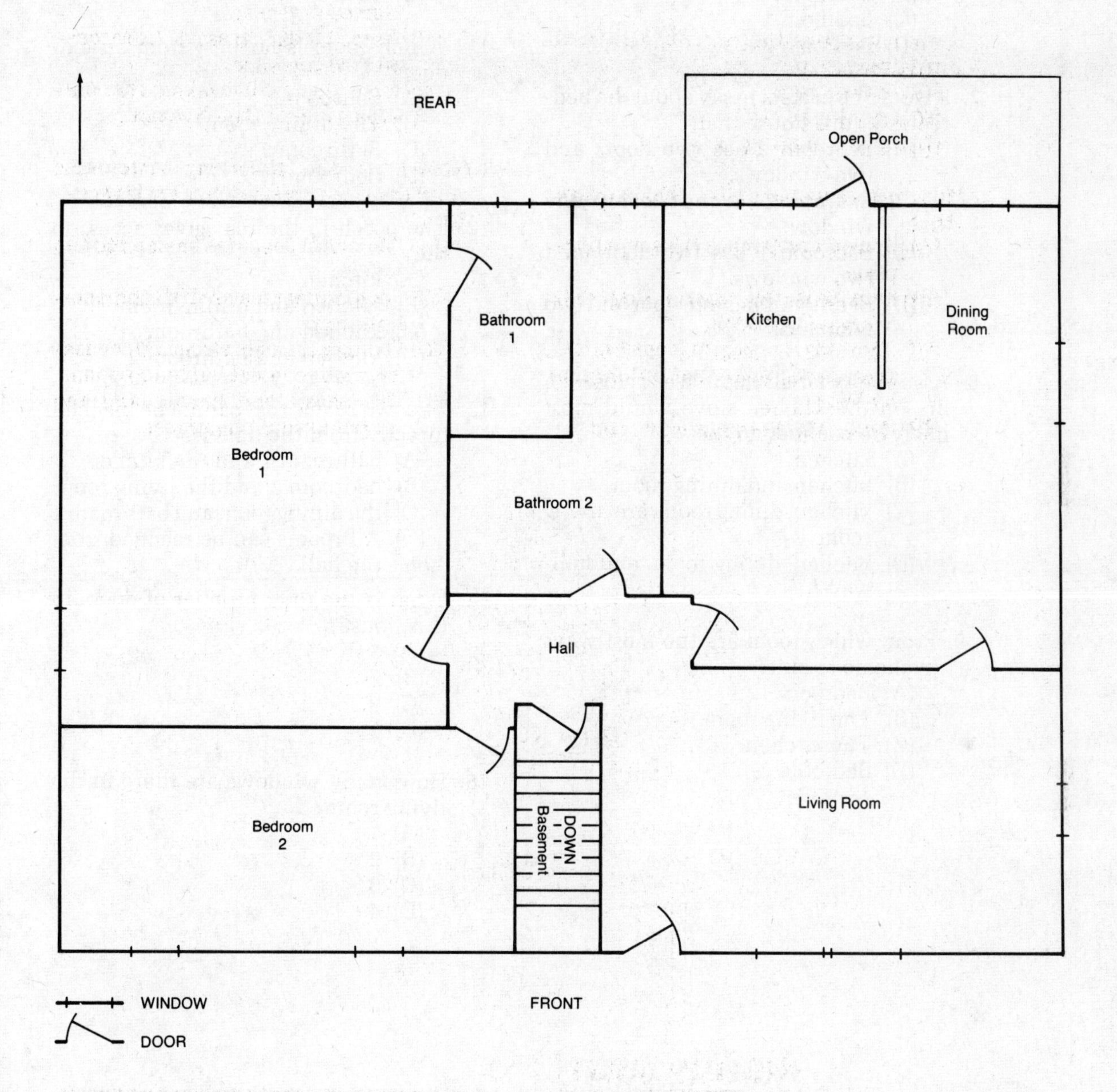

Use a full 5 minutes (no less). Then turn the page and answer the questions. Do not refer to the illustration again.

6. A fire coming up the stairs from the basement would most likely cut off passage to the exit doors for all of the following *except*
   (A) The kitchen
   (B) Bathroom 2
   (C) Bedroom 1
   (D) Bedroom 2

7. It would be correct to say about the bedrooms in this house that:
   (A) Bedroom 2 has two doors and two windows.
   (B) Bedroom 2 has one door and one window.
   (C) Bedroom 1 has two doors and two windows.
   (D) Bedroom 1 has one door and two windows.

8. Assuming that all doors were closed, a fire on the kitchen stove would most likely be confined to the
   (A) kitchen
   (B) kitchen and dining room
   (C) kitchen, dining room, and living room
   (D) kitchen, dining room, and hallway

9. From which room are the most ways available to exit?
   (A) Bedroom 1.
   (B) The living room.
   (C) The kitchen.
   (D) Bedroom 2.

10. How many windows are there for possible escape?
    (A) 9
    (B) 10
    (C) 11
    (D) 12

11. Which room is farthest from an exit door?
    (A) Bedroom 2.
    (B) The dining room.
    (C) Bathroom 1.
    (D) Bedroom 1.

12. The porch in the rear gives access to the
    (A) kitchen
    (B) kitchen and dining room
    (C) kitchen and bathroom 1
    (D) kitchen and bathroom 2

13. The two rooms that cannot be reached directly from the hall are
    (A) bathroom 1 and the kitchen
    (B) bedroom 2 and the living room
    (C) the dining room and bathroom 1
    (D) All rooms can be reached from the hall.

14. What is the total number of doors in this house?
    (A) 5
    (B) 7
    (C) 9
    (D) 11

15. How many windows are there in the living room?
    (A) 1
    (B) 2
    (C) 3
    (D) 4

GROUP THREE

Directions: Answer questions 16 through 20 based on the illustration on this page. You are permitted 5 minutes to study and commit to memory as much as you can of the illustration. You are *not* permitted to make any written notes during the 5 minutes you are studying the illustration.

After 5 minutes, stop studying the illustration and answer the questions without referring to the illustration. For each question choose the one *best* answer—(A), (B), (C), or (D)—and write the corresponding letter on your answer paper next to the number of the question.

Now start your 5 minutes on a clock and begin.

Sprinkler System

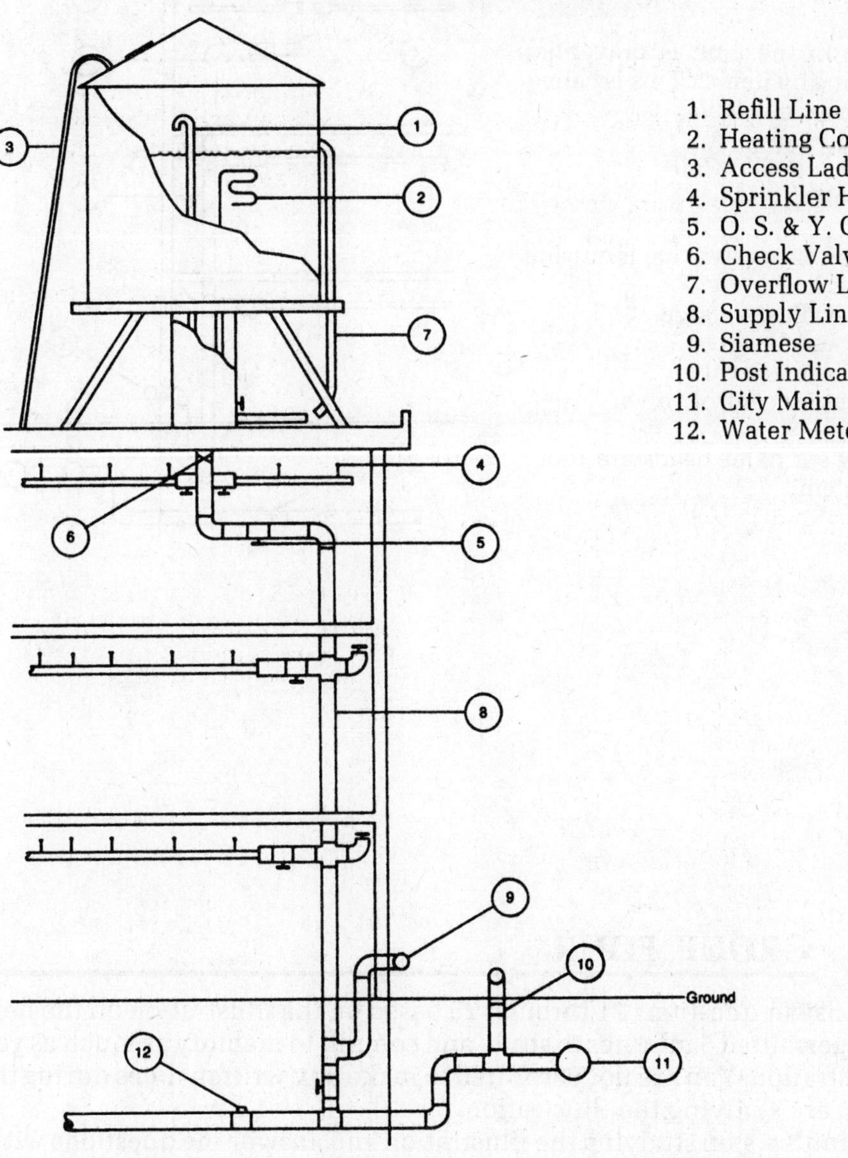

1. Refill Line
2. Heating Coil
3. Access Ladder
4. Sprinkler Head =
5. O. S. & Y. Control Valve
6. Check Valve
7. Overflow Line
8. Supply Line to System
9. Siamese
10. Post Indicator Valve
11. City Main
12. Water Meter

Use a full 5 minutes (No less): Then turn the page and answer the questions. Do not refer to the illustration again.

16. Item 9 is a
    (A) drain valve
    (B) post indicator valve
    (C) supply line
    (D) siamese

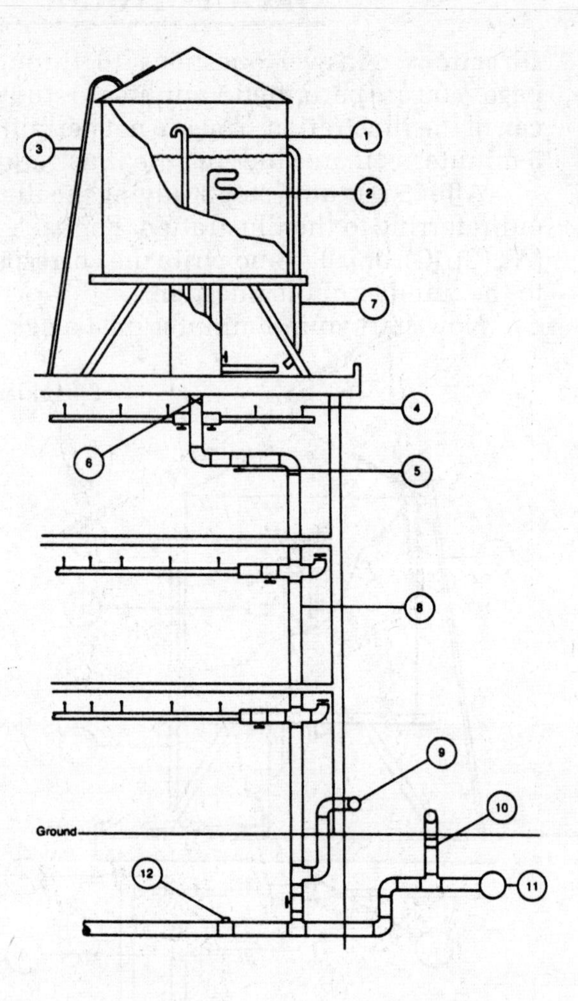

17. The water is prevented from flowing back into the tank through the supply line by the use of a one-way check valve. By what number is the check valve indicated?
    (A) 4
    (B) 6
    (C) 8
    (D) 12

18. The water in the tank is prevented from freezing by item 2. This is called
    (A) an overflow valve
    (B) a refill valve
    (C) an O.S. and Y. Coil
    (D) a heating coil

19. Which is the only valve that is outside the building?
    (A) Post indicator valve.
    (B) O.S. and Y. valve.
    (C) Water meter valve.
    (D) Sprinkler control valve.

20. How many sprinkler heads are there on this system?
    (A) 8
    (B) 10
    (C) 12
    (D) 15

GROUP FOUR

Directions: Answer questions 21 through 25 based on the illustration on the next page. You are permitted 5 minutes to study and commit to memory as much as you can of the illustration. You are *not* permitted to make any written notes during the 5 minutes you are studying the illustration.

After 5 minutes, stop studying the illustration and answer the questions without referring to the illustration. For each question, choose the one *best* answer—(A), (B), (C), or (D)—and write the corresponding letter on your answer paper next to the number of the question.

Now start your 5 minutes on a clock and begin.

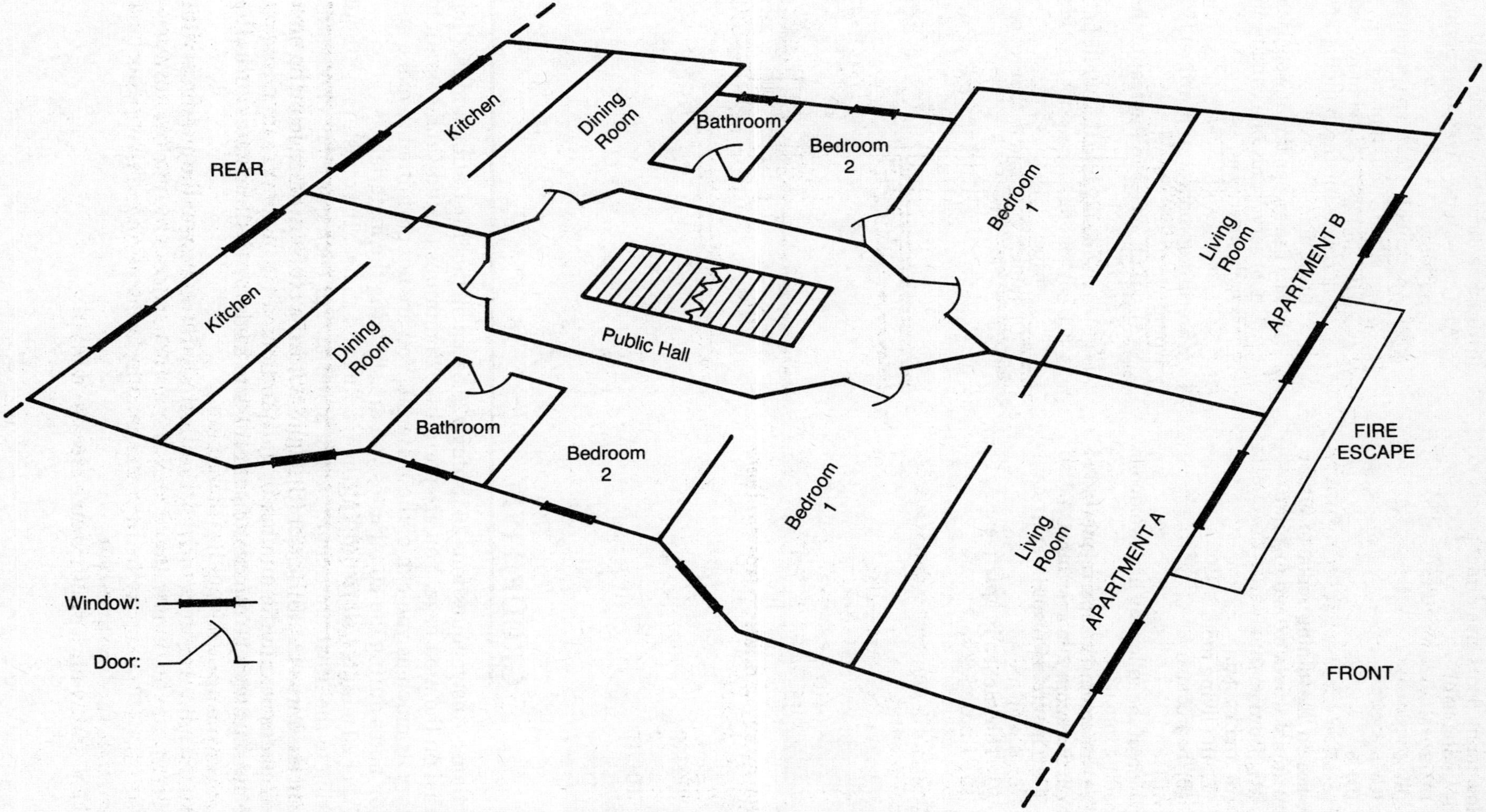

REAR

Kitchen

Dining Room

Bathroom

Bedroom 2

Bedroom 1

Living Room

APARTMENT B

Public Hall

Kitchen

Dining Room

Bathroom

Bedroom 2

Bedroom 1

Living Room

APARTMENT A

FIRE ESCAPE

FRONT

Window:

Door:

Use a full five minutes (no less). Then turn the page and answer the questions. Do not refer to the illustration again.

21. How many doors give direct access to the public hall?
    (A) 3
    (B) 4
    (C) 5
    (D) 6

22. A fire in the dining room of either apartment would extend *first* into
    (A) bedroom 2
    (B) the kitchen
    (C) the living room
    (D) bedroom 1

23. Which of the following has two doors leading into it?
    (A) Bedroom 1 in apartment B.
    (B) Bedroom 2 in apartment A.
    (C) The dining room in apartment B.
    (D) The kitchen in apartment A.

24. It would be *incorrect* to state that:
    (A) There are two ways out of this building.
    (B) Apartment A and apartment B are slightly different.
    (C) The number of windows on both sides of the building are the same.
    (D) The fire escape landing on each floor serves only one window in each apartment.

25. A fire in bedroom 2 of apartment A would
    (A) cut off access to the rear of the apartment
    (B) extend only into the public hall
    (C) cut off all access to the fire escape
    (D) remain confined to that room

GROUP FIVE

Directions: Answer questions 26 through 29 based on the illustration and reading material on the next page. You are permitted 20 minutes to study and commit to memory as much as you can of the material. You are *not* permitted to make any written notes during the 20 minutes you are studying the material.

After 20 minutes, stop studying the reading and answer the questions without referring to the illustration and reading material. For each question, choose the one *best* answer—(A), (B), (C), or (D)—and write the corresponding letter on your answer paper next to the number of the question.

Now start your 20 minutes on a clock and begin.

DIVISION OF FIRE PREVENTION
FIRE DEPARTMENT

F.P. Directive 1-79 (Revised) September 10, 1980

BY VIRTUE OF THE AUTHORITY VESTED IN ME UNDER SECTION 1105 OF the New York City Charter, I hereby promulgate Criteria for Official Type of "No Smoking" Signs. (Filed with the City Clerk on November 3, 1978). (Amendment filed March 31, 1980).

CRITERIA FOR OFFICIAL TYPE "NO SMOKING" SIGNS

1. The following criteria shall be used for printing of OFFICIAL type Fire Department "NO SMOKING" signs.

2. Permission is granted to anyone who desires to print the new type sign, in strict conformance to the criteria, for use or sale.

3. Signs shall be used only where required by provision of law or Fire Department regulation.

4. Existing official type signs may be used until stock is depleted.

5. Specific Criteria:

a. Pictorial Description and Layout:

NO SMOKING
IN THESE PREMISES
UNDER PENALTY OF FINE OR IMPRISONMENT, OR BOTH BY ORDER OF THE FIRE COMMISSIONER

b. Overall sign size: 10"x14" (or 11"x15")

c. Symbol: International "NO SMOKING" symbol (3¾"x¼" Cigarette and Smoke in Black; 5/16"x5/16" Ash in Red; 6½" Diameter x ¾" Circle with ¾" Cross Bar in Red).

d. Legends:

Legends:	Letter Size—Capitals:	Color:
NO SMOKING	2" Min.	Red
IN THESE PREMISES	1" or ¾" min.	Black
UNDER PENALTY OF FINE OR IMPRISONMENT, OR BOTH, BY ORDER OF THE FIRE COMMISSIONER	¼" min.	Black

e. Material: Heavy stock cardboard or other durable material.

6. In Nursing Homes, Hospitals, Sanitoria, Convelescent Homes, Homes for the Aged or for chronic patients, or portions of buildings used for such purpose, as set forth in C19-165.4 Administrative Code, certain areas may be approved by the Fire Commissioner as locations where smoking is permitted. In all other portions of such premises, "NO SMOKING" signs shall be posted, and shall conform to the criteria set forth in Section 5 of these regulations, except that the statement "EXCEPT IN DESIGNATED LOCATIONS" shall immediately follow the legends "NO SMOKING IN THESE PREMISES". The additional legend "EXCEPT IN DESIGNATED LOCATIONS" shall be in BLACK CAPITAL LETTERS at least ½" high letter size.

AUGUSTUS A. BEEKMAN
Fire Commissioner.

Use a full 20 minutes (no less). Then turn the page and answer the questions. Do not refer to this material again.

26. It would be accurate to say about official "NO SMOKING" signs that:
    (A) Only the circle and cross bar are red.
    (B) The pictorial design is optional.
    (C) They state that the penalty for smoking is always a fine.
    (D) The cigarette and smoke are colored black.

27. The new official "NO SMOKING" signs may be
    (A) used on any premises
    (B) printed only by licensed, approved printers
    (C) printed only on durable material
    (D) printed in either the new or the old style

28. For certain premises the writing on the "NO SMOKING" sign may be modified to read
    (A) "Except in Designated Locations"
    (B) "No Smoking Here"
    (C) "Smoking Permitted Here"
    (D) None of the above; the sign cannot be altered.

29. For official "NO SMOKING" signs, there are
    (A) four acceptable colors
    (B) two acceptable sizes
    (C) only two acceptable types of materials
    (D) three acceptable types of signs

GROUP SIX

Directions: Answer questions 30 through 34 based on the written material and illustration on the next page. You are permitted 20 minutes to study and to commit to memory as much as you can of the material. You are *not* permitted to make any written notes during the 20 minutes you are reading the material.

After 20 minutes, stop reading the material and answer the questions without referring to the written material and illustration. For each question, choose the one *best* answer—(A), (B), (C), or (D)—and write the corresponding letter on your answer paper next to the number of the question.

Now start your 20 minutes on a clock and begin.

The Axe

The axe is one of the primary tools used by fire fighters and its construction, its use, and maintenance must be understood.

The "4 C's" of the axe, a common tool, are as follows:

- Construction
- Care
- Carry
- Chopping

Construction

The head is made of steel and has two usable surfaces. The flat surface is used for pounding, and the double-edged surface, for cutting. The handle is most often made of wood (hickory is common) and fits into a slot in the head. The cutting edge of the head has a tempered tip. The axe makes use of the principle of the inclined plane.

Care

Like all fire fighting tools, the axe must be properly maintained. Clean burrs from the head with a file and then wash with kerosene. Sharpen the axe with a slow-turning sandstone wheel, not on a grindstone. If a wheel is not available, use a file.

After the head is cleaned, wash the handle using unsoaped Brillo, and then put a light coat of oil on the head. Axe handles should not be painted; paint hides defects and could lead to a defective tool being used at a fire.

Carry

The best and safest way to carry the axe is by using two hands. One hand carries the head; the other, the handle at a point opposite the head. This method is not always available, however, for there are times when the second hand is needed for other tasks. When carrying the axe in one hand, use that hand to engulf the head, with the cutting edge facing forward. This will allow the axe handle to fall at your side and will keep you in control of the blade and handle movement.

Chopping

Hold the axe in a comfortable, balanced position with one hand at the grip and the other at the shoulder. Chop in a circular motion, making many small cuts rather than attempting to cut through on each swing. Before you begin, clean the area around the spot to be cut and look around to make sure that it is safe to swing the axe. When cutting, the blade should enter the surface at about a 60 degree angle. When cutting roofs, cut the sheathing first, then go back and cut the roof boards. Do not cut on top of roof joists or in the center of the bay. Attempt to cut at a point 2 inches from the joist, on the inside of the hole to be made.

Use a full 20 minutes (no less). Then turn the page and answer the questions. Do not refer to this material again.

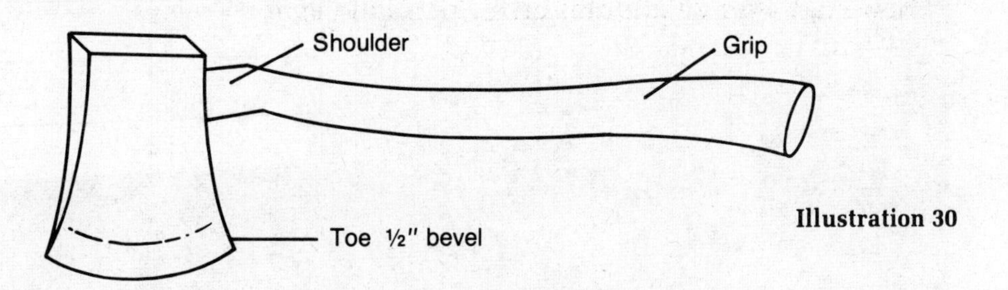

Illustration 30

30. Which of the following is not one of the "4 Cs" of the axe?
    (A) Construction
    (B) Care
    (C) Cleaning
    (D) Carry
31. After the handle is washed, you should
    (A) put a clean coat of paint on it.
    (B) varnish it lightly.
    (C) put a light coat of oil on it.
    (D) check it for defects.
32. The best technique to cut a roof using an axe is to
    (A) use a circular motion, taking small cuts.
    (B) cut all the material at once.
    (C) use the mechanical advantage of the wedge by taking a full overhead swing.
    (D) cut in the center of the bay.
33. What is the best angle at which to cut a roof?
    (A) 30 degrees
    (B) 60 degrees
    (C) 75 degrees
    (D) 90 degrees
34. The best way to carry an axe is
    (A) head in hand.
    (B) handle in hand.
    (C) head or handle in hand.
    (D) head and handle in hands.

GROUP SEVEN

35. Your commanding officer sends you to the back of a burning house and asks you to assess the situation from that location. When you arrive at the back of the house you could expect it to look similar to (see picture 35):
    (A)
    (B)
    (C)
    (D)

A
B
C
D

36. When operating from the basket of an elevated platform truck, you would be most correct if you thought the roof of the building looked like (see picture 36):
    (A)
    (B)
    (C)
    (D)

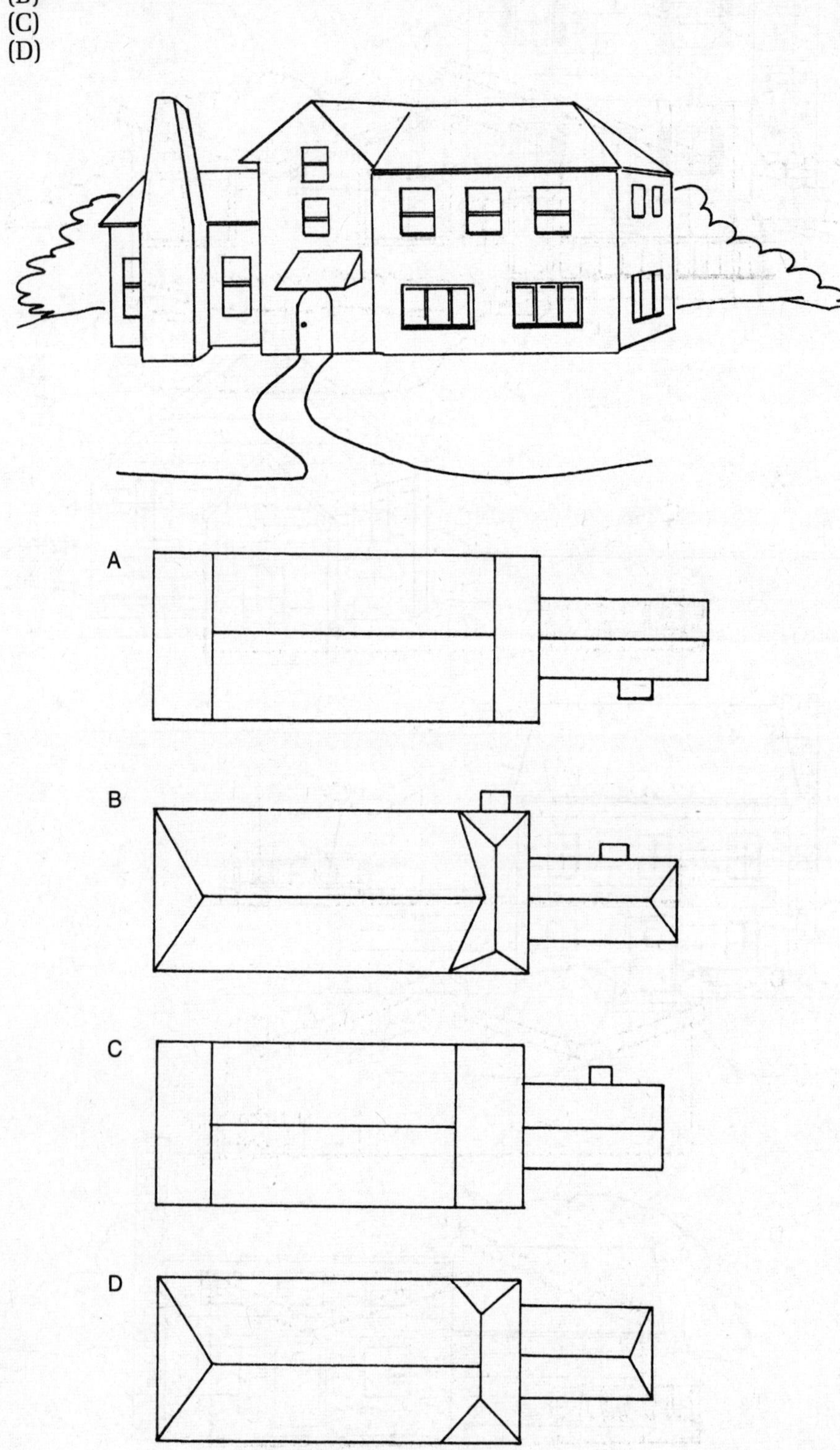

37. The illustration that most likely represents what the building would look like when viewed from the rear is:
    (A)
    (B)
    (C)
    (D)

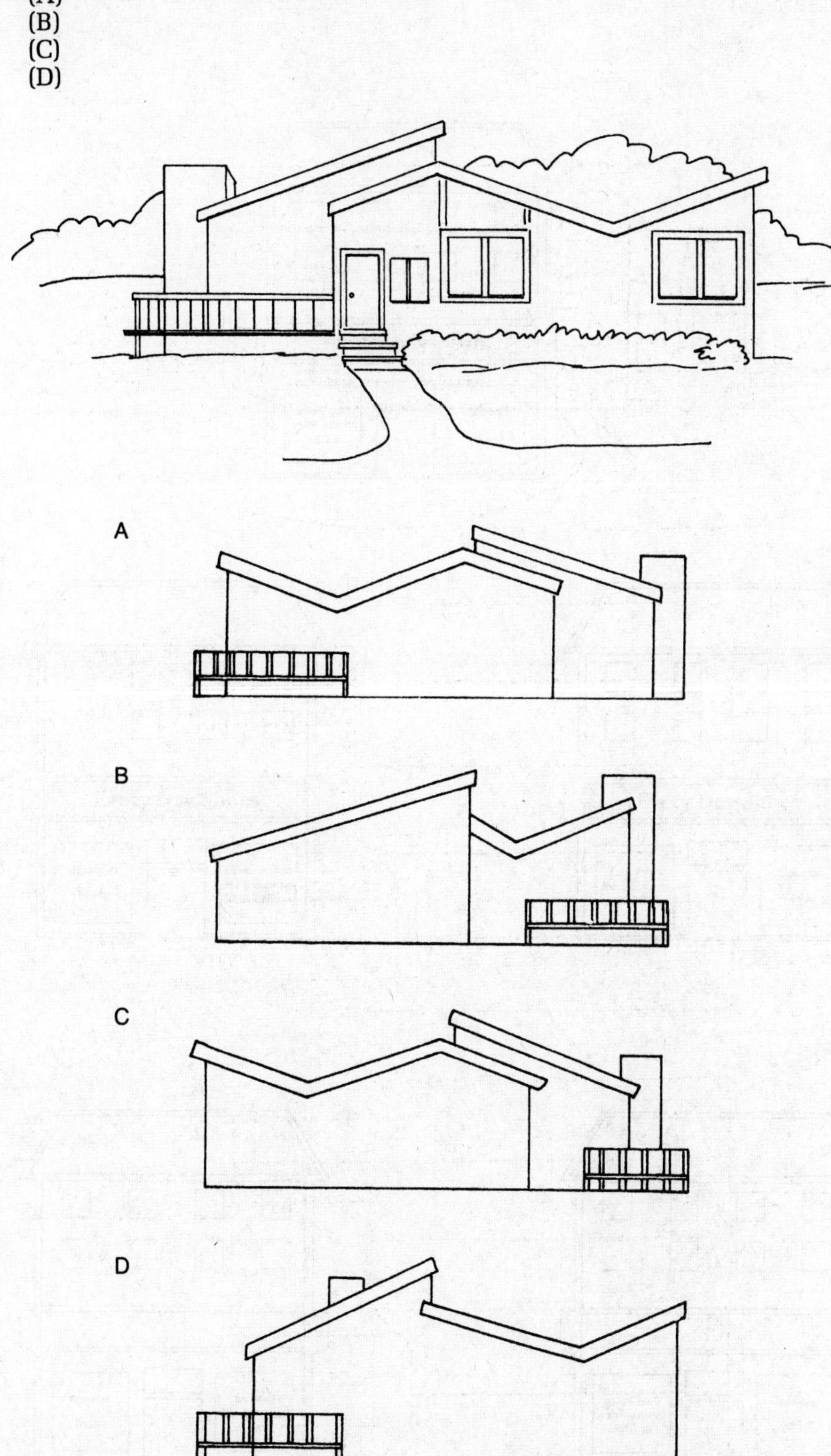

38. The illustration that depicts the image you would see if you were inside the shoe store and looking out is:
    (A)
    (B)
    (C)
    (D)

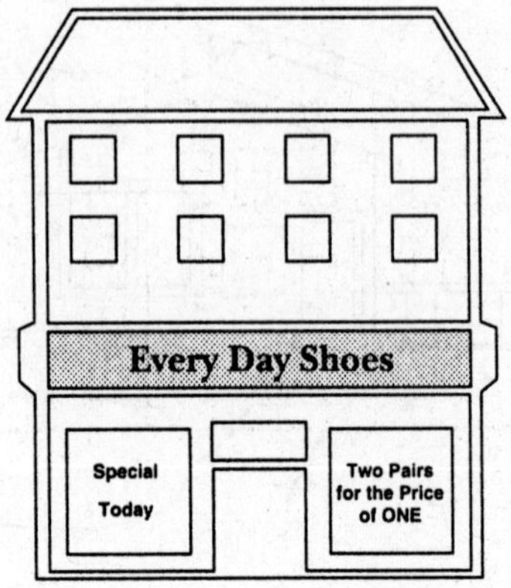

Every Day Shoes

Special Today

Two Pairs for the Price of ONE

A

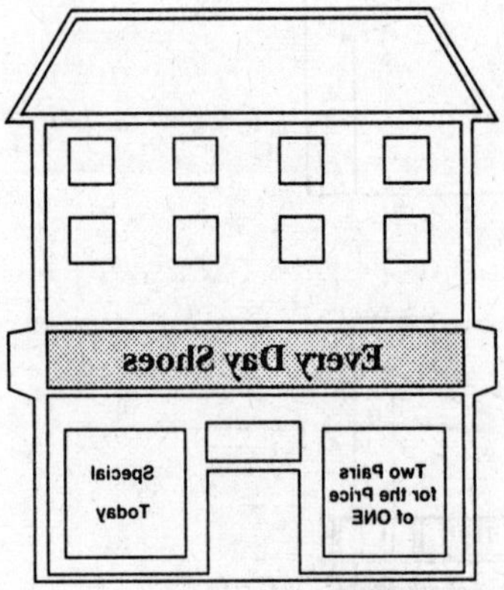

B

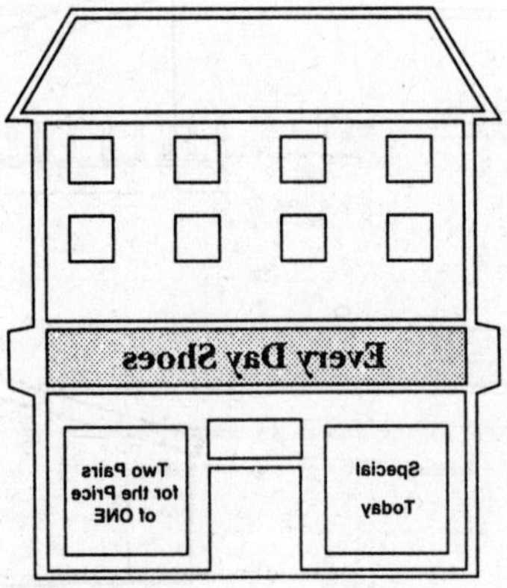

C

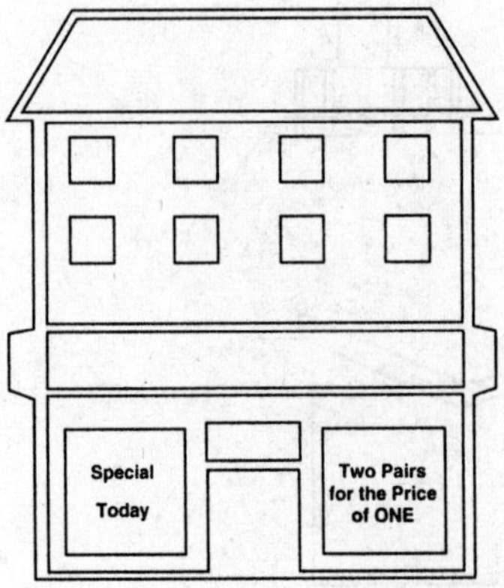

D

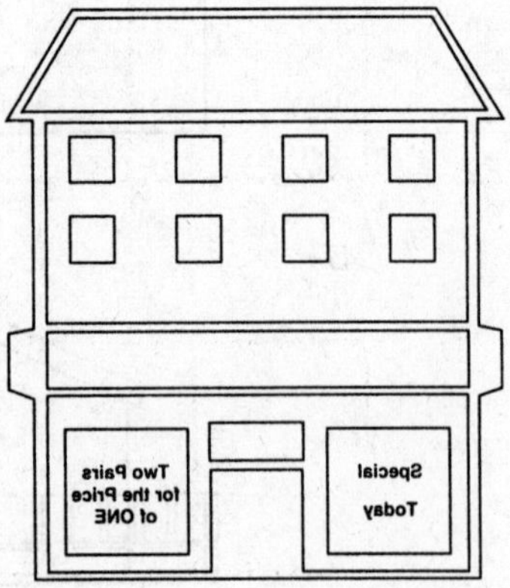

39. If you were sent to the rear of this row of houses, you would expect it to look like:
    (A)
    (B)
    (C)
    (D)

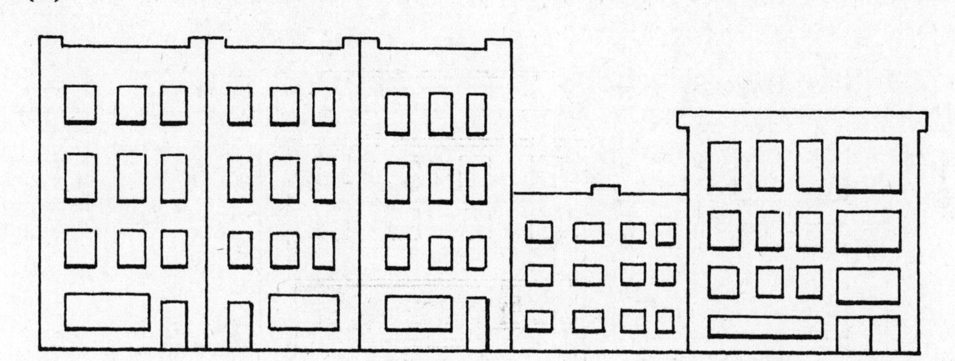

A

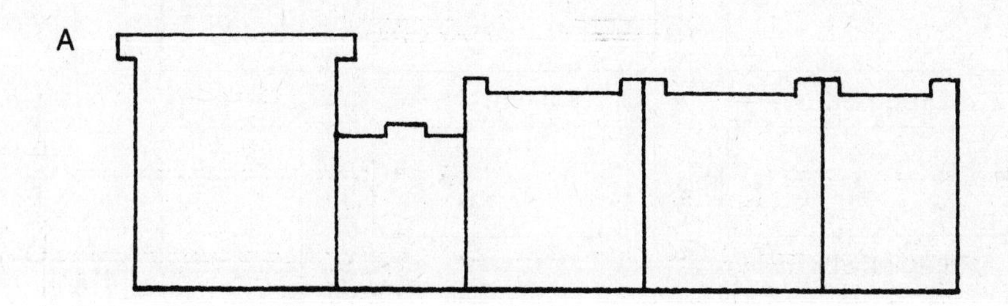

B

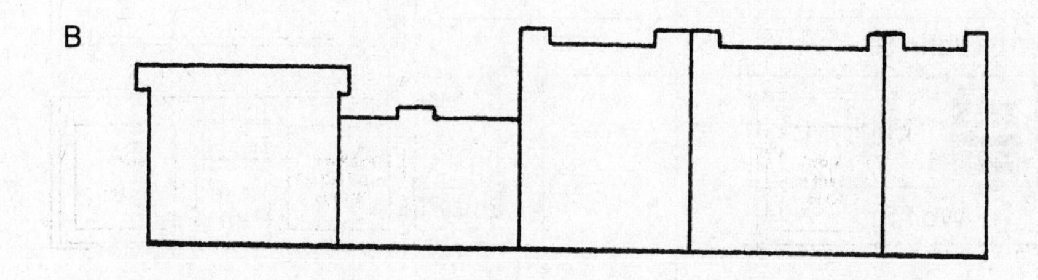

C

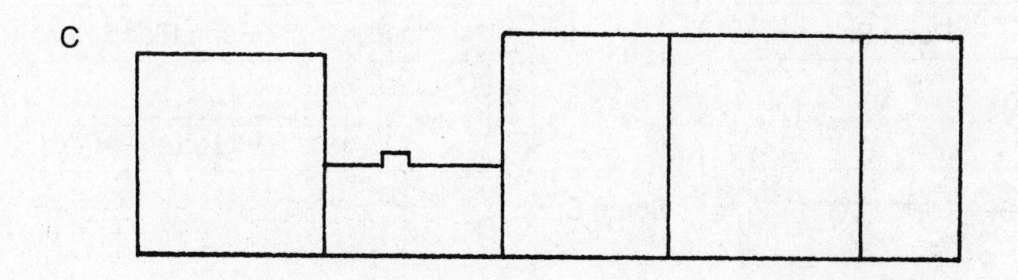

D

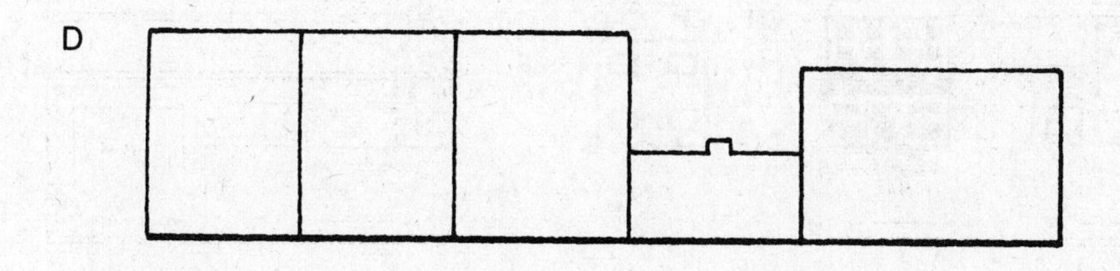

GROUP 8

Directions: Base your answers to questions 40 through 43 only on the information provided in the illustration below. All traffic laws, rules and requirements must be obeyed. The shortest, most direct route should be used.

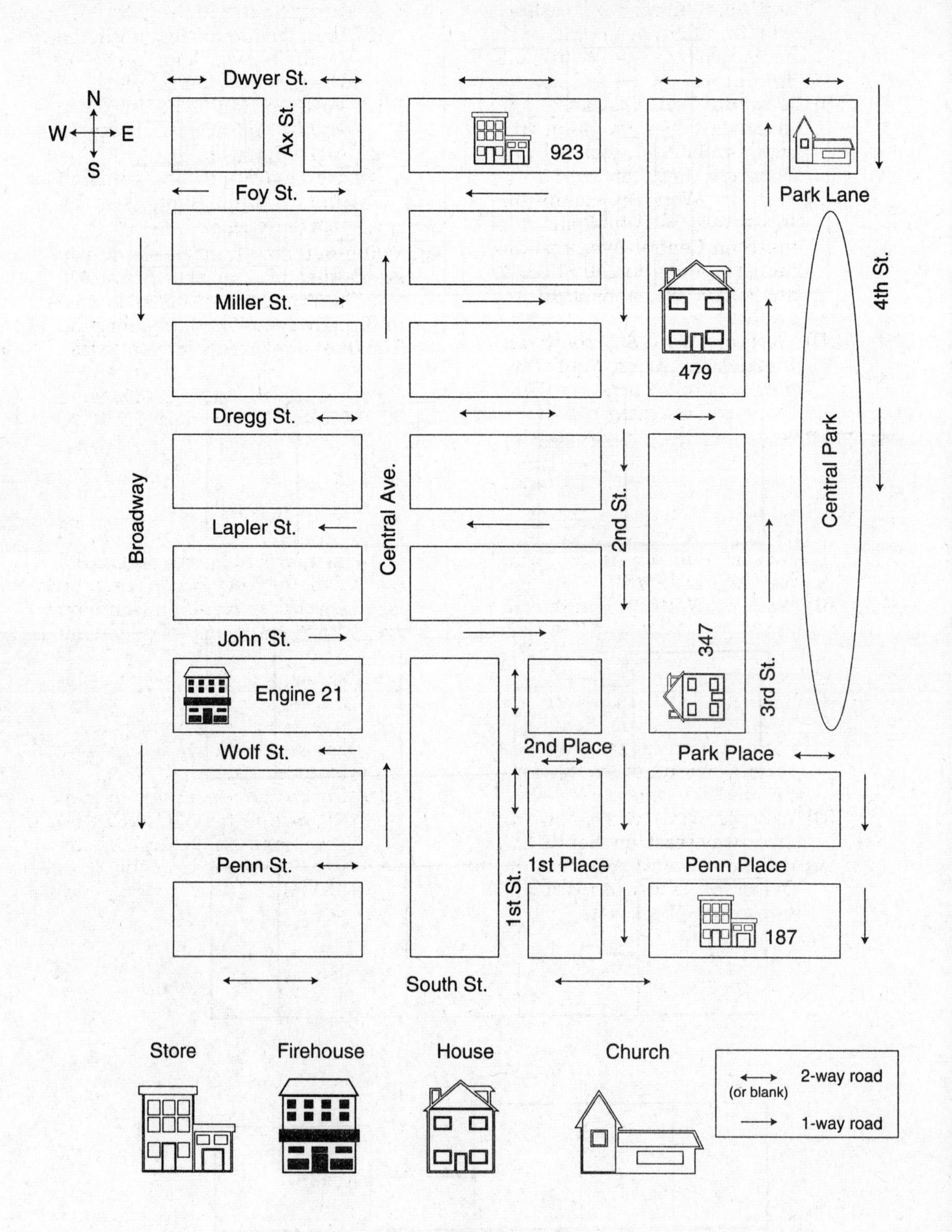

40. Engine 21 receives notification of a fire at 347 3rd Street between Park Place and Dregg Street. The most direct route for the driver to take is:
    (A) West on Wolf St., south on Broadway, east on Penn St., north on Central Ave., east on John St., south on 2nd St., east on Park Place, and north on 3rd St.
    (B) West on Wolf St., north on Broadway, west on John St., south on 2nd St., east on Park Place, and north on 3rd St.
    (C) East on Wolf St., south on Broadway, east on Penn St., north on Central Ave., east on John St., south on 2nd St., east on Park Place, and north on 3rd St.
    (D) West on Wolf St., south on Broadway, east on South St., north on 2nd St., east on Park Place, and north on 3rd St.
41. Firefighter Jones has been asked to give a fire safety talk at the Park Lane Church. The best route to take from the firehouse of Engine 21 is:
    (A) North on Broadway, east on Dwyer, south on 4th St., and west on Park Place.
    (B) West on Wolf St., south on Broadway, east on Central, north on 4th St., and West on Park Lane.
    (C) West on Wolf St., south on Broadway, east on Penn St., north on Central Ave., east on Dregg St., north on 3rd St., and east on Park Lane.
    (D) West on Wolf St., south on Broadway, east on South St., north on Central Ave., east on Dwyer St., south on 4th St., and west on Park Lane.
42. While inspecting 187 South St., a call for a person with breathing difficulty is received. You are to respond to 479 Dregg St. You should go:
    (A) West on South St., north on Central Ave., and west on Dregg St. to 479 Dregg St.
    (B) West on South St., north on Central Ave., and east on Dregg St. to 479 Dregg St.
    (C) West on South St., north on 2nd St., and west on Dregg St. to 479 Dregg St.
    (D) West on South St., north on 2nd St., and east on Dregg St. to 479 Dregg St.
43. While returning from an alarm, you are notified of a reported fire at 923 Foy St. Your fire apparatus is now on Broadway just north of Dregg St. The most direct route for you to take is:
    (A) Go south on Broadway to Lapler St., east on Lapler St. to Central Ave., north on Central Ave. to Foy St., and east to 923 Foy St.
    (B) Go south on Broadway to Dregg St., east on Dregg St. to Central Ave., north on Central Ave. to Dwyer St., east on Dwyer St. to 2nd St., south on 2nd St. to Foy St., then west on Foy St. to 923 Foy St.
    (C) Go south on Broadway to John St., east on John St. to Central Ave., north on Central Ave. to Foy St., and then east on Foy St. to 923 Foy St.
    (D) Go south on Broadway to John St., east on John St. to Central Ave., south on Central Ave. to Foy St., then east on Foy St. to 923 Foy St.

Directions: Base your answers to questions 44 through 48 only on the information provided in the illustration below. All traffic laws, rules and requirements must be obeyed. The shortest, most direct route should be used.

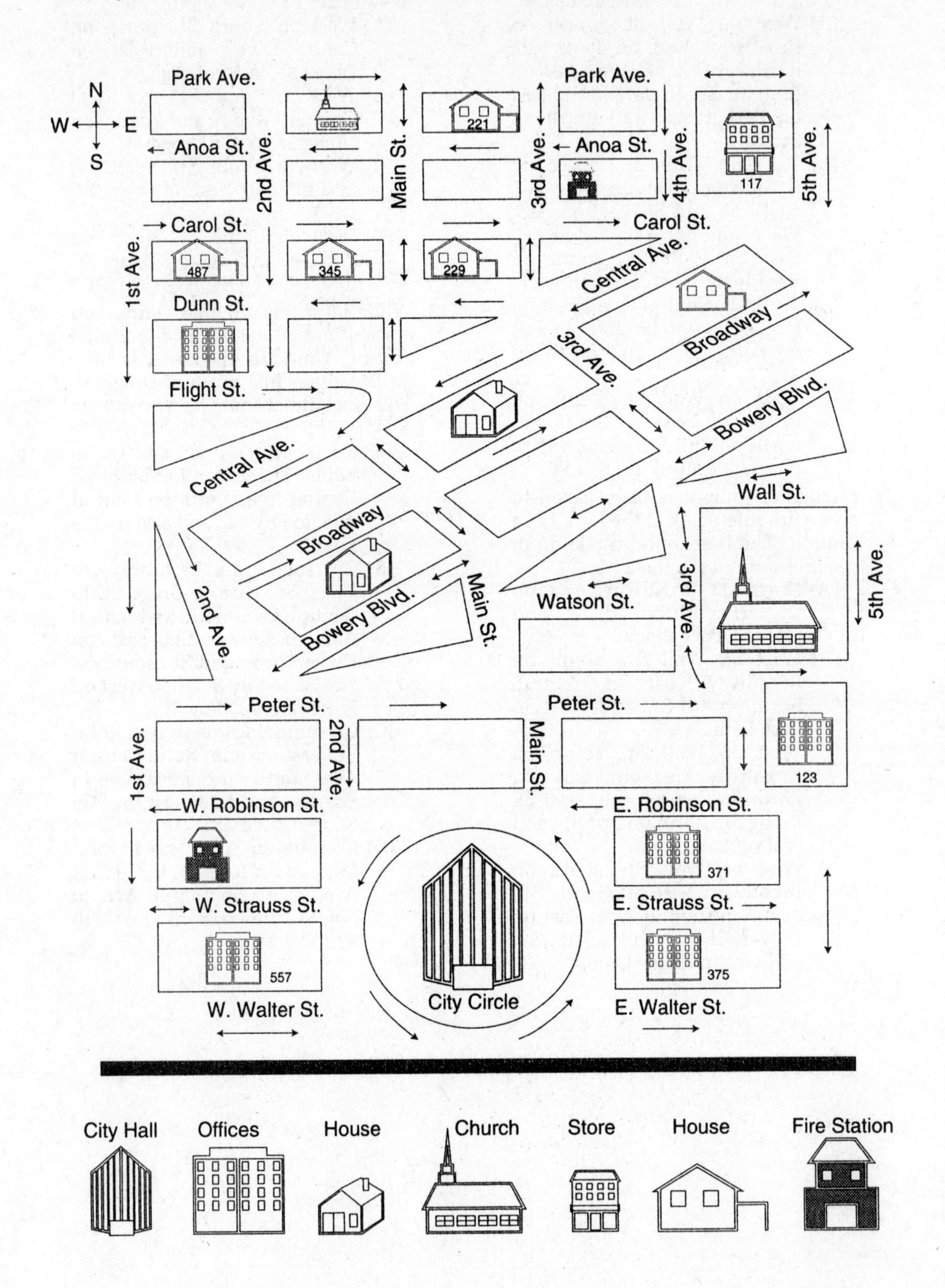

44. A firefighter assigned to the Carol St. firehouse is told to work in the Strauss St. firehouse for the next two weeks. The most direct route to the Strauss St. firehouse is:
   (A) East on Carol St. to Central Ave., Central Ave. to 2nd Ave. South, then east on Walter, north to Strauss, and east to the firehouse.
   (B) East on Carol to 5th Ave., south on 5th Ave. to E. Walters, west on E. Walters to 1st Ave., north on 1st Ave. to W. Strauss, and east to the firehouse.
   (C) East on Carol to Central Ave., southwest on Central to 1st Ave., south on 1st Ave. to W. Strauss St., and east on W. Strauss to the firehouse.
   (D) East on Carol to 5th Ave., south on 5th Ave. to Bowery Blvd., southwest on Bowery Blvd. to 1st Ave., south on 1st Ave. to W. Strauss St., and then east on Strauss to the firehouse.

45. The fire company assigned to inspect 123 E. Robinson St. parked its apparatus in front of the building. While conducting the inspection, the fire company is told there is a fire in a church on 2nd Ave. at Park Ave. The best way for the fire company to get there is:
   (A) Go north on 5th Ave. to Park Ave., west on Park to 2nd Ave., and then south on 2nd to the church.
   (B) Go west on Robinson to 1st Ave., north on 1st Ave. to Park Ave., east on Park to 2nd Ave., and then south on 2nd Ave. to the church.
   (C) Go north on 3rd Ave. to Anoa St., west on Anoa to 2nd Ave., and then north on 2nd Ave. to the church.
   (D) Go west on Robinson to Main St., north on Main to Park Ave., west on Park Ave. to 2nd Ave., and finally south on 2nd Ave. to the church.

46. The business district in this town is located on the
   (A) east side.
   (B) south side.
   (C) north side.
   (D) City Center.

47. It would be most correct to state that "4th Ave. . . ."
   (A) runs from Park Ave. to Central Ave.
   (B) is a two-way street.
   (C) is two blocks long.
   (D) runs east to west.

48. A man living in a house on Dunn St. at 1st Ave. wants to make a personal complaint about what he believes may be a local fire hazard. By car, the closest place to report the fire hazard is
   (A) the W. Strauss St. firehouse.
   (B) the Carol St. firehouse.
   (C) City Hall.
   (D) the office on E. Walter St.

Answer Key and Explanations

ANSWER KEY

1. B	8. B	15. C	22. B	29. B	36. C	43. B
2. D	9. B	16. D	23. A	30. C	37. C	44. C
3. B	10. D	17. B	24. C	31. D	38. D	45. D
4. B	11. C	18. D	25. A	32. A	39. B	46. B
5. C	12. B	19. A	26. D	33. B	40. A	47. C
6. A	13. C	20. D	27. C	34. D	41. C	48. A
7. C	14. C	21. B	28. A	35. D	42. B	

ANSWER EXPLANATIONS

GROUP ONE

1. **B** See items 10 and 11.
2. **D** A close look at the operating nut and to what it is connected reveals that this hydrant is in the closed position. In this position the operating nut will turn in only one direction—to the right, thus lowering the main valve and allowing water to flow in.
3. **B** Choice A—item 7 is related to allowing water into the hydrant; item 9 helps to let it out. Choice C—item 13 restricts the travel of the main valve. Choice D—item 8 is the standpipe of the hydrant.
4. **B** To answer this question, first read the statement and then trace the paths of the numbered items.
5. **C** The operating nut is connected to the main valve rod, which in turn is connected to the main valve. Choice A—the chain connects the cap to the barrel. Choice B—the bonnet does not turn; it is bolted into position. Choice D—the base is not identified, and the bonnet and barrel are held together by bolts.

GROUP TWO

6. **A** The kitchen has an exit door to the rear. Passage can be made through the dining room and then the living room.
7. **C** A door to the hall and one to the bathroom, and a window on the west side and one on the rear.
8. **B** These two adjoining rooms have an open passageway, which would permit the fire to extend rapidly.
9. **B** The living room has three windows, two doors, and access to the hall, which leads to four other rooms.
10. **D** If you count all the windows, the answer is 12. The living room has two windows which are adjacent to each other.
11. **C** From this room you must pass through bedroom 1, then the hall, and finally the kitchen or living room.
12. **B** The kitchen has a door and a window leading to the porch; the dining room has one window leading to the porch.
13. **C** To reach these two rooms you must first pass through another room.
14. **C** There are seven interior doors and two exterior doors.
15. **C** There are one side window and two front windows.

GROUP THREE

16. **D** This is attached to the supply line riser and is used to pump water into it.
17. **B** If you did not remember this, you should be able to reason it out. The check valve prevents the water from flowing back and forth. We would want water from the tank to flow into the system, but we wouldn't want water from the supply line to flow into the tank. Therefore the point where they meet should have a check valve.
18. **D** This is indicated as item 2.
19. **A** The post indication valve is located in the street and is used to shut off the water supply from the city main. Choices B and D—the O.S. and Y. valve is the sprinkler control valve.
20. **D** There are five on each of the three floors.

GROUP FOUR

21. **B** Two doors in each apartment lead to the public hall.
22. **B** The kitchen is immediately adjoining the dining room; all the other rooms are separated from it by a hall passage.
23. **A** This room has a door leading from the public hall and an interior door leading to the center hall.
24. **C** The shaft on apartment A's side is diamond shaped with four windows. The shaft on apartment B's side is square and contains only two windows. Choice A—there are two possible exits, the stairs and the fire escape. *Note:* The question referred to the building, not the apartments. Choice B—the shafts and the number of doors are different. Choice D—the fire escape serves one window in each apartment.
25. **A** This room is located on the center hall and would not allow passage through it. Choice B—the fire would extend into the hall and bedroom 1 (if the door was open). Choice C—it should be possible to use the fire escape from the rooms adjoining it. Choice D—the fire would not be confined because there is no dividing wall or door to stop it.

GROUP FIVE

26. **D** See section 5.c. Choice A—the ash is also red. Choice B—the picture is required. Choice C—the penalty can be a fine, imprisonment, or both.
27. **C** See section 5.e. Choice A—official signs may be used only in places where required by law or fire department regulation. Choice B—the signs may be printed by anyone. Choice C—the new style must be used for new signs; the old signs may be used until the supply is exhausted. Choice D—this number is not known. We know about the new sign and we know that there are existing official type signs, but we don't know how many.
28. **A** See section 6. Permitting smoking in these controlled areas reduces the temptation to smoke in rooms where dangerous gases used for medical purposes may be present.
29. **B** Signs may be either 10" x 14" or 11" x 15". Choice A—there are only two colors, red and black. Choice C—a number of "other" durable materials are available.

GROUP SIX

30. **C** Cleaning is only one part of care; sharpening is the other.
31. **D** Once the handle is washed, you have finished except to check for defects. It should not be painted (A), varnished (B), or oiled (C).
32. **A** This correct answer is found in the second sentence of the section entitled "Chopping." Choice D—the cut should be made 2 inches from the joist (the last sentence under "Chopping").
33. **B** The 60 degree angle is mentioned under "chopping."
34. **D** The passage states that the best method is to use two hands, and then explains how to hold the axe. Choice D—is the only selection that meets this criterion.

GROUP SEVEN

35. **D** Choice D is the correct image. Choices A and B are images from a side view, and Choice C is an image from the front view.
36. **C** Choice C is the correct image. Choice A is incorrect; it does not show the chimney and the canopy over the front door. Choice B shows the canopy and the chimney on opposite sides of the house and does not show the peaked roofs. Choice D shows the chimney and canopy on the back of the house, and does not show the peaked roofs.
37. **C** Choice C is the correct image. Choice A shows the porch on the wrong side. Choice B shows the single slanted roof portion of the house as the larger of the two sections, but the original drawing shows the multipeaked roof as the larger portion of the house. Choice D shows the chimney in the center of the single slanted roof, but the original illustration shows it at the end of the house.
38. **D** Choice D is the correct image. Choices A and B are incorrect. You would not be able to see the "Every Day Shoes" sign from inside the store. Items that can be seen from the inside of buildings are those that appear on glass or can be seen through the glass, such as a street pole, mail box, canopy, and writing on the canopy. Choice C is incorrect because the writing on the windows would now appear on the opposite side and in reverse. Items that were on your left when looking in from the outside will appear on your right when looking out from the inside.
39. **B** Choice B is the correct image. Choice A is incorrect because the single building at the end is not larger than the group of three buildings. Choices C and D are incorrect because they do not show the projections on the side (cornice) and above (parapet walls) of the roof line which should be visible from the rear.

GROUP EIGHT

40. **A** Choices B, C, and D are incorrect because they violate the traffic laws. Although these routes may be shorter, the directions are specific about not violating the traffic laws. A normally unrecognized portion of these questions is the testing of your ability to understand and follow orders and directions.
41. **C** This is the shortest, most direct route. Choice B is incorrect because it fails to tell you how to get from Broadway to Central Ave. Choice A violates the traffic law. Choice D is a good route, but it is longer.
42. **B** This is the shortest, most direct route. Choice A is incorrect be-

cause going west on Dregg St. is heading away from 479 Dregg St. Choices C and D violate the traffic law.

43. **B** Choice A and C violate the traffic law. Choice D takes you in the wrong direction when you reach Central Ave.

GROUP NINE

44. **C** This is the correct path. Choice A violates the traffic laws and takes you in the wrong direction on Walter St. Choice B violates the traffic law at E. Walters St. Choice D violates the traffic law at the intersection of Bowery Blvd. and Peter St., and it omits going through Peter St. to reach 1st Ave.

45. **D** This is the most direct, lawful route. Choice A would require a violation of the traffic law on E. Robinson St., Choice B violates the traffic law at 1st Ave., and Choice C violates the traffic law at 2nd Ave.

46. **B** The area where the largest number of offices are and where City Hall is located would be considered the business district. The north and center of the city are residential, having a large number of houses, stores, and churches. The east and west sides are mixed residential and business.

47. **C** This is correct. Choice A is incorrect because 4th Ave. runs from Park Ave. to Carol St., Choice B is incorrect because it is a one-way street, and Choice D is incorrect because it runs north to south.

48. **A** This is correct. This is a two-part question: Where do you personally report a local fire hazard, and how do you get there? Choice B is not direct and is substantially longer than Choice A. Choices C and D are incorrect because the complaint should be reported to the local firehouse first.

CHAPTER 7

Understanding Reading and Verbal Comprehension Questions

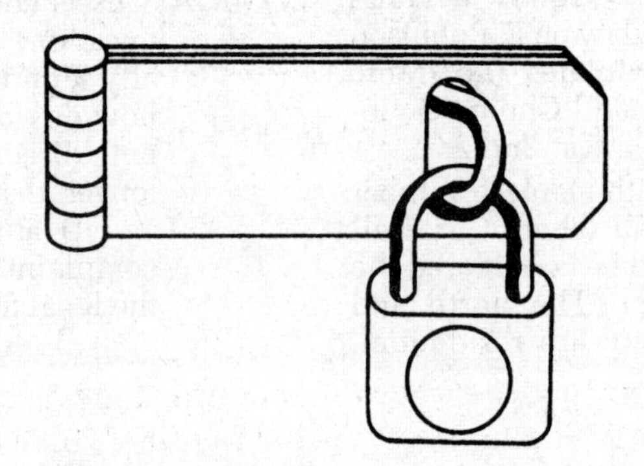

Because so much of the Firefighter Examination requires the ability to understand written material, you must take extra time and care with this chapter. Don't be in a hurry to move on to the next group of questions; first make sure that you have the reading skills necessary to answer correctly any questions you will be asked.

Firefighters must have the ability to understand one another and the public. This skill is referred to as verbal comprehension. Good verbal comprehension means that firefighters can interpret what another individual is saying. A good vocabulary is part of this skill but not all of it.

As a firefighter, you must be able to listen to a description of what is happening and understand what it means to you. This skill is tested in two different ways during the fire department entrance examination: orally and written. The oral interview tests your ability to hear a question, interpret what is being asked, and effectively respond. This activity was described in Chapter 2.

The second manner in which communication skills are tested is through written materials. Using written materials to test for verbal skills is not the best technique; however, in those situations where the examiner will be unable to hold an effective oral interview, written materials are often substituted. The written materials for the verbal comprehension sections are very similar to the reading comprehension materials. The questions are slightly different because the candidate is asked to come to a conclusion based on the information provided by the passage, to put items in the passage into some order, or to express the original idea in a clear and precise format.

Reading ability can be improved with patience and practice, in the same way you learn to type or to hit a baseball. The brain is a collection of nerve cells; how the brain remembers is not precisely known. However, the brain is like a muscle, which you must exercise and train to get into shape. Reading is a skill that requires exercising of the brain—the more you do it, the more you practice, the more you challenge yourself, the greater the results you achieve. In this chapter, you will find general hints for improving your overall reading skill, as well as specific strategies for answering reading comprehension questions. Spend time learning this material, and even after you have mastered it don't hesitate to go back and review.

How to Concentrate

How often have you "read" something by looking at the words without concentrating on their meaning or the author's intent? How many times have you "read" something and then discovered when you finished that you had no idea of what you had "read"? The problem of letting your mind wander is the biggest roadblock to overcome if you want to become a good reader and test taker. Jumping ahead to see what's coming or glancing back to make sure you read what you thought you read are probably the second and third most common factors in poor reading and test taking. The kind of reading required to be a successful test taker can be learned; it is the kind that calls for short periods of total concentration. This skill is developed by using a system—a system that allows your mind to create mental pictures of what you are reading and then to ask questions about them.

A good way to develop your ability to concentrate is to work with short passages from newspapers or magazines. Choose an article that interests you, read it carefully, and then put it aside. Now try to write down the key points of the article without looking at it again. Start small, that is, with short, easy articles. As you improve, increase the length of the articles and also try more difficult materials. You will become more proficient with practice. Your key to success is to concentrate and focus on what you are reading, while avoiding the distractors around you.

Increasing Your Vocabulary

Concentration will not help you if the reading passage contains a significant number of words that you don't understand. Fortunately, some simple guidelines and a little work can help you improve your vocabulary dramatically.

1. When you are reading a book, magazine, or other material that belongs to you and you come across a word you don't fully understand, write the word in the margin. If the reading material is not your own, write the word on a card and indicate where you read it. In this way you can look the word up at your convenience and then go back to the passage to see how it was used.

There are two good techniques for increasing your vocabulary. One uses a system of 3 × 5 cards; the second, a notebook. The card system works well because it allows you to carry a small stack of cards with a significant number of words on them. The cards take up almost no room, can be easily carried in a pocket or purse, and are readily available for review when you have free time. Put the words on one side of the card and the definitions on the reverse side, and carry a few blank cards to ensure that you have a place to enter any new words you come across. When you write the words and definitions, organize the material so that it can be read by flipping the cards from the bottom up. This will allow you to read through the cards faster, to put the cards you are finished with in the back of the pile, and almost always to keep your place.

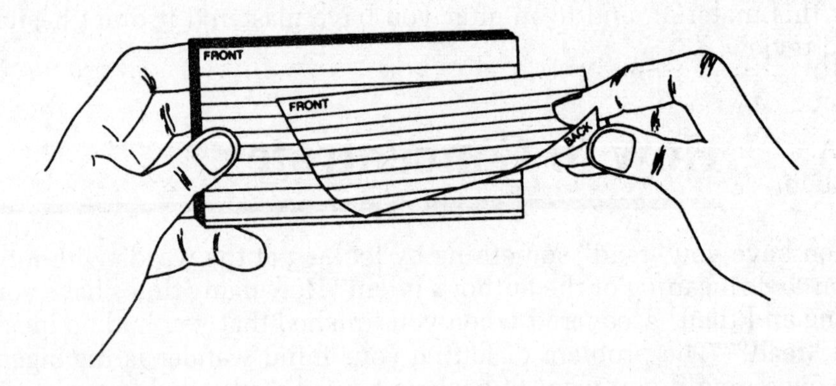

Be sure to carry the cards with the words that are giving you the most trouble, and as you review them, mark the words you are still not sure of. The next time you take a stack of cards to review, take only the ones with marked words. You should review all the words periodically.

The notebook has the advantage of being permanent and potentially much more comprehensive. If you choose this method, on each page make two columns, one about 1½-inches and the second occupying the rest of the page. In the first column put the word, and in the second the definition. In this way you can practice by going from definition to word, or word to definition.

2. As soon as possible, look up the meaning of the word in Chapter 5 if it is fire-related or in the dictionary. Write a brief but accurate definition on the back of the index card or in the notebook.
3. Return to the material where you read the word, and make sure you now understand its meaning. If the material is your own, write the meaning near the word. Make sure you fully understand the use of the word in the sentence.
4. Try to use each new word in your everyday conversation. You will make a few mistakes in the beginning, but they will only help to reinforce your learning.
5. Keep a separate list of any words you misuse, and review them periodically until you are certain you have mastered them.
6. Ask a friend to test you on the meanings of these words, and spend more time on the words you still don't know.

Strategies for Handling Reading and Verbal Comprehension

1. READ THE DIRECTIONS. Today's examinations contain general directions at the beginning of the examination, along with specific directions preceding each different type of question. Be sure to *read all instructions* carefully before doing the questions.

   The most important direction given for reading and verbal comprehension questions is to answer the questions based solely on the information provided. Never introduce other knowledge you possess into a reading comprehension question. You may be better informed or have a different opinion from the author; however, the examiners are not looking for this. They are testing you on your ability to read and extract information from the materials given.

2. SCAN THE CONTENTS OF THE PASSAGE. Get an idea of what the reading is about—an overview or survey of the passage. You can do this by skimming and looking for signposts that indicate the content. A good rule is to read the first sentence of each paragraph and then the entire last paragraph. But do this rapidly, just to get an idea of what is contained in the passage.

3. READ THE STEM OF EACH QUESTION PERTAINING TO THE PASSAGE. After scanning the passage, go to the questions and read the stem of each question, which contains the information that precedes the answer choices. Understand what is to be tested for in the passage. Also, scan some of the choices, but do this quickly. By knowing what you will be asked, you can make the best use of your reading time.

4. READ THE PARAGRAPH CAREFULLY. Now that you know what the paragraph is about, and what type of information you need to answer the questions, read the paragraph *very carefully.* This is the time to put that concentration practice to work, the time to concentrate exclusively on the material in the paragraph. While you are reading the material, note the key facts and mark them according to strategy 5.

5. UNDERLINE OR CIRCLE KEY WORDS OR PHRASES. WRITE KEY FACTS IN THE MARGIN. Unless the directions prohibit doing so, all good multiple-choice test takers use their pencil to underline key items in the reading passage and to write notations in the margin.

   What kinds of key words or phrases should you underscore or circle?

   - Transitional words and terms that signal a change in thought. Examples: *therefore, however, nevertheless, on the other hand, yet, but*
   - Absolute words (usually a wrong choice as answers; they are generally too comprehensive and are difficult to defend). Examples: *never, always, only, none, all, any, nothing, everyone, nobody, everybody*
   - Limiting words (usually a correct choice as answers). Examples: *usually, generally, few, occasionally, sometimes, some, possible, many, often*

How should you indicate key facts? Place an asterisk in the margin next to important information, write short notes with arrows directed to the portion of the passage to which you are referring, or bracket the portion that must stand out.

Important information includes *all items covered in the questions* and, in fire-related materials, such points as the following:

- Time of occurrence
- Dangerous items
- Location of fire
- Number of exits
- Statements of or about key people
- Fire protection equipment
- Orders or directions

A little restraint is also needed. You must keep in mind the purpose of all this marking and noting—to help you understand the passage and retrieve information quickly. Too many marks or notes can be confusing and will obscure, rather than emphasize, important items.

6. ASK YOURSELF QUESTIONS WHILE READING THE PARAGRAPH. Your ability to understand the passage is increased significantly when you pause occasionally and ask yourself questions about what you are reading, for example:
   - What is the main idea of the passage as a whole?
   - What is the main purpose of each paragraph in the passage?
   - What action is required?
   - Who is directly affected?
   - What could have been done to increase safety?
7. DEVELOP A MENTAL PICTURE OF WHAT YOU ARE READING. Many people find that the best way to understand and enjoy a novel is to project themselves into the story, and then try to share the characters' experiences. The good test taker uses this technique when working with reading and verbal comprehension questions. Use your imagination to develop a mental picture or impression of what you are reading. If you practice this with your everyday reading material, such as newspapers and magazines, you will find that it helps you to retain what you read.
8. ANSWER THE QUESTIONS, using the strategies outlined in Chapter 3. Remember that every answer is contained somewhere in the reading passage; review the passage quickly to verify your choice and to eliminate the wrong answers. If you have noted key facts in the margin of the test booklet, you should be able to locate and review the appropriate section rapidly.

Practice Exercises

Much of the initial training that firefighters receive is from classroom instruction or from drill site simulations, which are of a hands-on nature; however, a great deal must also be learned from training bulletins, manuals, and guides. This kind of training requires the ability to read with understanding and to be able to apply what has been learned to solve a problem. The following series of exercise questions was developed to allow you to test your ability to read and apply knowledge. You can expect similar questions on the official examination.

Directions: For each group of questions, take 20 minutes to read the passage and answer the questions that follow. After reading the passage carefully, choose for each question the one best answer—(A), (B), (C), or (D)—and write the corresponding letter on your answer paper next to the number of the question.

GROUP ONE

Read the following passage and then answer questions 1 through 10 solely on the basis of information in the passage.

The All-Purpose Saw

The all-purpose saw, which has recently been introduced into the fire service, facilitates cutting operations, speeds ventilation, and generally provides greater operational efficiency at fires and emergencies.

This saw is powered by a high-speed gasoline engine which drives a 12-inch circular blade. The blade's maximum depth of cut is 4 inches. The two-stroke engine develops between 7 and 8 horsepower at a blade speed of 6,000 RPM. The combination of high horsepower, great speed, and use of proper blade enables the user to cut virtually any material encountered at fire and emergency operations.

The saw comes equipped with three cutting blades: a carbide-tip blade, a steel-cutting blade, and a concrete-cutting blade.

The carbide-tip blade is specially designed for cutting through gravel- and tar-covered roofs, wood flooring, light sheet-metal coverings, and other, similar substances. The steel-cutting blade is an aluminum oxide abrasive blade used for cutting through various types of steel found in automobile wrecks, bars on windows, and similar objects. The concrete-cutting blade is a high-speed silicon carbide abrasive blade used for cutting through concrete and other masonry products.

For maximum safety and greatest operational efficiency only sharp blades approved by the manufacturer should be used.

1. The steel-cutting blade is
   (A) made of aluminum.
   (B) used to cut windows.
   (C) an abrasive-type blade.
   (D) designed for cutting light sheet metal.
2. For which of the following would the fire service all-purpose saw NOT be used?
   (A) Making cutting operations easier.
   (B) Cutting a cord of firewood.
   (C) Making fire operations more efficient.
   (D) Speeding up the ventilation process.
3. The cut of the all-purpose saw is limited to a depth of
   (A) 2 inches.
   (B) 4 inches.
   (C) 6 inches.
   (D) 12 inches.
4. To cut a wood floor the best blade would be the
   (A) aluminum blade.
   (B) silicon carbide blade.
   (C) steel-cutting blade.
   (D) carbide-tip blade.
5. To breach a hole in a cinder-block wall the most effective tool is an all-purpose saw with a
   (A) high-speed abrasive blade.
   (B) carbide blade.
   (C) aluminum oxide blade.
   (D) silicon carbide blade.
6. The best saw blade for cutting a ventilation hole in a roof is
   (A) a silicon blade.
   (B) a carbide-tip blade.
   (C) an oxide blade.
   (D) none of the above.

7. The all-purpose saw is highly effective because of
   (A) its high speed.
   (B) the versatility of the saw blades
   (C) its high horsepower.
   (D) the combination of high speed, high horsepower, and versatility.
8. A firefighter alert for safety would use
   (A) the saw only on the ground.
   (B) only sharp blades.
   (C) a back-up firefighter to watch the operation.
   (D) only the carbide-tip blade.
9. Which of the following would be INCORRECT to say about the all-purpose saw?
   (A) It has been used by the fire service for many years.
   (B) It comes equipped with three types of blades.
   (C) It can cut almost any material.
   (D) It can be used at fires or emergencies.
10. The horsepower rating of the all-purpose saw is approximately
   (A) 3.5 horsepower.
   (B) 4.5 horsepower.
   (C) 7.5 horsepower.
   (D) 6,000 horsepower.

GROUP TWO

Read the following passage and then answer the questions 11 through 15 solely on the basis of information in the passage.

Lifting

Sprains and strains continue to account for a large percentage of the time lost to injuries. In a significant number of cases the circumstances indicate that a modification in the method of lifting might have prevented the injury. Lifting is so much a part of everyday living that most of us don't think about how we do it. Unfortunately, it is often done incorrectly, resulting in pulled muscles, disc injuries, or painful hernias. There are some simple, basic steps for safe lifting that everyone should be aware of and should use, not just on duty, but for off-the-job tasks as well.

Here are seven basic steps for safe lifting, to be applied whenever possible:

- If the object to be lifted is too heavy, get help.
- Part your feet comfortably, with one alongside the object and one behind it.
- Keep your back straight, nearly vertical.
- Keep elbows and arms close to the body.
- Grip the object with both hands, using the whole hand.
- Tuck your chin in.
- Keep your body weight directly over your feet.

11. It would be INCORRECT to lift a relatively heavy pail of debris with
   (A) elbows and arms close to the body.
   (B) chin tucked in.
   (C) body weight over the object.
   (D) back straight.
12. If you are a firefighter confronted with the need to lift an extremely heavy object, you would be acting correctly if you
   (A) advised your superior that the object could not be lifted.
   (B) requested help from your superior.
   (C) tucked in your chin, gritted your teeth, and lifted the object yourself.
   (D) convinced yourself the object would not be too heavy if you used the proper lifting technique.

13. INCORRECT lifting
    (A) leads to headaches.
    (B) seldom occurs.
    (C) can result in a pulled muscle.
    (D) cannot be avoided.
14. It would be INCORRECT to say about the process of lifting that:
    (A) There are seven basic steps for safe lifting.
    (B) Many injuries incurred while lifting could be avoided.
    (C) Lifting is seldom required outside our jobs.
    (D) Injuries incurred while lifting are responsible for much lost time.
15. Poor lifting methods lead to sprains and strains. These injuries
    (A) occur only when on duty.
    (B) are an unavoidable hazard of the firefighter's job.
    (C) usually occur at night.
    (D) account for a large percentage of the time lost to injuries.

GROUP THREE

Read the following passage and then answer questions 16 through 22 solely on the basis of information contained in the passage.

"Calculated Risk"

Probably the most overworked, the most deceptive, and the least understood term heard in discussions involving safety is "calculated risk." Each of the two words has a definite meaning in itself; used in combination and applied to familiar problems, however, the words usually imply a certain confusion, both in the mind of the person using them and in regard to the particular situation to which they are being applied.

Firefighters work in a hostile environment and are often faced with emergencies that prompt risk-tasking. How "calculated" can their actions be if time has not been taken to "calculate"? Without an estimate of the situation, the "risk" may be nothing but a foolhardy act that exposes firefighters to injury and thus complicates, rather than expedites, the accomplishment of the task. With reference to fire fighting, safety is a relative condition. Firefighters are obliged to accept something less than an absolute degree of safety, and this is understood as a condition of the job. We must approach with caution, however, the inclination to solve each difficult problem with the words "a calculated risk" unless the situation has really been evaluated.

It becomes fairly obvious from questioning individuals who have stated "I took a calculated risk" that in most cases no calculations, mental or otherwise, were ever made. In fact, it is not unusual to discover that the speaker had not the vaguest notion whether the chance for an accident was one in ten or one in a million.

Statements regarding calculated risk can be meaningful only when the individual concerned

1. Knows the job.
2. Applies that knowledge to the task.
3. Has sufficient experience to make an intelligent evaluation of the situation.

If these criteria are not met, the outcome will fall entirely to "chance." Firefighters must not allow chance to become a way of life. They cannot afford to dash about blindly, depending on Lady Luck for protection. The answer lies in disciplining oneself to size up a situation rapidly and to make moves based on job knowledge and know-how. Only then can a firefighter truthfully say, "I took a calculated risk."

16. The term "calculated risk" is
    (A) an infrequently used expression.
    (B) not part of a firefighter's vocabulary.
    (C) the best understood term in the fire service.
    (D) one of the most overworked and most deceptive phrases used in connection with safety.
17. A firefighter can assume that:
    (A) Chance is a way of life.
    (B) Lady Luck and calculated risk are synonymous.
    (C) Most people make a conscious effort to estimate a calculated risk.
    (D) Even though the risk was calculated, it may still cause injury.
18. The main point of this passage is:
    (A) Firefighters have to take risks.
    (B) Injuries can be reduced by clear thinking.
    (C) Following the rules for making a calculated risk will remove the threat of chance.
    (D) The firefighter's safety is the most important consideration in fire fighting.
19. To make a meaningful calculated risk one must
    (A) discipline oneself to size up a situation rapidly and make moves based on job knowledge and know-how.
    (B) develop the fire fighting skills that allows chance to become a way of life.
    (C) study the rules for probability theory and statistical analysis so that one can accurately determine the chance of an accident.
    (D) avoid working in areas that present unusual hazard potential.
20. Firefighters work in a hostile physical environment and are often faced with emergencies. It would be accurate to state that:
    (A) Firefighters are always required to take risks.
    (B) Only some firefighters should take risks.
    (C) Firefighters should evaluate the probability of success before taking a risk.
    (D) The calculation of risk is desirable but is time consuming, and firefighters often don't have time to do it.
21. It would be correct to say about "calculated risk" that:
    (A) Each of the two words has the same meaning.
    (B) Risk cannot be calculated.
    (C) The term is badly misunderstood.
    (D) The term is confusing.
22. It would be INCORRECT for a fire officer to believe that:
    (A) Firefighters must accept something less than an absolute degree of safety.
    (B) Risk may be taken without estimating the situation.
    (C) Safety is a relative condition where fire fighting is concerned.
    (D) Self-discipline is an essential component of fire fighting.

GROUP FOUR

Read the following passage and then answer questions 23 through 30 solely on the basis of information contained in the passage.

Heat Transmission

Fires can spread by three means of thermal (heat) transmission: conduction, convection, and radiation.

Conduction occurs in all materials to some degree. It is the transmission of heat through a material without any visible motion of the matter. An accepted explanation of this phenomenon is that thermally disturbed molecules transmit heat energy by agitating adjoining molecules. The ability to transmit this heat energy varies with the specific material.

- Solids are the best conductors.
- Liquids are poor conductors.
- Gases are relative nonconductors.

Of the solids, metals are the best conductors. It may be noted that the better thermal conductors are also usually the better electrical conductors.

Example: In the common radiator, heat delivered by a fluid is conducted from the inner to the outer surface of the radiator to heat space.

Convection is a common phenomenon. Here heat is transmitted by the automatic circulation of the fluid (liquid or gas) involved. Upon receiving heat, the fluid becomes less dense and tends to rise. A colder fluid, being heavier, falls to a lower level.

Example: Hot water in a furnace is heated and, because of temperature and density differential, rises to radiators above. There the heat is transmitted to the metal of the radiator. The liquid, being cooled, then seeks the lower (furnace) level, where the cycle is repeated.

For another example of convection let's look at the same radiator that conducted the heat to its outer surface. The surrounding air absorbs the heat from the radiator, becomes lighter, rises, and is then replaced by cooler and more dense air.

Radiation, a major cause of fire extension in large fires in built-up areas, is a phenomenon whereby heat energy is transmitted through space. All matter (other than that at absolute zero) emits radiation when thermally agitated. The degree of radiation varies with the material and the agitation. The wavelength of the heat waves varies from the shortest, ultraviolet, to the longest, infrared. The heat waves travel only in straight lines through space, without heating the space concerned, until they reach a body capable of absorbing the energy.

Example: Sun or heat energy from fire, passing through a window, heats material in the room.

In summation, it may be briefly stated that, as far as firemanics is concerned, the associations are as follows:

Conduction—solids.
Convection—fluids (liquids or gases).
Radiation—space.

23. Which of the following statement is correct according to the passage?
    (A) Metals are the best conductors.
    (B) Convection is the same as conduction.
    (C) Heat radiation is a major cause of cancer.
    (D) Conduction occurs only in liquids.
24. In accordance with the phenomenon of convection, at a fire heat should
    (A) disperse rapidly.
    (B) seek lower levels.
    (C) increase geometrically.
    (D) rise until stopped by a barrier.
25. Heat is transmitted through space by
    (A) conduction.
    (B) convection.
    (C) radiation.
    (D) none of the above.
26. The thermal transmission of fires occurs by all of the following EXCEPT
    (A) association.
    (B) conduction.
    (C) radiation.
    (D) convection.
27. Which of the following is correct in regard to the wavelengths of heat waves?
    (A) The shortest is ultrascopic.
    (B) The longest is infrared.
    (C) The shortest is ultrared.
    (D) The longest is infraviolet.
28. Which of the following statements is accurate?
    (A) Conduction occurs in all materials.
    (B) Liquids are good conductors.
    (C) Convection of heated air is rare.
    (D) All fluids are liquids.
29. Heat transmission is a constant problem. Fires spread by
    (A) conduction only.
    (B) thermal transmission.
    (C) the hot water in a furnace.
    (D) the sun.
30. Which of the following statements is INCORRECT?
    (A) The sun's rays are a form of convected heat energy.
    (B) Water in a pipe will rise.
    (C) The air is coolest close to the floor.
    (D) Radiation travels in straight lines.

GROUP FIVE

Many home owners and apartment dwellers concerned about their possessions and their personal safety install sophisticated locking devices on the entrances of their domiciles. These locking devices are effective at keeping out the unwanted; however, they often create prison-like conditions for the home owner or apartment occupant.

When a fire occurs, firefighters must gain rapid entry to a building to rescue people inside and to extinguish the fire. Elaborate locking devices have often proved burdensome to firefighters. However, the recent development of powerful hydraulic tools are allowing the firefighters to get to occupants faster and easier.

The term "Jaws" was given to a powerful pushing and pulling tool. This tool is able to exert more than 10,000 pounds of force and can rapidly pry open the most difficult obstruction. The Jaws of Life has been used as an automobile extrication tool very successfully, and has proven to be an important part of the firefighter's arsenal. This powerful rescue system is composed of a hydraulic spreading device, a power unit, high grade hydraulic hose, spread locking device and pulling jaws, cutting jaws, rams, hand pump, and chains with hooks.

The Jaws of Life is heavy and cumbersome. Heavy tools can not be carried easily or rapidly to remote locations. This is a major disadvantage of "Jaws." Another disadvantage to this tool is the need for a gasoline generator to supply power to operate the tool. The most common and preferred method for using the Jaws of Life is with the power generator; however, the hand pump should be used in areas where an explosive atmosphere exists. The hand pump is effective as a back-up solution should the gasoline pump become inoperative.

Another new tool, the Hand-held Hydraulic Press, weighs less than 30 pounds, is easily transported, and is operational with the use of a hand pump. This tool is similar to a hydraulic automobile jack. The hand-held hydraulic tool can exert a force of more than 4,000 pounds. This tool exerts a pushing force between the outer door jamb and the door; thus, it can only work on inward-opening doors. The jaws of the tool are tapped into place with a maul or the back of a 6-pound ax. The hand pump operates quickly to spread the jaws and open the door. The hand-held hydraulic press is not intended as an automobile extrication tool and should not be used for this type of rescue.

When hydraulic tools are not immediately available, the firefighter must use hand tools, such as a pry bar or ax. Some fire companies use a special tool to remove locking devices and open the door in the conventional method. This is known as the "through the lock" method. A tool known as the "K-tool" is hammered behind the rim of the lock, and then a pry-bar is used to force the locking cylinder out of the door. This allows the firefighter to get access to the interior of the locking device, which can then be easily opened. The problem the firefighter has is that sometimes occupants will put three or four locks on a door or will put recessed rims on the lock, making it impossible to get the "K-tool" into position.

31. Which of the following is most accurate?
    (A) Apartment dwellers almost always have three or four locks on their entrance doors.
    (B) The "Jaws of Life" is effective on automobiles using the "through the lock" method.
    (C) Sophisticated locking devices keep the occupants in a prison-like living condition.
    (D) Recessed rims on door locks make it possible to get the K-tool into position.
32. The tool a firefighter would most likely choose to force entry into an apartment would be a
    (A) maul or ax.
    (B) maul or ax and the K-tool.
    (C) maul or ax and hand-held hydraulic press.
    (D) maul or ax and the Jaws of Life.
33. It would be LEAST correct to state that:
    (A) The hand pump is not intended to be used as an automobile extrication tool.
    (B) The hand pump and the Jaws of Life are effective in an explosive atmosphere.
    (C) The hand-held hydraulic press is powered by the hand pump.
    (D) The "Jaws" with the hand pump exerts a force up to 4,000 pounds.
34. Which hydraulic tool would be used at an automobile accident?
    (A) Jaws of Life
    (B) automobile jack
    (C) K-tool
    (D) hand-held hydraulic press
35. A firefighter talking to a group of citizens would be most correct in saying that:
    (A) Of all the forcible entry tools, the Jaws of Life is the heaviest.
    (B) The K-tool is used for recessed rim locks.
    (C) The hand-held hydraulic press exerts 10,000 pounds of pressure.
    (D) Sophisticated locking devices have never been a problem for firefighters.
36. The Jaws of Life are for
    (A) exerting a force up to 4,000 pounds.
    (B) pushing and pulling.
    (C) rapidly forcing doors.
    (D) pulling trains apart.
37. Which of the following is not part of the Jaws of Life system?
    (A) spreading device
    (B) maul
    (C) chains
    (D) ram

Answer Key and Explanations

ANSWER KEY

1. C	7. D	13. C	19. A	25. C	31. C	36. B
2. B	8. B	14. C	20. C	26. A	32. C	37. B
3. B	9. A	15. D	21. C	27. B	33. D	
4. D	10. C	16. D	22. B	28. A	34. A	
5. D	11. C	17. D	23. A	29. B	35. A	
6. B	12. B	18. B	24. D	30. A		

ANSWER EXPLANATIONS

GROUP ONE

1. **C** The steel-cutting blade is described in paragraph four as an aluminum oxide abrasive blade. (A common practice of examiners is to cut off part of a statement, just enough to make the statement correct yet incomplete.) Choice A—this blade is not made of aluminum; aluminum oxide is a composition material. Choice B—the steel-cutting blade is used on window bars, not windows. Choice D—for light sheet metal the carbide-tip blade is used.
2. **B** This use is not mentioned in the passage. The saw could cut firewood, but the fire service does not officially use the saw for this purpose. The other choices are all specifically mentioned in the passage.
3. **B** This information is found in the second sentence of the second paragraph.
4. **D** Cutting wood is the primary use of the carbide-tip blade. See the opening sentence of the fourth paragraph.
5. **D** See the last sentence in the fourth paragraph. Choices A and B are other examples of how an examiner may cut off part of a statement; this time, however, the choices are made too broad and therefore incorrect.
6. **B** This is part of the opening statement in paragraph four.
7. **D** This statement is given in the second paragraph, last sentence. The other choices are incorrect because they are incomplete.
8. **B** This information is in the last paragraph. Choice A—this is impractical; the saw must be used on roofs and on window bars, which are not always on the ground floor. Choice C—this would be a good technique, but you are directed to choose your answer according to information in the passage. Choice D—this would limit the versatility of the saw and also is not justified by the passage.
9. **A** The opening statement of the passage tells us that the saw is a recent introduction into the fire service. The other choices are all true statements, according to the passage.
10. **C** This is found in the second paragraph (between 7 and 8 horsepower).

GROUP TWO

11. **C** The last basic rule states that the body weight should be kept over the feet, not over the object.
12. **B** By asking for help, you would have informed your officer of the problem and would be able to avoid injury (see the first basic rule). Choice A—this implies that you would not attempt to solve the problem of lifting the heavy object.

In multiple- choice questions you are required to select the best answer. Choice B, being justified by the passage, is superior to choice A. Choice C—this is an unsafe act. Choice D—the stem indicates the object is extremely heavy; therefore it should not be lifted without assistance.

13. **C** This statement is found in the fourth sentence. The other choices are contradicted by the passage.

14. **C** In the third sentence we are told that lifting is so much a part of everyday living that most of us don't think about how we do it. The other choices are all correct statements according to the passage.

15. **D** This information is found in the opening sentence of the paragraph. Choice A—the last sentence of the first paragraph states that safe practices are for use both on and off duty. Choice B—see the second sentence of the first paragraph. Choice C—there is no mention in the passage about when lifting injuries occur; you should not infer that they usually occur at night or during the day.

GROUP THREE

16. **D** This is the opening statement of the passage.

17. **D** Computing the chance of success helps to determine whether a firefighter will or will not take the risk. Even though this is done, there is still a chance of injury. Choice A—chance is complete randomness, with no thinking or interaction on the individual's part. Obviously, this does not have to be a way of life; people can make decisions, and these decisions do affect the outcome of events. Choice B—Lady Luck is chance; it is not calculated risk. Choice C—most people do not make a conscious effort to calculate the degree of risk (see paragraph three).

18. **B** The thrust of this passage is that injuries can be reduced by thinking about whether to take a particular risk. Choice A—firefighters must take risks, but this is not the *main* point of the passage; it is only a subpoint. Choice C—following the rules for a calculated risk will allow the decision maker to have more data on which to base the decision but will not rule out chance entirely. Choice D—a firefighter's safety is important; however, personal safety must often be overlooked to extinguish a dangerous fire or to save another's life (see paragraph two).

19. **A** See the last paragraph. Choice B—chance should not be a way of life. Choice C—these theories have very little application on the fire ground. Estimating most "calculated risks" requires other knowledge and skills. Choice D—this is impractical in regard to fire fighting.

20. **C** This is the theme of the passage. Choice A—firefighters are required to take risks but not *always*—an absolute term. Choice B—only some firefighters take a risk at any one given time, but all firefighters are presented with the challenge of having to take risks at some time. Choice D—the fact is that they do have time and must take it (see paragraph two).

21. **C** The first sentence tells us that "calculated risk" is a "least understood" term. "Badly misunderstood" is another way of saying the same thing. Choice A—each of the words has a definite meaning in itself, and these meanings are different. To "calculate" you must make a computation. "Risk" implies taking a chance which may lead to failure. Choice B—risk can be calculated, maybe not as well as we would wish, but to some degree. Choice D—the term itself is not confusing; it is improper use of the term that has led to confusion (see the first paragraph).

22. **B** In the second paragraph the author states "without an estimate of the situation, the 'risk' may be nothing but a foolhardy act. . . ." The other statements are all true according to the passage.

GROUP FOUR

23. **A** Solids are the best conductors; of the solids, metals are the best. Choice B—conduction is the transmitting of heat by agitation of adjoining molecules. In convection heat is transmitted through a circulation process. Choice C—this is not mentioned in the passage and therefore is not an acceptable choice (see the stem of the question). Choice D—conduction occurs in all materials to some degree (see paragraph two).
24. **D** Heat is transmitted to the fluid (gas or liquid). The fluid then becomes lighter and rises; however, it can rise only if there is room to do so. If a ceiling or roof has no opening, heat will collect there until it finds or burns a hole through it. There is no justification in the passage for any other choice.
25. **C** As stated in the passage, radiation travels through space in straight lines without heating the space until it finds a body capable of absorbing the heat.
26. **A** Most questions on entrance examinations are looking for correct answers; however, you must also be prepared for the opposite. You must also watch out for the word EXCEPT; it is the key to finding the correct answer since it changes the direction of the stem. There is no mention of thermal transmission by "association" in the passage.
27. **B** This practice of jumbling parts of words requires some patience on your part. You must find the section in the passage that applies and then mark each answer as correct or incorrect. There is only one correct statement.
28. **A** This is found in the opening sentence of the second paragraph. Choice B—liquids are poor conductors. Choice C—the passage states that convection is a common phenomenon. Choice D—fluids may be *liquids* or *gases*.
29. **B** "Thermal transmission" is a synonym for "heat transmission" (see the opening statement).
30. **A** The sun's rays are radiation waves. Choices B and C—these are correct; see the example under convection. Choice D—this is correct; see the paragraph on radiation.

GROUP FIVE

31. **C** This is stated in the first paragraph. Choice A is incorrect; to be correct, change "almost always" to "sometimes." Choice B is incorrect. A technique for distracting the candidate is to combine a true statement with an unrelated statement that makes the answer incorrect. Choice D is incorrect; it should read "impossible."
32. **C** The message the writer is trying to give is the difficulty firefighters have in gaining access to the occupancy and the new tools that are now used by firefighters. The hand-held hydraulic press rapidly and effectively accomplishes the mission, is lighter and quicker than the Jaws of Life, and much more effective than the K-tool or other hand-held tools. It therefore is the most likely choice of firefighters. Choice A would be the last choice of firefighters, Choice B is better than A, but not as good as C. Choice D is incorrect. The maul is not used with the Jaws of Life.
33. **D** The third paragraph states the Jaws of Life exerts a force up to 10,000 pounds. Choice A is a correct statement; the hand pump is a power source for tools. Choice B and C are correct.
34. **A** The third paragraph tells the reader this tool has been used very successfully on automobile extrication. Choice B is incorrect. The automobile jack, although conceptually the same as the hand-held

hydraulic press, should not be used. Choice C is incorrect. The K-tool is used for door locks. Choice D is incorrect. The last sentence of the fourth paragraph tells the reader not to use this tool for automobile extrication.

35. **A** The third paragraph tells the reader the Jaws of Life is heavy, and a heavy tool can not be rapidly carried. The passage also tells the reader that the hand-held hydraulic tool is easily transported; thus, the reader can deduce that the hand-held hydraulic tool is lighter than the Jaws of Life. The maul and ax weigh about 6 pounds. Choice B is incorrect. The K-tool is not effective on recessed rim locks. Choice C is incorrect. The hand-held hydraulic press exerts a force of 4,000 pounds. Choice D is incorrect. Sophisticated locks have been a problem.

36. **B** This is stated in the third sentence. Choice A is incorrect. It should read 10,000 pounds. Choice C is incorrect. The tool can easily force doors but cannot be put into place and made operational rapidly. Choice D is incorrect. There is no mention of trains in the passage, and the instructions to the candidate say to use only the information in the passage. An answer may be true but not correct. You are being tested on your ability to understand what you read and your ability to follow written directions.

37. **B** The parts of the Jaws of Life system are listed in paragraph three; the maul is not listed as one of them.

CHAPTER 8

Decision Making, Reasoning, and Problem Solving

This part of the examination measures your ability to use logic and common sense in the solving of problems commonly confronted by firefighters.

The questions will generally relate to situations that require you either to take some action, to explain why an action has or would be taken, or to interpret what the action implies.

Understanding the Decision and Reasoning Problem

Decision making, reasoning, and problem solving are closely related subjects. The search to understand how they differ has led to many excellent theories; unfortunately it has also led to more new and challenging questions. For our purposes let us not worry about a theory; instead let us use the application of such a theory to our best advantage.

There are two types of reasoning questions: deductive, using general statements to get to a specific conclusion, and inductive, using specific statements to get to a general conclusion.

Five Steps for Handling Problem-Solving Questions

The steps necessary to arrive at the best solution for a question of this type are as follows:

1. *Identify* the PROBLEM.
2. *Identify* the POSSIBLE solutions.
3. *Select* the BEST SOLUTION.
4. *Eliminate* the OTHER SOLUTIONS.
5. *Make* the DECISION.

Now let's look at each of these steps in greater detail.

1. Identify the problem: The stem of the question should reveal this. Read the question carefully, and clearly identify what is being asked. What is the special problem or point of the question? Is there a specific order you must follow? Are there any special instructions you must follow?
2. Identify the possible solutions: Read ALL the choices quickly to identify the one that best solves the problem.
3. Select the best solution and then defend it: Answer the questions "Why does this solve the problem?" "Does it solve the problem with the least damage and or danger (keeping in mind that damage can be both physical and mental)?" "Will it solve the problem within a reasonable time?" "Have you answered all the parts of the question? Are they in the correct order?"
4. Eliminate choices that:
   - are contradictory to what is required,
   - call for unnecessary actions or risks (some risk may be necessary),
   - require that an order be disobeyed,
   - insult, disregard, interfere with, or hazard a citizen,
   - benefit you alone,
   - result in failure to obey a law,
   - only partially solve the problem, or
   - call for action that you are not authorized to take or are not in a position to implement.
5. Now make the decision and record your answer:

Decision-making problems can cover a wide variety of areas. For this reason there are no absolute rules, only guides. The ideal answer will (1) solve the problem (2) with a high degree of safety (3) in the least amount of time (4) with the least cost to the citizens and the fire service.

Practice Exercises

Directions: Read each question carefully, and apply the five steps outlined in the preceding section. Select the one *best* answer—(A), (B), (C), or (D)—and write the corresponding letter on your answer paper next to the number of the question.

GROUP ONE

1. A firefighter performing a fire prevention inspection finds a maintenance person using a match to check for the source of a suspected natural gas leak. The FIRST action the firefighter should take is to
   (A) tell the person how dangerous it is to get so close to the gas with a match.
   (B) explain that the area should be well ventilated before lighting the match.
   (C) tell the maintenance person to extinguish the match immediately.
   (D) issue a written order to the person in charge to have the match extinguished and to prevent this from happening again.
2. A firefighter attempting to extinguish a fire in a portable electric heater disconnected the main electrical supply to the building. The firefighter's action was

(A) correct—this guarantees that no electricity will flow into the heater
(B) incorrect—disconnecting the plug would accomplish the same result and be faster
(C) correct—this action is needed before pulling the plug out
(D) incorrect—the fire should always be extinguished first

3. When the fire department responds to a fire in a U.S. mailbox, it uses an extinguishing agent other than water. The primary reason for this action is that:
   (A) Water applied to a fire in a confined mailbox will create an explosive hazard.
   (B) Water damage to the unburnt mail will make it undeliverable.
   (C) Approved chemical agents are more effective than water.
   (D) The delay caused by stretching a hose line would allow it to grow.

4. The fire fighter outside the building in illustration 4
   (A) is not really needed
   (B) should be equipped with a self-contained breathing apparatus
   (C) is checking the hose for leaks and defects
   (D) should be inside the door so he can watch the more senior fire fighters and learn from them

Illustration 4

5. The purpose of the blocks in illustration 5 is to
   (A) keep the hose from sliding
   (B) keep the hose dry
   (C) allow cars to pass over the hose
   (D) slow down the stretching of the hose line

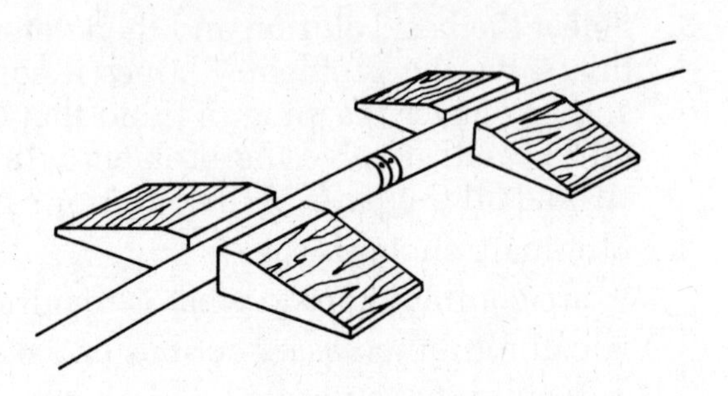

Illustration 5

6. The function of the wooden block and rubber tube on the hoses in illustration 6 is to
   (A) protect against chafing when the hose rubs on the ground
   (B) indicate which hose line is bigger
   (C) keep the hose kinked
   (D) Identify the hose connected to the hydrant.

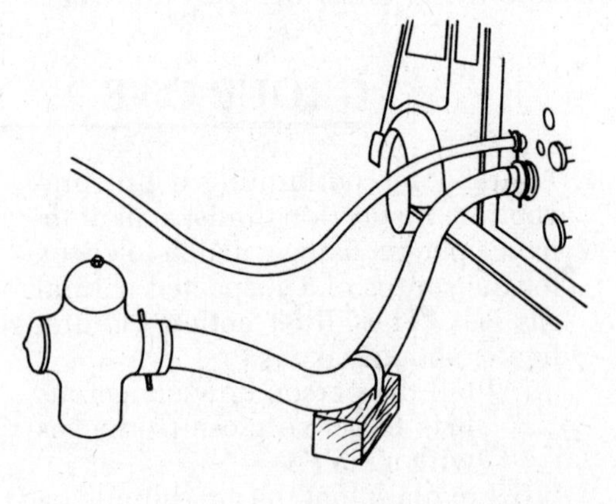

Illustration 6

7. Fire fighters often tie ladders as shown in illustration 7. This is done to
   (A) prevent them from being stolen
   (B) prevent them from touching the fence
   (C) prevent unauthorized persons from using them
   (D) Prevent them from slipping

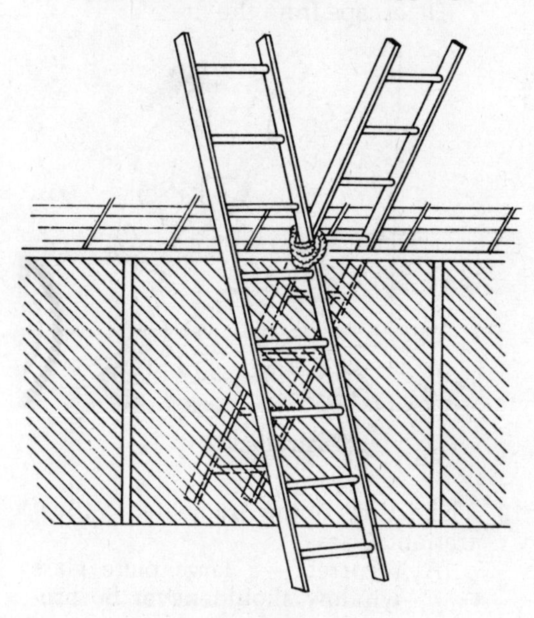

Illustration 7

8. In illustration 8 the fire fighter is
   (A) attempting to get up the ladder while the woman comes down
   (B) trying to take the child away from the woman.
   (C) preventing the woman from falling while descending the ladder
   (D) testing how much weight each of the ladder rungs will hold

Illustration 8

9. The correct action to take when your clothing is on fire is to
   (A) drop and roll
   (B) beat out the fire with your hands
   (C) run fast to blow out the fire
   (D) yell for help

GROUP TWO

10. The pike pole is being used in illustration 10 to
    (A) pull the ceiling down to expose the fire
    (B) feel for the hot spots in the ceiling
    (C) make a hole in the ceiling to allow the smoke to rise to the floor above.
    (D) provide a pole to slide to safety on

Illustration 10

11. The fire fighter on the ground in illustration 11 is
    (A) controlling the speed at which the fire appliance is raised
    (B) pulling the appliance down
    (C) keeping the appliance from going into a window
    (D) keeping the appliance from rubbing against the building

Illustration 11

12. The fire fighter in illustration 12 is attempting to
    (A) ventilate the apartment
    (B) test his new axe
    (C) find hidden fire
    (D) escape from the fire

Illustration 12

13. The actions of the fire fighter in illustration 13 are
    (A) incorrect—a large plate glass window should never be broken
    (B) correct—at a fire, plate glass should always be broken
    (C) incorrect—the window should be broken at the bottom so that glass doesn't slide down the pole
    (D) correct—breaking the window in this way tends to push the glass into the store and allows the heat to escape safely above the fire fighter

Illustration 13

14. Fire drills are held frequently in schools to
    (A) teach the students to be quiet
    (B) make the students familiar with fire department regulations
    (C) train the students to leave the school in a quick and orderly manner
    (D) demonstrate and enforce the need for discipline

15. Many fires have been started by a combination of heating equipment and combustible materials. The most likely reason that fires of this nature increase when the weather turns cold is that:
    (A) More heating units are in operation.
    (B) Heating equipment tends to break down more quickly in the cold.
    (C) Flammable liquids are stored near the heating equipment.
    (D) Cold air often clogs the burner of the heating unit.

16. Upon arrival at a reported fire in a factory, fire fighters find a very heavy smoke condition. A fire fighter sent to check the rear of the building meets a person who states that he is the driver of a flat-bed tractor-trailer truck parked at the factory's loading platform and that the smoke is being created by a fire in cardboard boxes loaded in the truck. The FIRST action the fire fighter should take is to
    (A) tell the driver to drive the truck away from the building
    (B) tell the driver to disconnect the tractor and drive it away from the fire
    (C) tell the driver to go into the factory and alert the occupants
    (D) ask the driver to help the fire fighter stretch a hose line to extinguish the fire

17. Upon arriving at a serious automobile accident fire fighters find one driver unconscious in the front seat of her car. Of the following actions which would be LEAST appropriate for the fire fighters to take?
    (A) Stretch a hose line.
    (B) Pull the driver out of the car as fast as possible.
    (C) Set up a system of lights or flares to stop and divert oncoming traffic.
    (D) Stabilize the car by putting chocks under the wheels and disconnecting the battery.

18. A fire fighter would be correct if he concluded that the knot shown in illustration 18 is
    (A) designed not to slip
    (B) improperly tied
    (C) designed to be untied easily
    (D) a square knot

Illustration 18

19. Many people utilize an "octopus connection," as shown in illustration 19, to make maximum use of an electrical outlet. This is unsafe because
    (A) drawing excessive current from the outlet will overheat the wires in the wall
    (B) children may play with the plugs
    (C) the weight of the excess plugs tends to pull the device out, causing a short circuit
    (D) drawing excessive current from the outlet will overheat the wires in an appliance

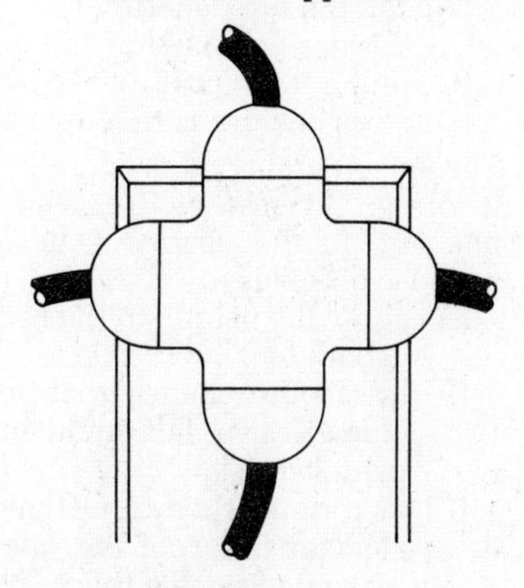

Illustration 19

GROUP THREE

20. While a person is lighting the oven in a gas stove, the match goes out. The proper action to take is to
    (A) quickly light another match and try again.
    (B) turn off the gas and wait awhile.
    (C) close the oven door and quickly light another match.
    (D) put one's finger over the gas jet and use a cigarette lighter to light the oven.
21. If a car goes into a skid, the driver should turn the wheels into the skid to retain control. The drawing that correctly illustrates this concept is

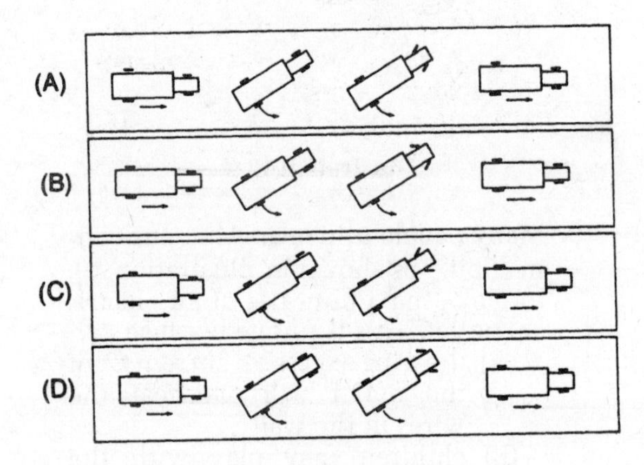

22. An electric iron left lying on an ironing board has started a smoldering fire. The correct FIRST action to take is to
    (A) call the fire department.
    (B) throw a pot of cold water on it.
    (C) pull out the plug.
    (D) cover the fire with a towel.
23. Theater programs often include this warning: "In case of fire, walk, do not run, to the nearest exit." The purpose of this is to
    (A) allow the people in the back to get out first.
    (B) lessen the chance that someone will trip, fall, and cause a panicky rush.
    (C) let the audience know that theaters are fireproof and safe.
    (D) Make it clear that there is little danger of fire in a theater.
24. When a small fire occurred on a kitchen table, the housewife threw a heavy scatter mat on it to extinguish it. This action was
    (A) correct—the mat will cut off the oxygen supply.
    (B) incorrect—the mat will spread the fire to the curtains.
    (C) correct—the mat will cool the fire.
    (D) incorrect—the woman should have run from the apartment and called the fire department.
25. While operating the nozzle of a hose line, a firefighter is told by the officer to "shut the nozzle." Which of the following is the best reason for closing the nozzle slowly?
    (A) The sudden stopping of the water pressure may break the hose.
    (B) Slowly closing the nozzle will reduce water damage.
    (C) Slow closing allows air into the hose.
    (D) The nozzle can't be closed fast.
26. A firefighter has just climbed a ladder and is about to search for a lost child in a heavily smoked-filled apartment. The most appropriate action to take before entering the window is to
    (A) make sure there is another way out.
    (B) use a tool or a foot to test the stability of the floor.
    (C) tie the ladder to a radiator.
    (D) put the fire helmet and flashlight by the window as a guide to getting out.
27. A hose line has been run between two rungs of a ladder. This would be considered
    (A) incorrect—the ladder can no longer be moved and used elsewhere.
    (B) correct—the weight of the hose is supported by the ladder.
    (C) incorrect—the weight of the hose will damage the ladder.
    (D) correct—the weight of the hose holds the ladder in place.

28. After a major fire the lieutenant left one firefighter at the scene to
    (A) watch for any rekindling of the fire and extinguish it.
    (B) prevent anyone from removing property.
    (C) search for lost valuables and tools.
    (D) assist the fire marshals in their search for evidence of arson.
29. At a fairly large fire in a factory, a firefighter discovers a number of compressed gas cylinders that are severely exposed by the fire. The most critical hazard in this situation is:
    (A) The valuable gas and cylinders may be damaged.
    (B) The cylinders may explode.
    (C) The toxic fumes from the compressed gas may be released.
    (D) The flammable compressed gas may be released and intensify the fire.
30. The most likely purpose of the hose line shown in illustration 30 is to
    (A) safeguard the occupants as they evacuate the building.
    (B) provide protective vapor barrier.
    (C) cool the surface of the exposed building.
    (D) extinguish the fire in that building before fighting the fire in the other building.

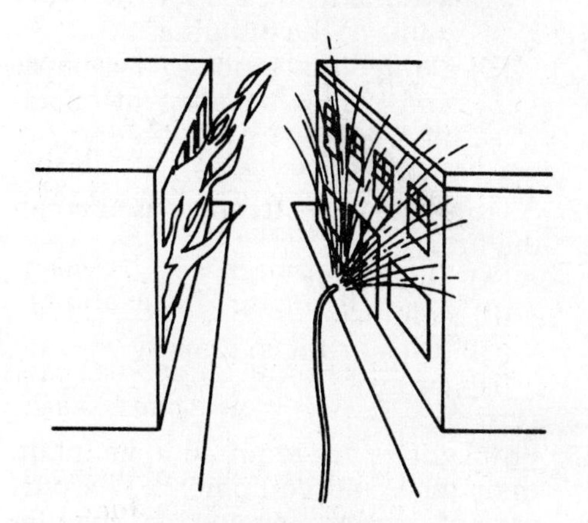

Illustration 30

31. Which of the following is the primary purpose of fire department inspections of private property?
    (A) To ensure that fire department equipment is not tampered with.
    (B) To train firefighters in private building-construction techniques.
    (C) To find and punish citizens who violate the fire prevention laws.
    (D) To have conditions that create fire hazards corrected.
32. During welding operations, sparks that could start a fire are given off. To ensure maximum safety, welding should be
    (A) done only outdoors.
    (B) limited to fire-resistant buildings.
    (C) preceded by a general clean-up and removal of combustible materials.
    (D) followed by a thorough soaking of all materials in the area.
33. The poor visibility in buildings on fire is
    (A) not serious—most people's eyes adjust quickly to the change in light.
    (B) serious—smoke obscures the light and acts as an irritant to the eyes.
    (C) not serious—the light from the fire is bright enough to show the exits.
    (D) serious—smoke causes breathing difficulties that often lead to serious injury.
34. Firefighters try to limit the amount of water they use at fires. The primary reason for this is to
    (A) limit the use of expensive water.
    (B) prevent collapse of the building.
    (C) avoid damage to the electrical and heating systems.
    (D) limit damage to furnishings and personal belongings.

GROUP FOUR

35. A sprinkler system can be supplied with water from either a single source or from two or more sources. Firefighters prefer multisource systems because
    (A) the chance of water failure is reduced.
    (B) installation is less expensive.
    (C) the multisource system works faster.

(D) the multisource system puts out more fires

36. When a fire is burning freely, it produces sparks and sometimes brands, that is, large pieces of burning materials which can be carried away from the fire by the air currents. The danger of sparks and brands is greatest from a fire
(A) at a gasoline station
(B) in a lumber yard
(C) in an electrical distribution plant
(D) at an open parking garage

37. While on the way home, a person hears and then sees a bell marked "sprinkler," and notices water coming from a pipe about 10 feet away. It is 2 A.M., the building is closed and locked, and no fire alarm box is in sight. The person knows, however, that there is a firehouse three blocks away. Given these conditions, the proper action is to
(A) write down the address, and then go home and call the police department
(B) search around the neighborhood for the nearest fire alarm box
(C) go to the firehouse and notify the man on housewatch
(D) do nothing; the water is probably just an overflow

38. The safest way to descend a fire escape ladder is backwards, with the front of your body facing inward, toward the steps, close to the handrail, testing each step with your foot before applying your full weight. The most likely reason for this method is that:
(A) If a step is missing or broken, you will be less likely to fall.
(B) It is not necessary to use the handrail because the steps are less worn near the wall.
(C) Fire escape steps are less likely to rust near the wall.
(D) The correct procedure is to walk with the front of your body facing away from the steps.

39. At a fire that burned for a long time but has now been extinguished, a fire fighter is told to go to the fourth floor to retrieve several tools. His FIRST action should be to
(A) refuse to go
(B) go but retrieve only the tools that he brought up
(C) check to see whether the floor is safe
(D) cut a hole in the floor to lower the tools through

40. "Elevators should not be used to exit from a building on fire." Which of the following choices is NOT likely to be the reason for this advice?
(A) The elevator shaft acts as a chimney, pulling smoke and heat up it.
(B) The elevator car may take the passenger directly to the fire floor.
(C) Fire fighters need the elevators to transport their equipment to the fire area.
(D) Occupants of the building may become trapped in the lobby while waiting for the elevator.

41. The primary reason for stretching the first hose line to the interior stairs of the building is that:
(A) The fire is almost always in this area.
(B) It is less difficult to do this than to stretch up ladders.
(C) The hose then protects the main avenue of escape for the occupants of the building.
(D) Doing this requires less hose and makes an efficient operation.

42. A large-size hose stream is used when the fire is
(A) almost out
(B) confined
(C) just starting to grow
(D) far away

43. Fire fighters are required to maintain their tools and equipment. The purpose of regular cleaning and maintenance of fire apparatus is to
(A) give the fire fighters something to do between fires
(B) identify defects in the apparatus and reduce the chance of deterioration
(C) make the apparatus and equipment look good to superiors
(D) teach discipline and compliance

44. Boundary lines around a fire area, called "fire lines," are normally set up by the police department to keep citizens out of the immediate fire fighting area. The most likely purpose of fire lines is to prevent
    (A) an arsonist from getting away
    (B) fire fighters from talking with the local people
    (C) citizens from interfering with the fire fighting operations
    (D) citizens from seeing fire fighters make an error

45. While responding to a fire, fire officers do not talk with the fire fighter driving the apparatus. The primary reason for this is to
    (A) allow the officer to watch how the fire fighter handles the apparatus
    (B) ensure that the chain of command is maintained
    (C) allow the officer to prepare mentally to fight the fire
    (D) ensure that the driver is able to concentrate on getting the apparatus to the fire safely

Answer Key and Explanations

ANSWER KEY

1. C	10. A	19. A	28. A	37. C
2. B	11. D	20. B	29. B	38. A
3. B	12. C	21. D	30. C	39. C
4. B	13. D	22. C	31. D	40. D
5. C	14. C	23. B	32. C	41. C
6. A	15. A	24. A	33. B	42. D
7. D	16. A	25. A	34. D	43. B
8. C	17. B	26. B	35. A	44. C
9. A	18. C	27. A	36. B	45. D

ANSWER EXPLANATIONS

GROUP ONE

1. **C** This eminently hazardous situation requires immediate corrective action. Choice A and D require additional time, which may allow the gas to explode. Choice B is incorrect; a match should not be used under any circumstances to detect a gas leak. There are gas detector machines for this purpose.

2. **B** Using the simplest, most direct, and most effective solution is the best method. Choices A and C are wrong because the method mentioned in the stem requires additional time and results in increased danger, as does choice D.

3. **B** The use of water will cause the ink on the mail to run and thus make it undeliverable. Choice A—water would turn to steam and vent safely out the mail slot. Choice C—this may be true in some cases but not necessarily in all. Choice D—there should be no delay; a portable extinguisher could be used.

4. **B** All fire fighters who will work in the proximity of the fire should be equipped with self-contained breathing apparatus, which allows them to go into the fire area without delay if needed. In this illustration the fire fighter is feeding additional hose into the crew who are attacking the fire, which is an important activity. There is no justification for choices A and C. Choice D makes the assumption that the fire fighter is not a senior fire fighter; nowhere in the question or illustration is this indicated. Answers not based on given facts should be avoided.

5. **C** When a car or truck drives over a hose, it can cut off the water supply

and leave the firefighter in a dangerous position; also, driving over hose may damage it. The blocks are not used for the purposes mentioned in choices A, B, and D.

6. **A** As the water flow is started and stopped, the hose will rub against the ground, thereby wearing a hole in it. There is no justification for the other choices.
7. **D** Tying the ladders increases the degree of safety when climbing over a high fence. The reasons given in choices A, B, and C make no sense in a fire situation.
8. **C** When the firefighter guides each step of the person, and keeps body contact, the person's fear is reduced and the chance of descending the ladder safely is increased. Choice A—this would be an unsafe act. Choice B—once the mother and child are on the ladder, it could be unsafe to take the child away from the mother; in addition, she would normally be reluctant to trust the safety of the child to anyone else. Choice D—ladders are tested with weights; a human being should not be subjected to the possibility of ladder failure and personal injury.
9. **A** The drop-and-roll technique has proven to be very successful for quickly extinguishing clothing fires. Choices B, C, and D would all lead to an increase in the volume of fire and in the chance of serious burns.

GROUP TWO

10. **A** The pike pole is used to extend the firefighter's reach. Choice B—a wooden pole is a poor conductor of heat. Choice C—such an action would lead only to a more serious problem on the floor above the fire. Choice D—this is obviously wrong and should be the first choice to be eliminated.
11. **D** The firefighter at the ground level holds the fire appliance away from the building so that it does not get damaged. Choices A and B—from illustration 11 you cannot determine whether the fire appliance is being raised or lowered. However, in either case the speed and force of raising or lowering are controlled by the firefighter on the roof, not the one on the ground. Choice C—illustration 11 shows the fire appliance being raised or lowered in the area between the windows. It was prepositioned in this way to prevent it from hitting the windows.
12. **C** Fire travels upward and extends itself in hidden voids between the studs and the wall coverings. There is no justification for choices A, B, or D.
13. **D** As the pole is swung, it will continue on its arc and will fall into the window, reducing the chance of glass sliding down the pole. This action will let the heat and smoke escape safely, while keeping the firefighter relatively safe. Choices A and B—a plate glass window should be broken only when absolutely necessary and when doing so will significantly improve the fire operation. Choice C—this procedure is not the preferred one because heat and smoke liberated at this point can severely injure the firefighter.
14. **C** The basic subject of the question is safety. Drills are designed to reinforce what has been learned, and a fire drill is concerned with safe escape from a burning building. The reasons in choices A, B, and D, although they may be desirable, are not concerned with safety and therefore are not the main consideration in fire drills.
15. **A** When the weather turns cold, heating units are turned on; however, items temporarily stored near the burner during the warm months may be forgotten about and not removed. The combination of sufficient heat and ignitable materials starts a fire. Choices B and D—it is not the weather that causes a breakdown of heating equipment; it is improper design, use, or maintenance of the equipment. Choice C—flammable and combustible liquids are routinely stored near

heating equipment but, if kept in proper storage tanks, are not a problem.

16. **A** The cardboard boxes are not in danger of exploding; therefore, it should be safe to move the truck. Moving the truck away from the building will reduce the chance that the fire will spread into the factory. Choice B—disconnecting the tractor would not move the trailer or the part of the truck that is on fire, nor would it help to reduce the chance of fire extending into the factory. Choice C—this task should be undertaken by the fire fighters. Choice D—the truck should be moved first; then additional fire fighters should be called to stretch hose.

17. **B** Pulling the driver out of the car could result in serious damage if she is suffering from injury. It is *not* necessary to rapidly remove the person from the car; a thorough medical examination should be undertaken before any attempt is made to move her. However, since there is no fire, the area around the automobile should be controlled and proper safety precautions taken (A, C, D).

18. **C** By pulling the rope at point C the knot can be easily and rapidly untied. This type of knot is used when something must be temporarily held in place and quickly released.

19. **A** The adapter, sometimes called an "octopus connection," allows several appliances to be connected to the same outlet. This is not a problem if the appliances are drawing a low amount of current and the sum of the currents of all the appliances is less than the carrying capacity of the wire. However, the chance of exceeding the carrying capacity of the wiring in the walls (not in an appliance, as in choice D) is great, and a fire may result. Choice B—children should not be permitted to play with *any* plugs. Choice C—These plugs are usually designed to stay in position. If they do become loose and slip out, this would disconnect the power to the various objects and reduce the electrical drain on the wires in the wall. In this case there would be no danger of a short circuit. Choice D—in this situation it is not the appliance or the appliance cord that is important, but rather the wire inside the wall and the wall outlet box. The "octopus" is pulling more current through this wire than it can handle. As a result the wire in the wall (not in the appliance) becomes very hot and often starts a fire.

GROUP THREE

20. **B** Waiting will allow the gas that has already escaped to dissipate. Choices A, C, and D—these actions may cause an explosion and fire, in fact, choice C will almost always result in an explosion.

21. **D** In choice D the front wheels are turned in the direction of the arrow indicated at the rear wheels. This action tends to straighten the vehicle and break the skid.

22. **C** This is a small fire and can usually be controlled without professional help, eliminating choice A. The *first* thing to do is to remove the live electricity so that there is no danger of shock. *Then* put water on the burning materials; this is the second, not the first, step, eliminating choice B. After the materials have cooled, soak them in a large pail or tub. Finally, notify the fire department that you had a fire but have already extinguished it. (*Note:* If the fire is large, that is, if the ironing board is free burning, *call the fire department immediately.*) Choice D—for this type of fire, smothering often is not effective; a cooling agent is required.

23. **B** Walking tends to ensure safe footing and to prevent tripping. When someone trips and falls, the escape path is blocked and other people become afraid that they will not get out; this often leads to pushing and then to panic. Choice A—theaters must provide enough exits for all occupants; in emergencies all exits, not just the ones in the rear, are

used. Choice C—"fireproof" is a poor term and has been dropped from the fire service vocabulary because it is very misleading. The correct term is "noncombustible." The fact that a building will not contribute combustible materials does not mean that the furnishings, scenery, and costumes won't burn. Choice D—there is often great danger of fires in theaters, and for this reason they are very heavily regulated by fire prevention laws.

24. A If a fire is small, it often can be extinguished by quickly covering it with a substantial mat, rug, or blanket. *However, this technique is limited to small, incipient fires.* Choice B—a heavy mat will tend to smother, not spread, a small fire. Choice C—this statement is incorrect because the mat does not provide any cooling. Choice D—*for small fires* proper action for quick extinguishment is important. If the fire grows beyond the small stage, the fire department must be notified.

25. A When flowing, water loses some of its pressure to friction. If, however, the flow is suddenly stopped, the pressure builds up fast and can burst the hose. Choice B—closing the nozzle would result in less water damage but the hose may still burst, as indicated in choice A. Choice C—once water is flowing and all the air is pushed out, air cannot get back into the system until the hose is disconnected and drained. Choice D—a nozzle can be closed fast; this leads to water hammer.

26. B This ensures the firefighter's safety; the floor may have been weakened by the fire or even removed during a renovation. Choice A—it is impossible to make sure that there is another way to get out. What the firefighter must do is make sure that no one moves the ladder and that the way the firefighter went in remains available to get out. Choice C—this is impracticable and may well be impossible—there may not be a radiator. Choice D—for the firefighter to put the helmet and flashlight by the window is an unsafe act; it deprives the firefighter of head protection, and the ability to see in dense, dark smoke.

27. A It should be possible to reposition ladders at fires as necessary. Hose lines should be stretched up ladders but should be lashed inside the building so as not to interfere with the ladders.

28. A At major fires small pockets of fire may continue to burn for long periods. A firefighter remains at the scene to extinguish these small fires. The reasons given in choices B, C, and D are not functions of the firefighter.

29. B The gas in the cylinder will seek to expand when heated. If it cannot, it will exert a strong force on the wall of the container, which will eventually explode like a rocket, sending flying fragments in many directions. Choice A—in comparison to the potential for death and destruction if B occurs, the cost of damage to the gas and cylinders is negligible. Choices C and D—there is no indication in the question that the gas is either toxic or flammable. Do not read into the question information that is not there.

30. C Heat energy from the fire often starts fires in other buildings. A protective stream of water will reduce the chance of damage to this building. Choice A—occupants would be directed to use the secondary means of egress. Choice B—a vapor barrier would not protect against radiant heat. Choice D—there is no indication of fire in the building at which the hose is directed.

31. D Fire prevention is a very appropriate name for what the fire department is trying to accomplish. If the inspections result in correcting a problem, thereby preventing a fire, the fire department has succeeded in its purpose. Choices A and B are important subcomponents of the fire prevention program. Choice C mentions punishment, but compliance is the object.

32. C Cleaning up and removing combustible materials will prevent a fire. Choices A and B are not always possible, and choice D may lead to serious unnecessary water damage to surrounding items.

33. **B** The smoke generated at a fire rapidly reduces visibility by blocking and scattering the light rays. It also acts as an irritant to the eyes, causing a burning and tearing sensation. Choice A—this is not necessarily so, particularly when smoke irritation is involved. Choice C—smoke rapidly obscures almost all natural light and most artificial light as well. Choice D—this is true but does not relate to the question asked.

34. **D** The cost of water damage at a fire can often exceed the cost of the damage done by the fire. Therefore firefighters are taught to use water effectively to extinguish the fire but to control and avoid improper or excessive use of it. Choice A—water is relatively inexpensive. Choice B—firefighters channel the water out of the building, thereby reducing the possibility of collapse. Choice C—at fires these systems are shut down to avoid damage and shock.

GROUP FOUR

35. **A** A second source of water supply ensures that water will reach the sprinkler heads when something happens to the original source. For example; a multisource system may consist of the city water main, the gravity tank on the roof of the building, and a pressure tank somewhere within the building. Installation of such a system is *not* less expensive (choice B), nor does the system work faster (choice C) or put out more fires (choice D).

36. **B** This is a good question to find out whether you can think about your surroundings. To answer the question you must first identify what substance would most likely be burning at each of the locations named and then determine whether that substance would produce sparks and brands. Wood burning in a lumber yard would present the greatest chance of this occurring.

37. **C** Two decisions are required in this problem: (1) should the condition be reported, and (2) if yes, then how? (1) The correct choice here is yes; report the condition (the water motor gong, or alarm bell, is installed on the outside of the building to alert passersby of the need to report that something is wrong). (2) It is almost always correct to choose the method that will most directly solve the problem. In this situation the problem is solved by going directly to the fire station. Choices A, B, and D will result in a delayed alarm and an increase in fire growth and damage.

38. **A** The use of a fire escape is dangerous; it should be, and often is, limited to emergency conditions. Safety in descending these stairs is very important. The stem describes the most appropriate procedure. If you choose choice D, you are fighting the question. To succeed in test taking, you must answer the question asked, not the question you would like to be asked.

39. **C** The decision involves personal safety. An officer who orders a firefighter into a building must think it is safe. However, the firefighter must personally check for safety when carrying out the order and, if conditions are not safe, report this back to the officer. Choice A—the firefighter should not refuse as a first action, but should first check the floor out. Choice B—if the floor is safe, the firefighter must retrieve *all* the tools or equipment the firefighter was sent to get. Choice D—cutting a hole would not be done; it would weaken the floor.

40. **D** While this is a possibility, it is the least likely of the choices offered. Lobbies are provided with directional signs to stairs and exits. All the other choices are valid reasons for not using the elevator.

41. **C** Positioning the hose at the interior stairs permits the occupants to escape safely and the firefighters to gain rapid access for search and ventilation. Keep in mind that most people will try to use the normal entrance route first; it is usually also the most reliable and quickest

way out. Choice A—this statement is not correct; most fires are not in the interior-stairs area. Choice B—this is not always correct; a ladder may be the easiest and quickest method to reach a fire on the second floor front, but it may result in pushing fire into the stairway, cutting off that area of escape. Choice D—in many cases more hose, not less, may be required.

42. **D** The larger the stream, the greater the distance water can be thrown. In choices A, B, and C a small stream should suffice.

43. **B** During the cleaning and maintenance process possible defects are identified and corrected, and harmful road tars and other chemicals that destroy apparatus are removed. The reasons given in choices A, C, and D are less important (D) or even trivial (A, C).

44. **C** The fire scene is often very dangerous; well-intentioned but untrained citizens often offer help or advice and wind up getting injured or becoming a nuisance. There is no justification for the other choices.

45. **D** When conversation is limited, the driver is better able to direct his full attention and effort to the task of getting the apparatus and other fire fighters to the scene safely. Responding to an alarm requires the full and undivided attention of the apparatus driver. Unnecessary conversation can lead to momentary distraction and an accident. Choices A, B, and C are related but are significantly less important.

CHAPTER 9

Understanding and Working with Tools

The work of a fire fighter requires the use of many and varied tools. This chapter reviews tool identification, understanding, use, and safety.

Fire fighters are not employed to build or repair; their use of tools is generally limited to the control of emergencies, extrication, and the opening up of building components for ventilation and overhaul. The types of tools they use are typical of the tools found in the carpentry, plumbing, and electrical trades. There are certain specialized tools that the fire service makes use of, but they have not been included because they require specialized knowledge that you will acquire after you have become a fire fighter.

Classification and Identification of Tools

There are two basic groups of tools: hand tools, for prying, cutting, drilling, holding, pounding, twisting, etc.; and power tools, for cutting, prying, turning, etc.

HAND TOOLS

HAMMERS

Hammers are used for pounding, shaping, and breaking by applying a striking force.

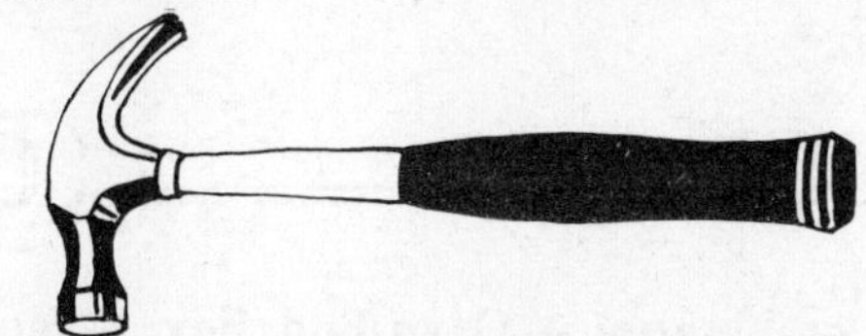

Nail Hammer (general-purpose hammer): used to hammer in nails and remove them.

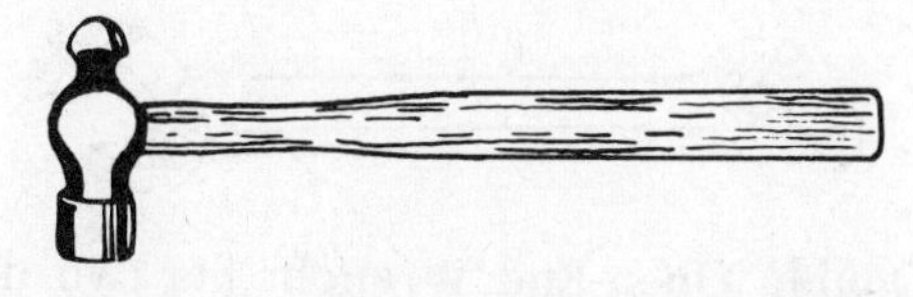

Ball Peen Hammer: used to shape and forge metals, set rivets, and drive a chisel.

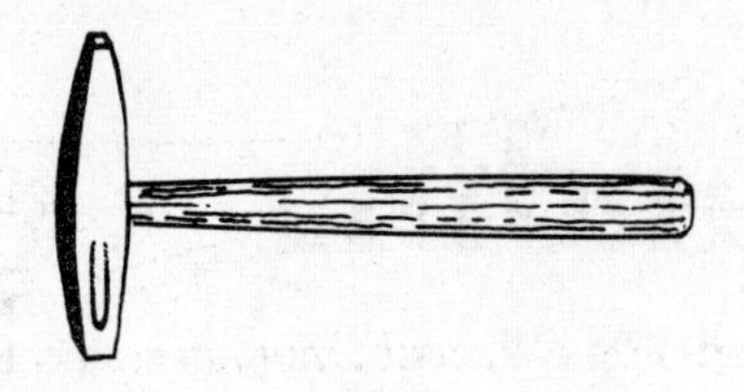

Tack Hammer (often magnetic): used to drive in small nails called tacks.

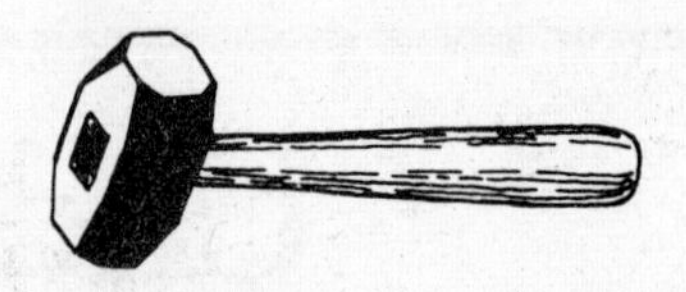

Hand Drilling Hammer: used for heavy work with star drills in masonry, with a wedge to split logs, with a chisel to cut rivets, etc.

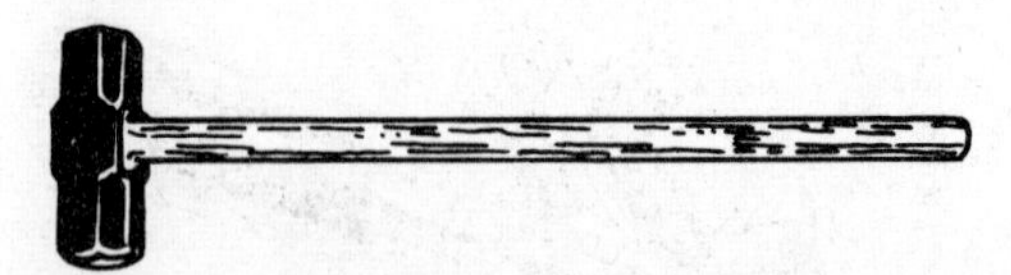

Sledge Hammer: used for heavy-duty work, pounding in or breaking. For light blows just let the head of the hammer fall; for heavy work swing the head.

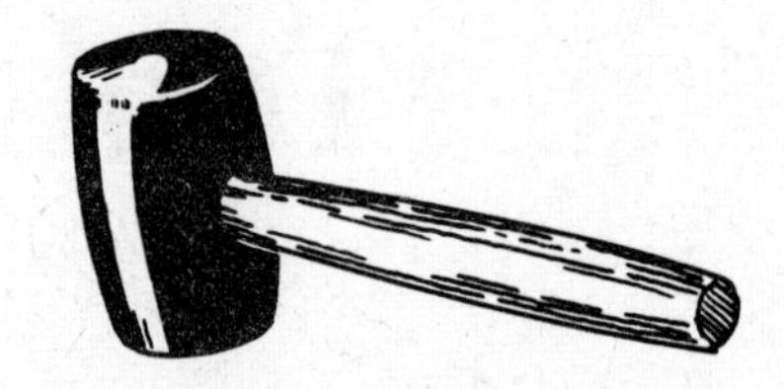

Mallet (rubber or wood): used to avoid damaging other tools or surfaces.

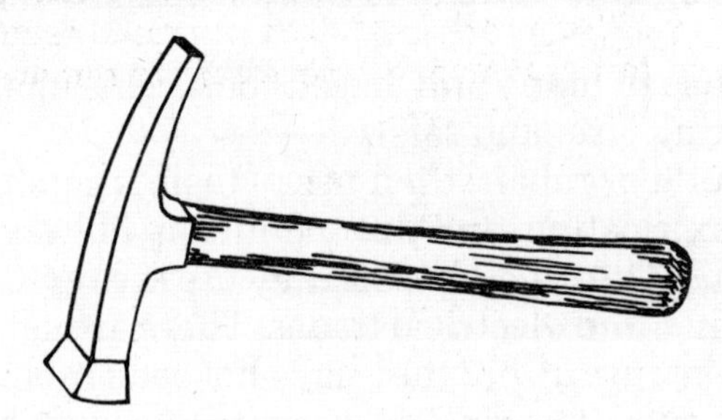

Mason's or Bricklayer's Hammer: used to drive cold chisels and to shape or trim brick and masonry.

WRENCHES

Wrenches are used to exert a twisting force on bolts, nuts, pipes, etc. There are three basic types: nonadjustable, adjustable, and special-use.

Nonadjustable Wrenches

All nonadjustable wrenches are used to tighten or loosen nuts and bolts. They come in a variety of sizes; the wrench size must be fitted correctly to the nut or bolt. Open-end wrenches are easy to use and can reach into difficult spaces but tend to slip off. Box wrenches do not slip off easily but are slow when operated in tight spaces.

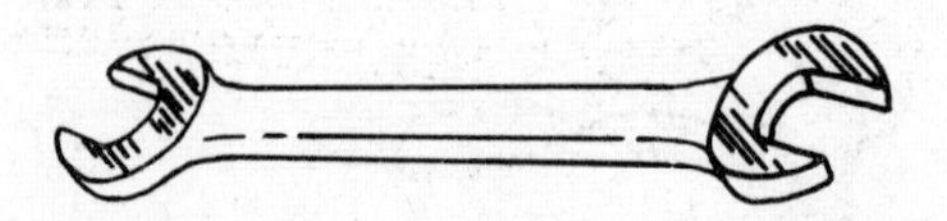

Double Open-End Wrench: fits two different sizes.

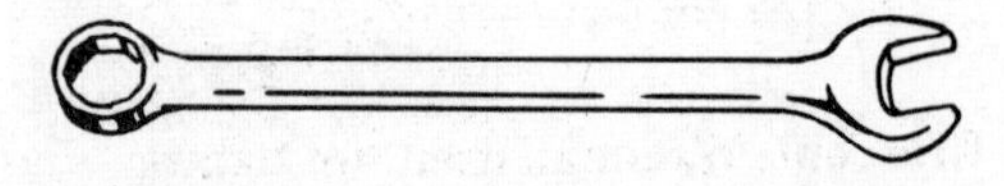

Combination—Open-End/Box Wrench: fits one size.

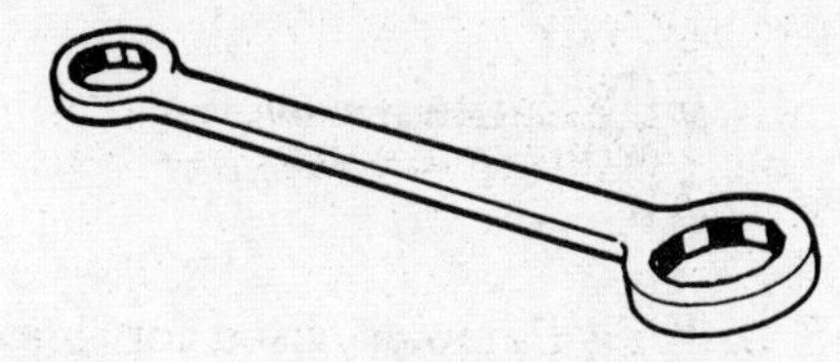

Double Box Wrench: fits two sizes.

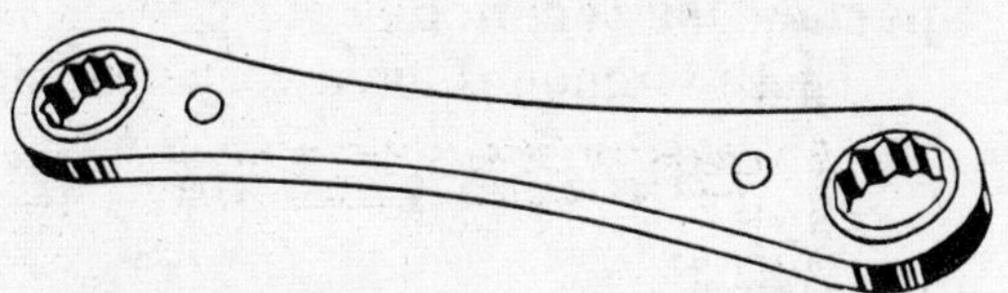

Ratchet Box Wrench: not necessary to remove on return stroke.

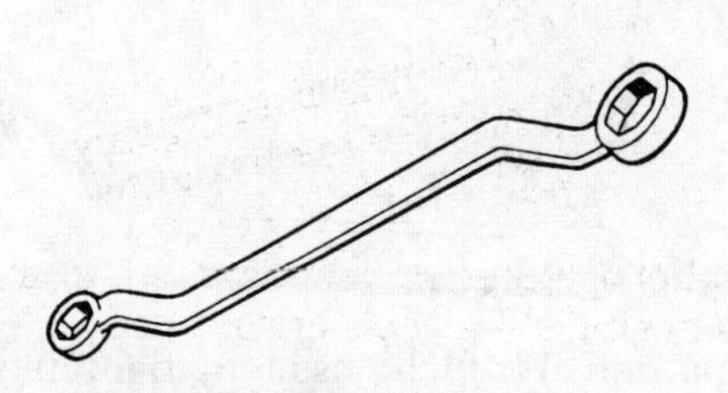

Offset Box Wrench: provides access to hidden nuts or bolts.

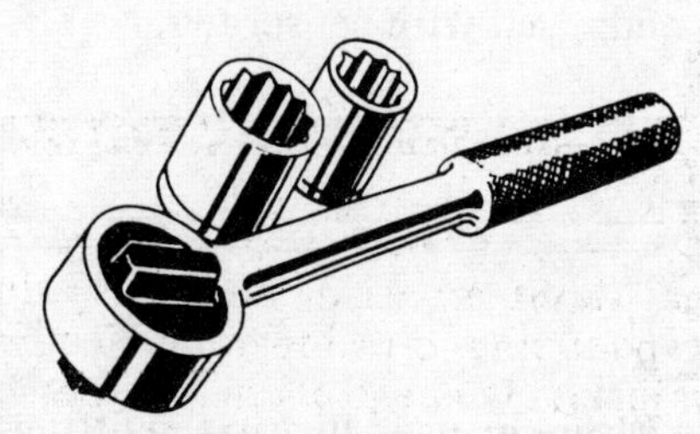

Ratchet Handle and Socket Wrench: provides a variety of sizes and rapid removal or tightening; not necessary to remove on return stroke.

Adjustable Wrenches

Adjustable wrenches allow the user to fit the wrench size to the object being turned.

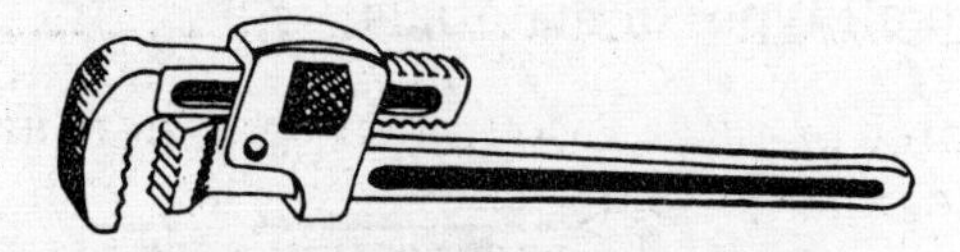

Pipe Wrench or Stillson Wrench: used on pipes and rods; should not be used on nuts or bolts.

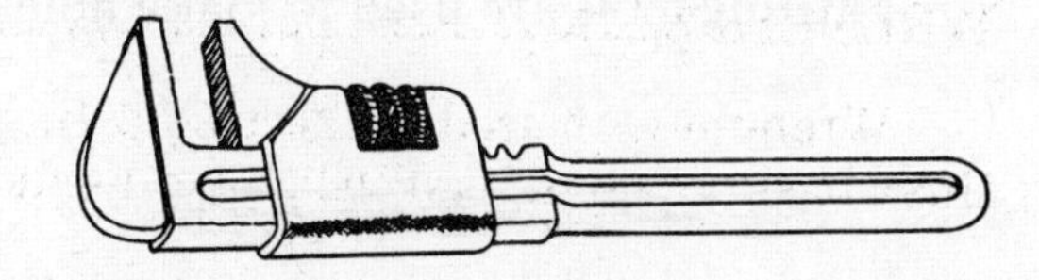

Monkey Wrench or Screw Wrench: used on nuts or bolts of any size.

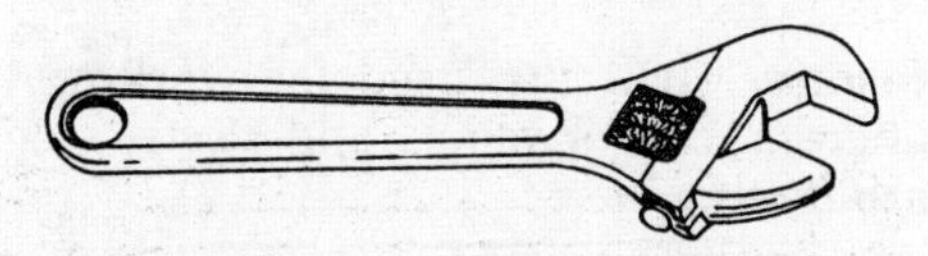

Crescent Wrench: used to tighten and loosen nuts and bolts; strongest and safest when pressure is applied to the side with the fixed jaw.

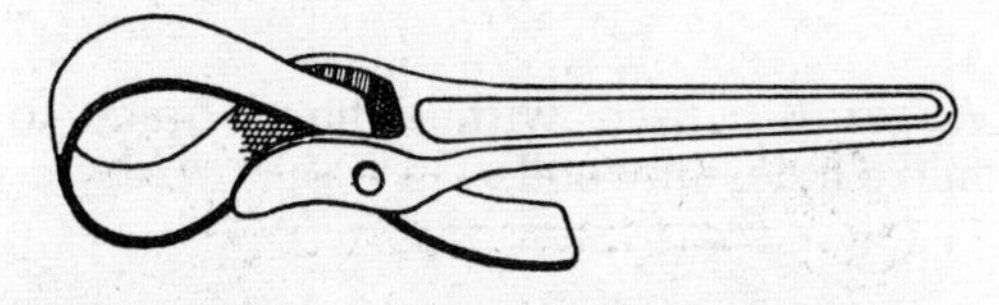

Strap Wrench: used to grip pipes or other objects whose surfaces should not be damaged.

Special-Use Wrenches

These wrenches have very specific, generally very limited, uses.

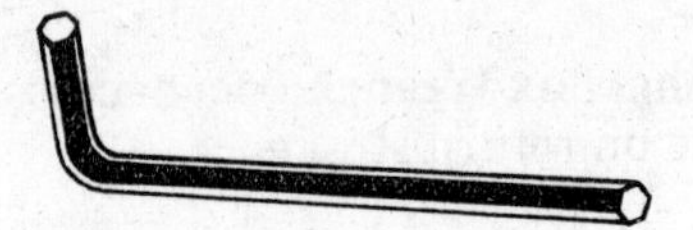

Allen Wrench (hex key or set screw wrench): is a hexagonal steel bar with a bent end; used with set screws.

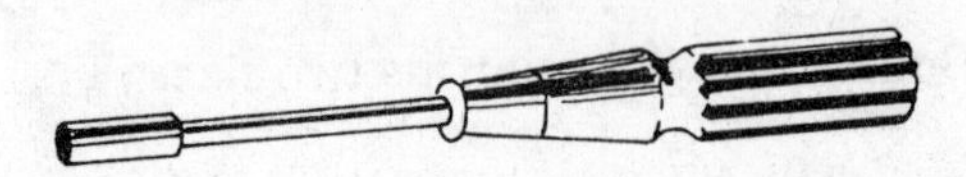

Nut Driver: looks similar to a screwdriver and is used on small nuts and bolts.

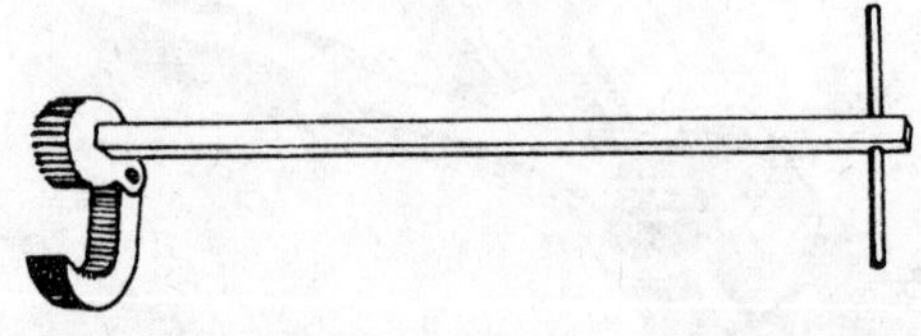

Basin Wrench: used to grip the fittings of sink basins where other wrenches will not fit.

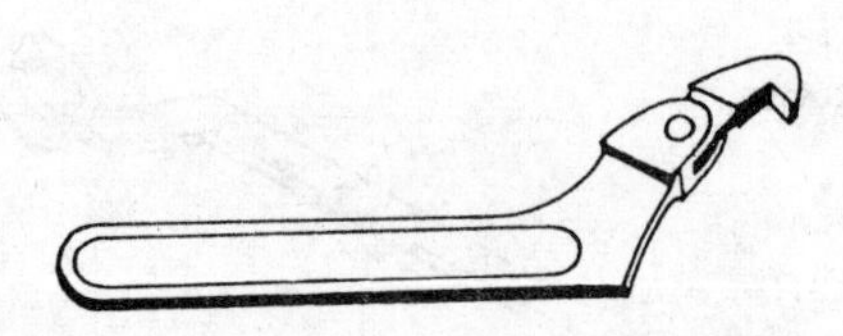

Spanner Wrench: used to tighten hose couplings and other special nuts or bolts.

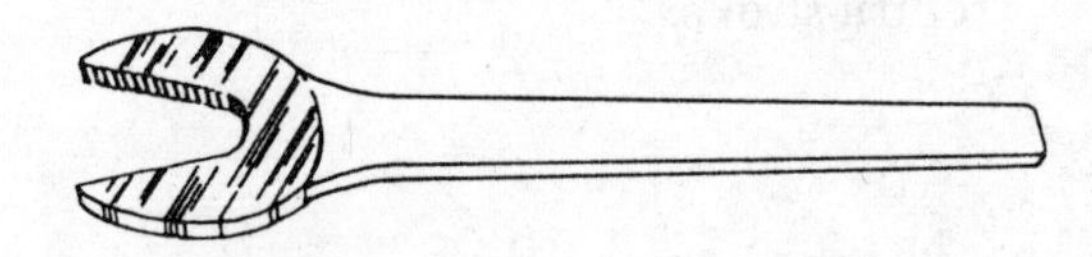

Crocodile or Bulldog Wrench: looks like an open-end wrench with one side of the bite smooth and the other serrated.

DRILLING TOOLS

Drilling tools are used to make holes in all types of materials.

High-Speed Twist Drill: used with a power drill for making holes in metal.

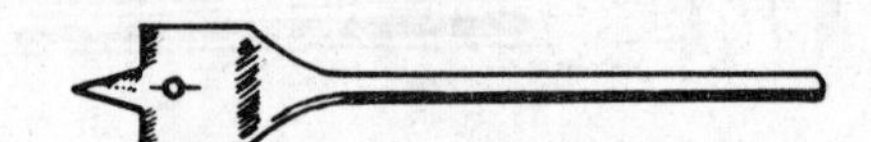

Power Wood Bit: used with a power drill to make holes in wood.

Auger Bit: used with a hand brace to make holes in wood.

Masonry Drill Bit: used for working in concrete, cinder block, and stone; has a carbide tip.

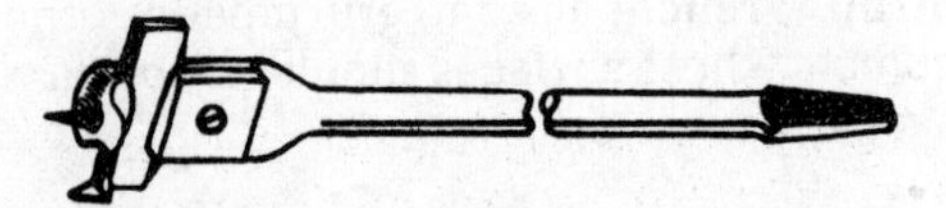

Expansive Bit: used with a hand brace to make holes of varying sizes.

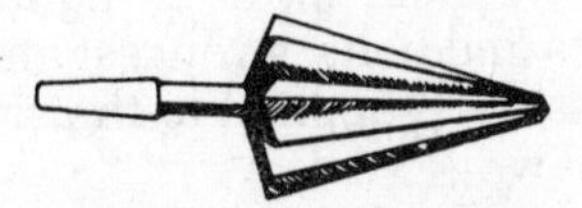

Reamers (are not drills): used to remove burrs from the inside of a pipe after it has been cut.

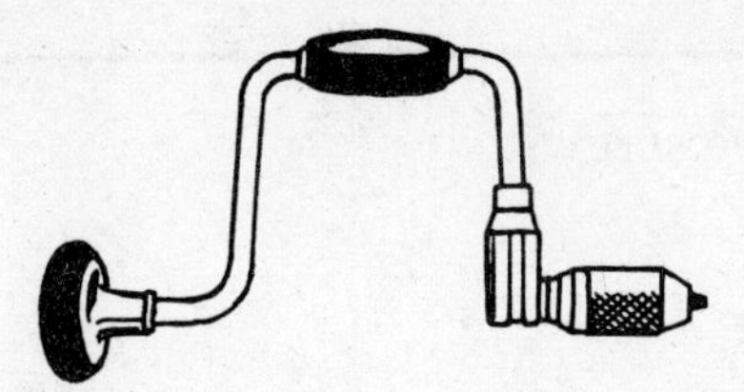

Hand Brace: used to drill large holes and also to increase the turning power of a screwdriver.

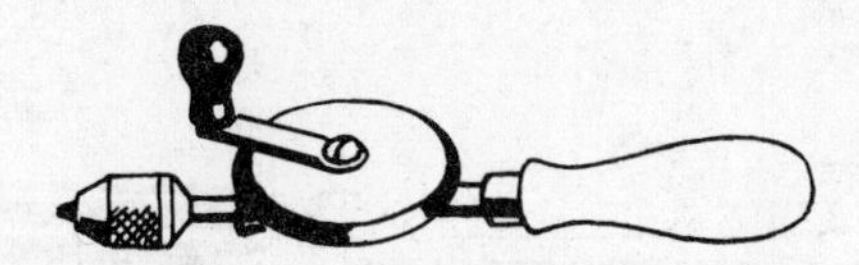

Hand Drill: used to drill small holes quickly in wood or metal.

SAWS

This group of tools is used to cut objects to size or to remove waste from materials.

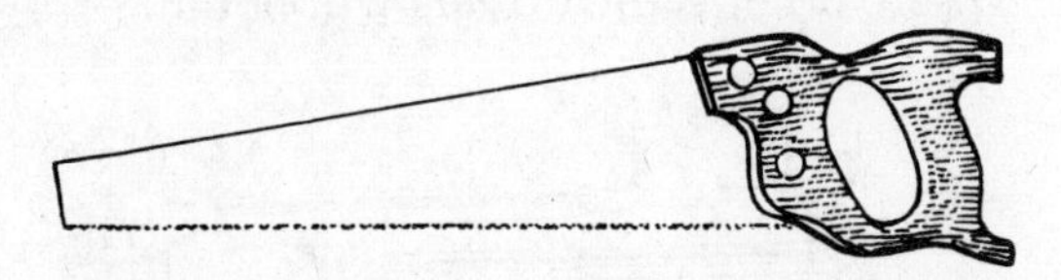

Hand Saw: used to cut wood to proper size.

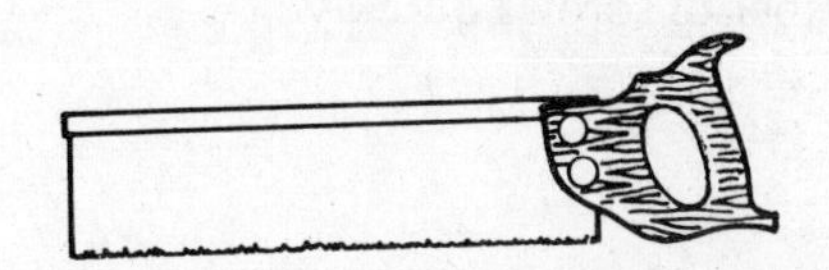

Back Saw: used to make fine cuts.

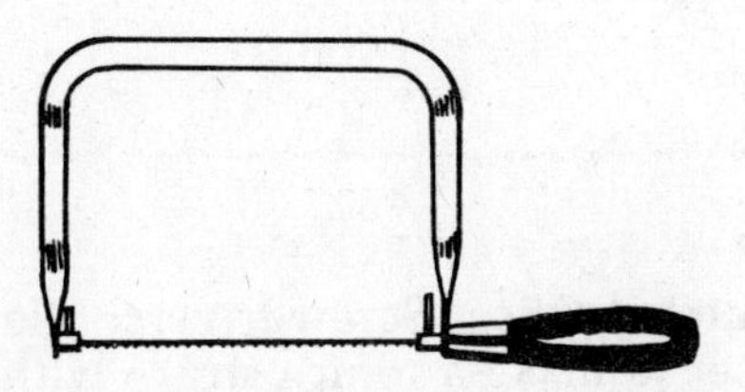

Coping Saw: used to cut curves and holes; depth of cut is limited by depth of throat.

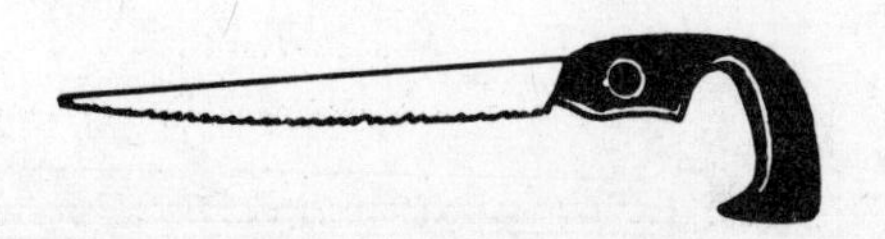

Keyhole Saw or Compass Saw: used to make cuts on the inside of a piece of work. A hole is drilled first, then the saw is inserted.

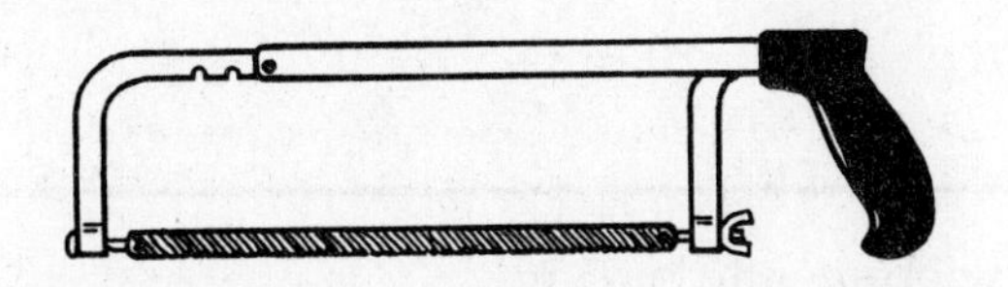

Hacksaw: used to cut metal and plastics.

Close Quarter Hacksaw: used where the frame of the hacksaw would obstruct cutting.

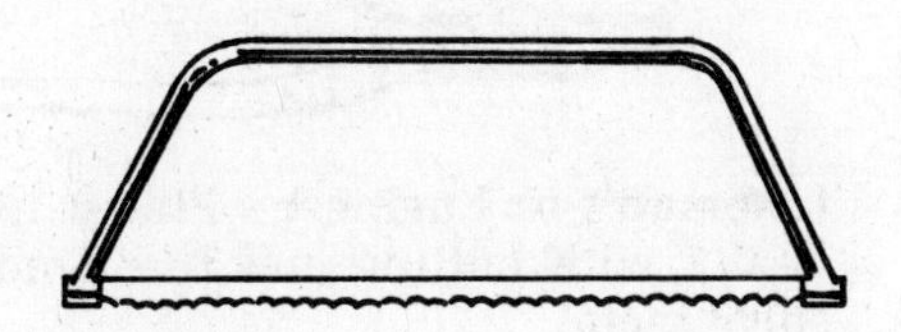

Tubular Frame Saw: used for very rough cutting such as trees and logs.

SCREWDRIVERS

These tools are used to drive and to remove screws.

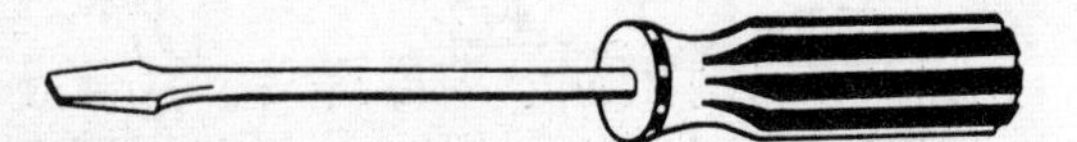

Standard Square-Blade Screwdriver: used with standard cut screws. Use the longer size where possible to make the work easier.

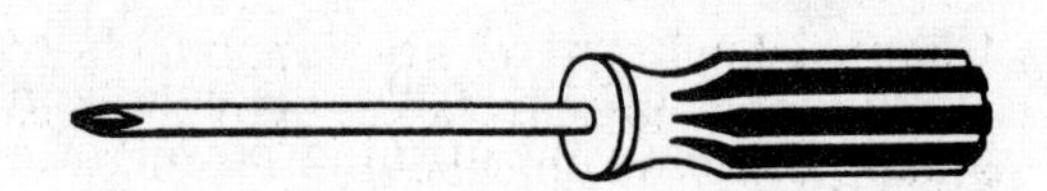

Phillips Head Screwdriver: used for special screws.

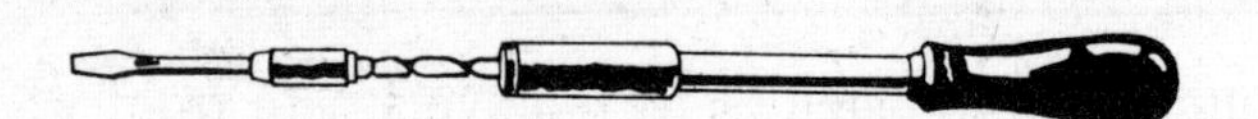

Ratchet Screw Driver; used to set a large number of screws quickly.

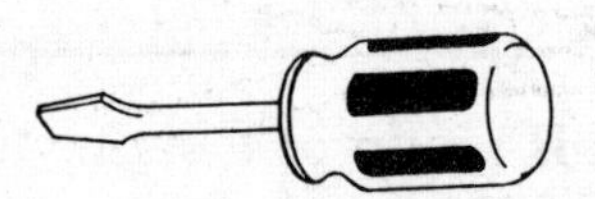

Stubby or Short Screwdriver: used where other screwdrivers will not fit.

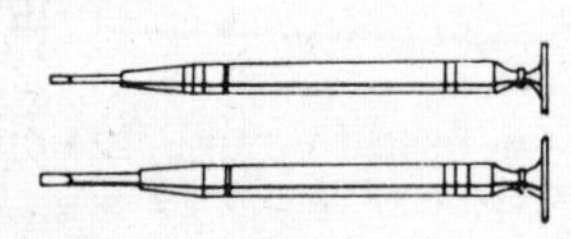

Jeweler's Screwdriver: used for very small, fine work.

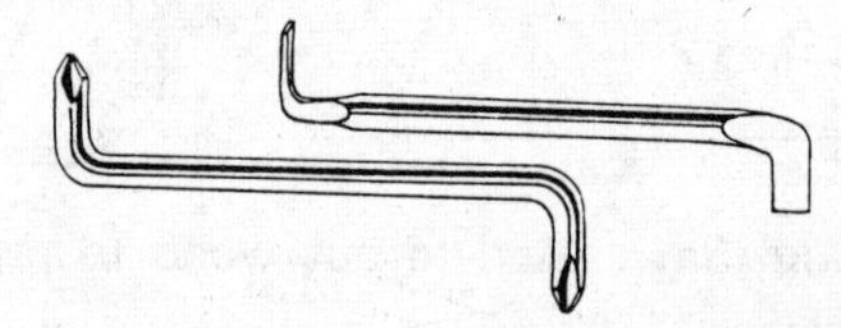

Offset Screwdriver: used to reach difficult spaces where straight screwdrivers will not fit.

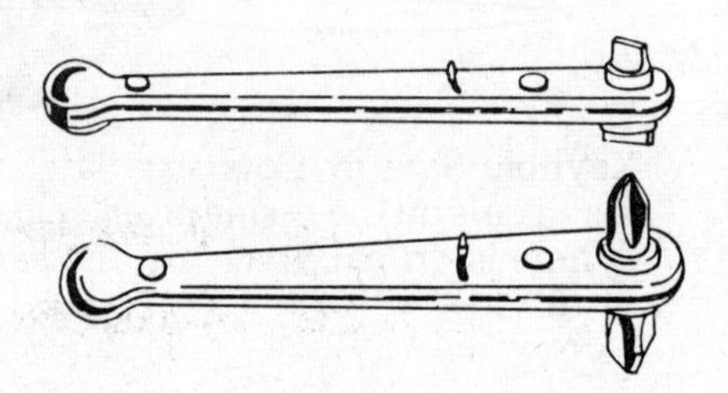

Ratchet Offset Screwdrivers: allows the user to make a return stroke without removing the screwdriver.

PLIERS

Pliers are used for holding, cutting, twisting, and prying.

Slip Joint Pliers: used to grip almost anything. The slip joint allows for two settings.

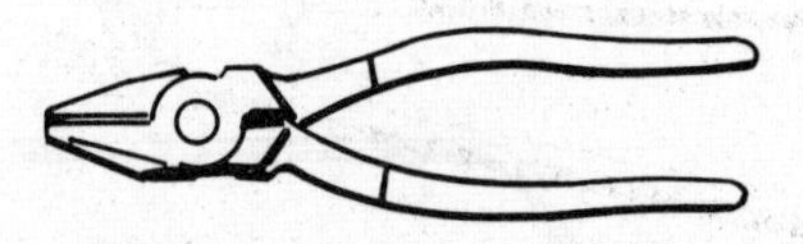

Lineman's or Engineer's Pliers: used for heavy wire cutting and the bending of sheet metal.

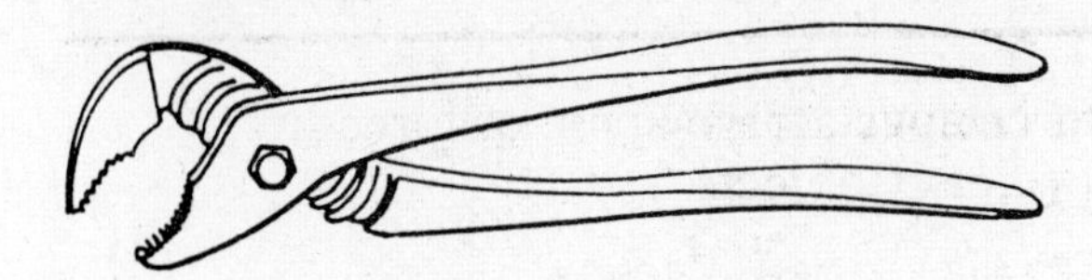

Channel Lock Pliers: are expandable to allow a larger gripping area; used on pipes and other round objects.

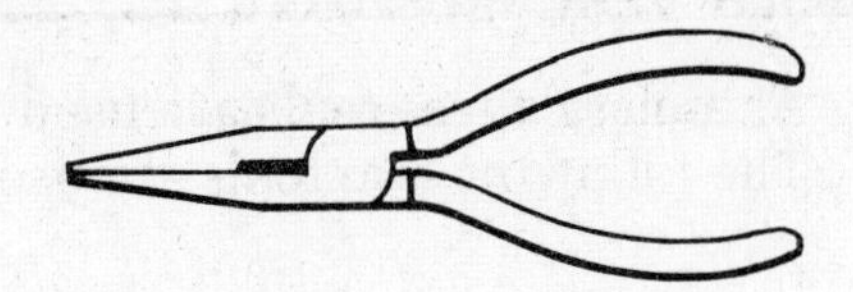

Long-Nose or Needle-Nose Pliers: used to shape wire or sheet metal and to reach inside an object.

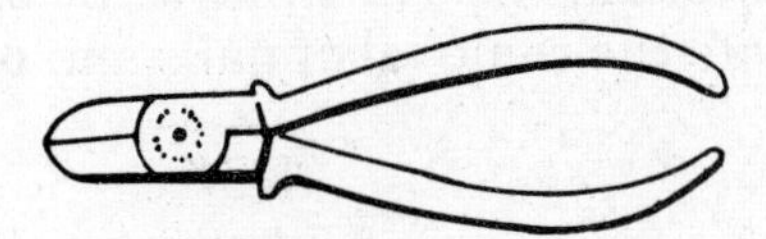

Diagonal Cutting Pliers: used to cut metal wire.

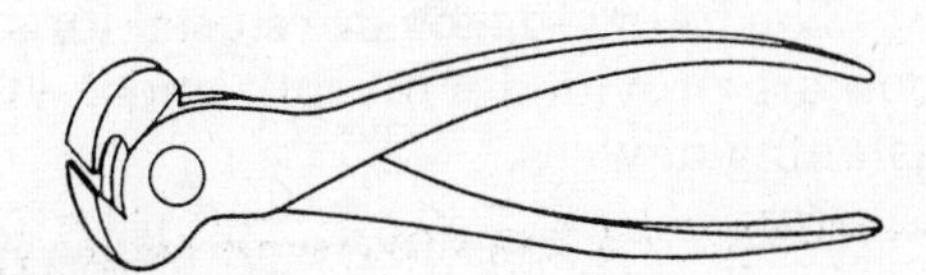

End Cutting Pliers: used to cut metal wire close to the surface.

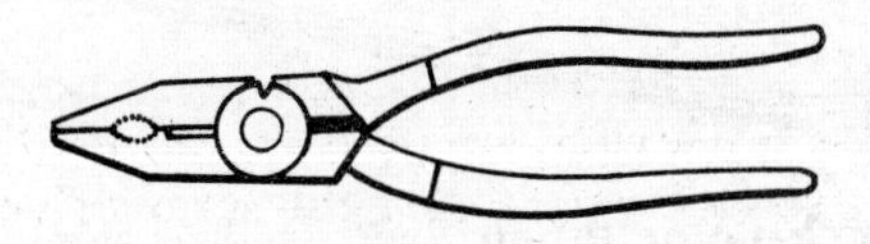

Electrician's Pliers: used to cut and strip wire.

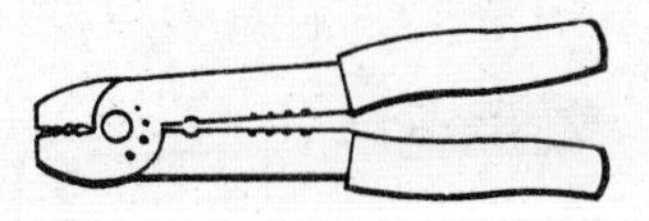

Electrician's Multipurpose Tool: used to strip wire, crimp wire or fittings, and cut small bolts.

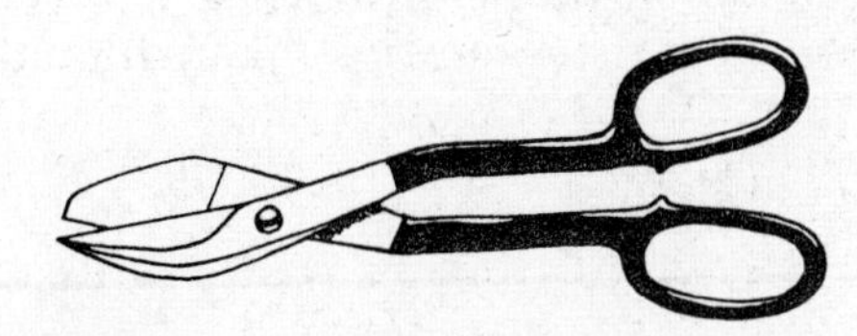

Tin Snips: used to cut sheet metal.

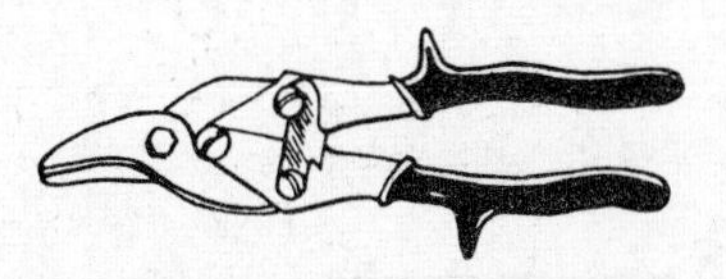

Duck Bill Snips: used to cut light sheet metal.

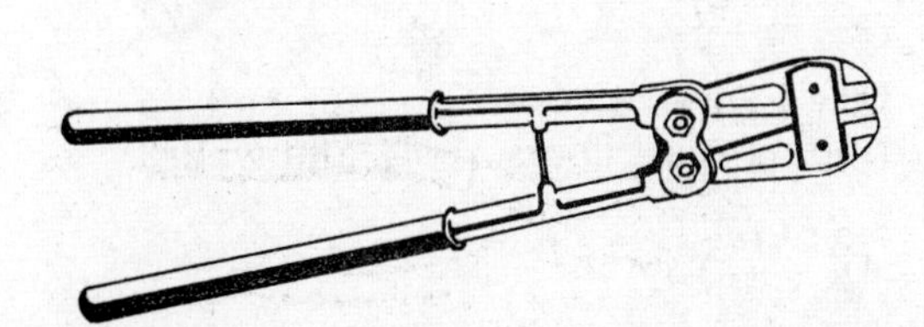

Bolt Cutter: used to cut steel rods, bolts, and locks.

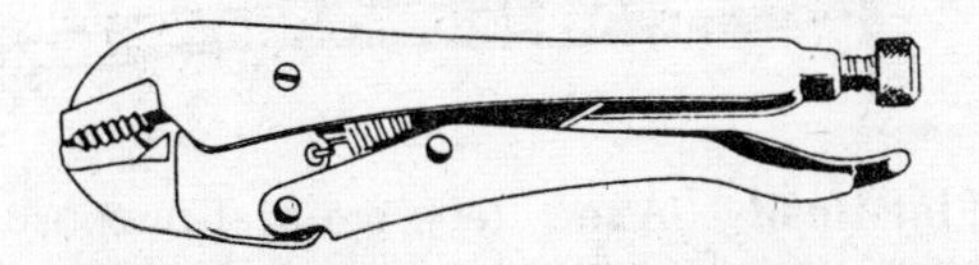

Combination Plier Wrench or Vice Grips (channel lock or vice grips): used to get a firm grip without having to apply a great amount of hand force; will hold the object firmly even after the fire fighter lets go; also used as a portable hand-held vice.

TOOLS FOR PRYING

As a class, these tools are used to force apart, force open, or force up. The following two tools are used for heavy objects.

Crow Bar: used as a lever to move or lift heavy objects.

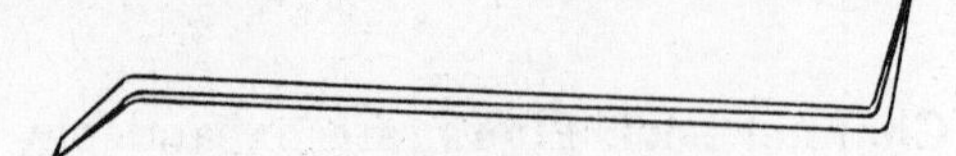

Stripping Bar (pry bar): used to pry apart and move heavy or well-secured objects.

The following tools are used for smaller, lighter objects. Note that some of the tools are provided with split ends, which facilitate the removal of nails and other fastening devices.

Jimmy Bar: used to pry lighter objects and remove fastening devices.

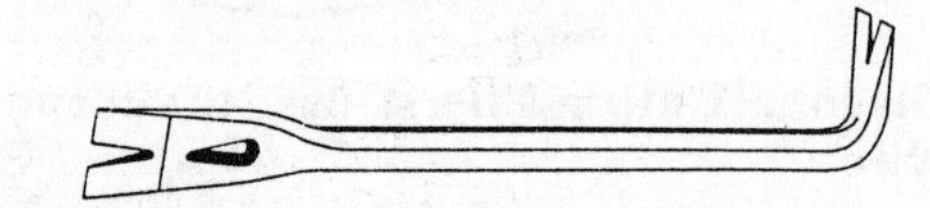

Offset Rip: used to pry light objects and remove fastening devices.

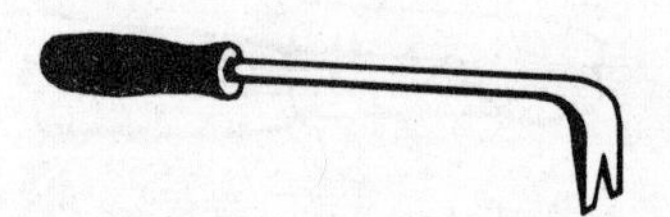

Nail Claw: used to remove nails rapidly.

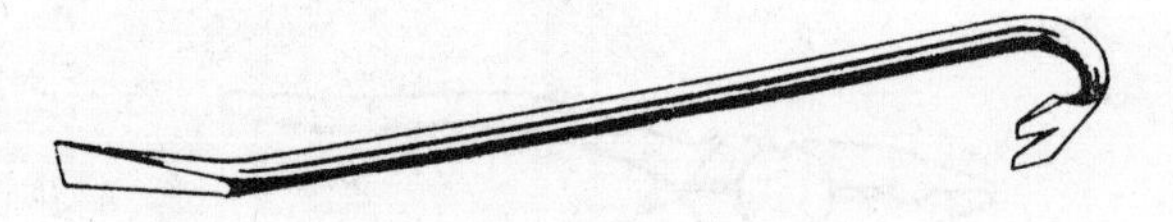

Wrecking or Pinch Bar: used to remove or demolish building components.

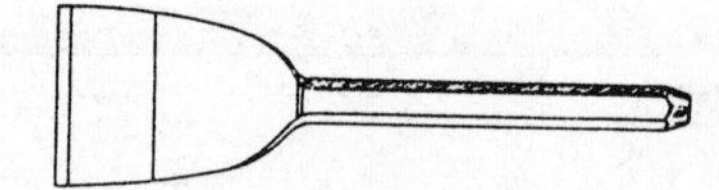

Electrician's Chisel: used to cut floor boards, molding, etc.

AXES

These tools are used to cut and hammer. They are for rough work.

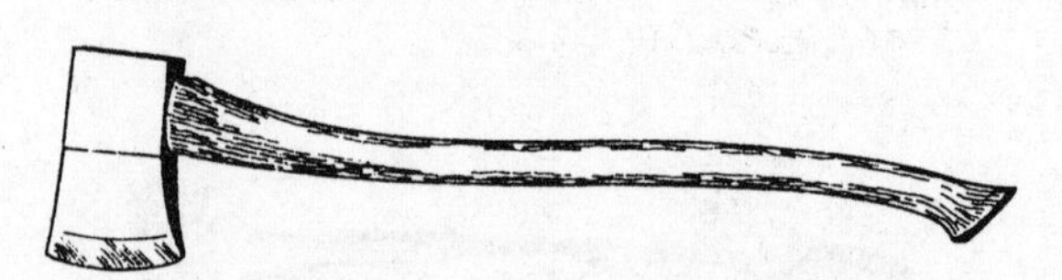

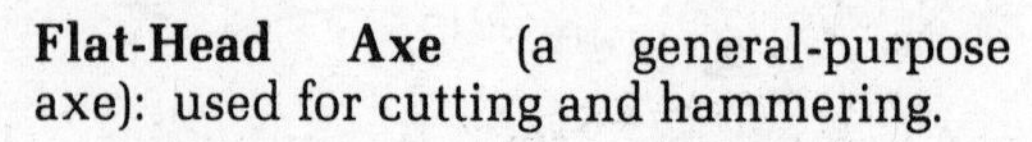

Flat-Head Axe (a general-purpose axe): used for cutting and hammering.

Fire Fighter's Axe: used for cutting and ripping apart floors, roofs, and walls.

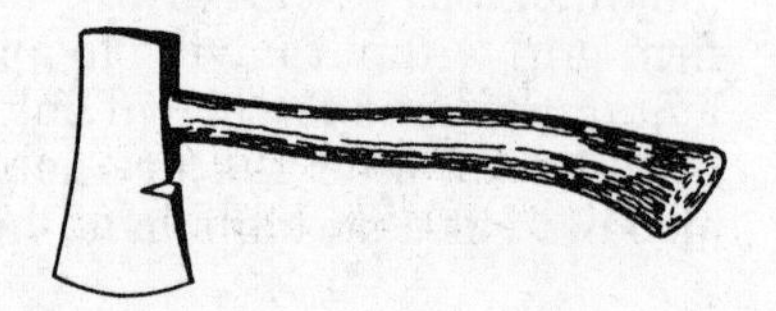

Half Axe: used to trim and nail light boards.

CLAMPS

Clamps are used to hold an object tightly while some other operation is performed.

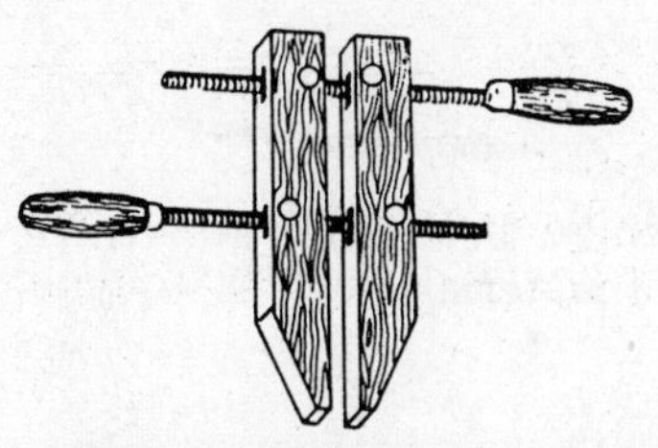

Adjustable Gluing Clamp: used to clamp wood and metal at almost any angle.

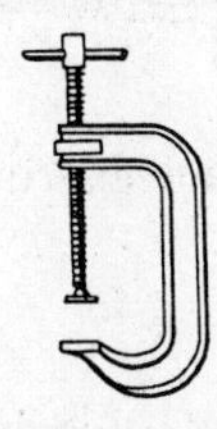

"C" Clamp: used to clamp wood or metal.

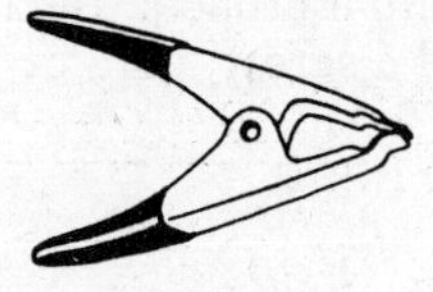

Spring Clamp: used to provide temporary light pressure.

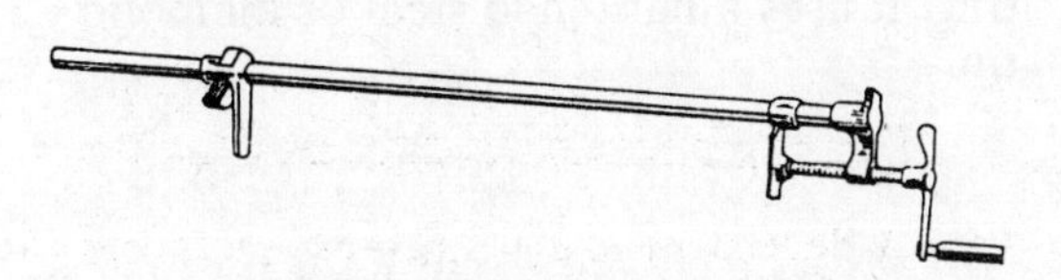

Pipe Clamp: used to hold large boards or objects together.

DIGGING TOOLS

These tools are used for digging holes and for moving debris or other materials.

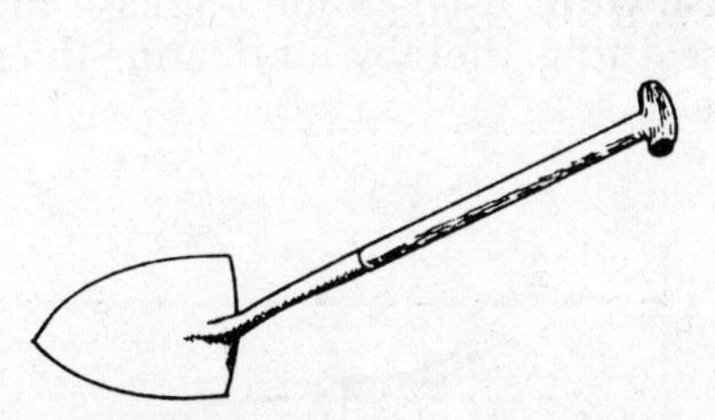

Round Shovel: used to dig holes and move materials.

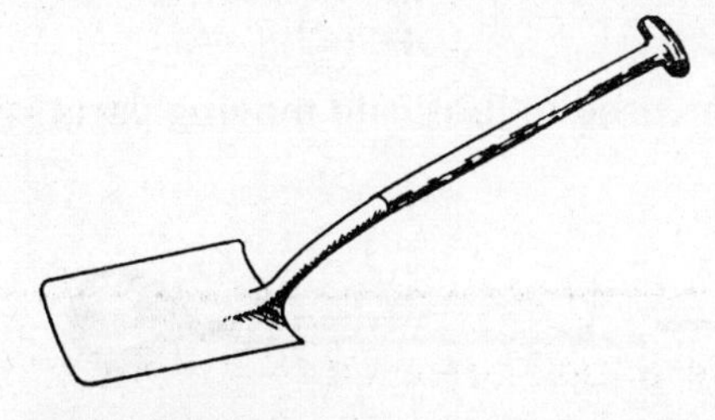

Broad Shovel: used to remove debris and other materials.

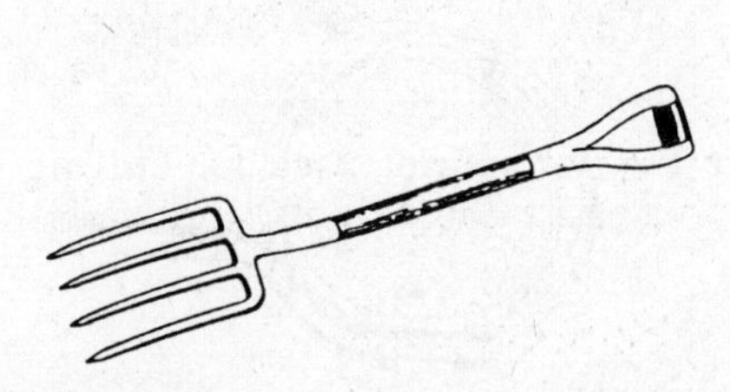

Digging Fork: used to break up earth for easy removal.

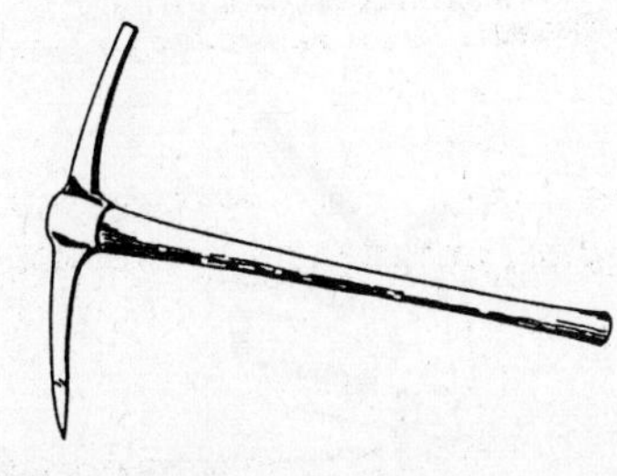

Pickaxe (a chisel-type tool): used to break up soft materials, asphalt, and earth.

MISCELLANEOUS HAND TOOLS

These tools have individual, specialized uses.

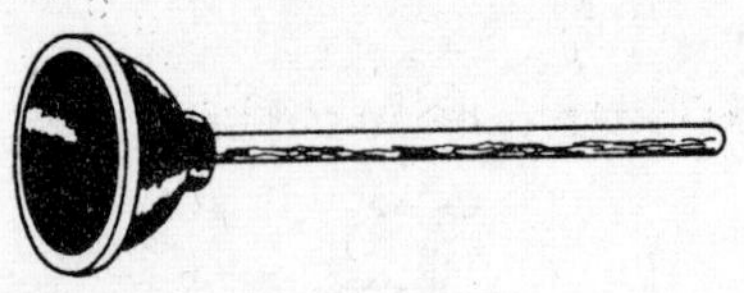

Plunger: used to free obstructions in sinks and toilets.

Utility Knife: used to cut and trim a wide variety of materials.

Glass Cutter: used to scribe glass for cutting. It uses a hardened steel or diamond tip.

Steel-Cutting or Cold Chisel: used to cut grooves, chip a broad surface, or shear off the head of a screw, bolt, or rivet.

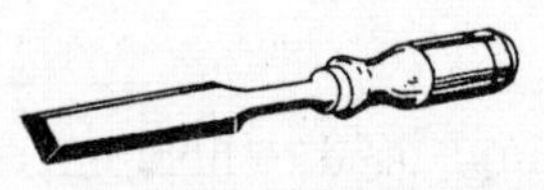

Wood Chisel: used to cut an edge or groove in wood; often used with a mallet.

Pocket Knife: used to cut, shape, and trim a wide variety of materials.

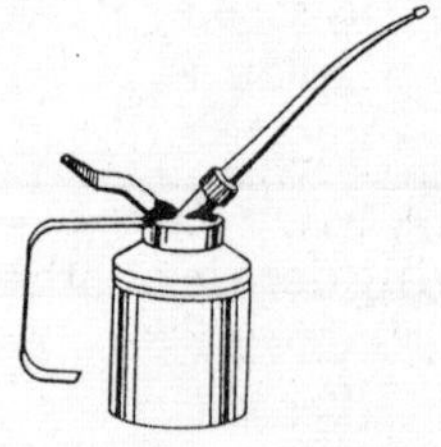

Oil Can: used to lubricate moving parts.

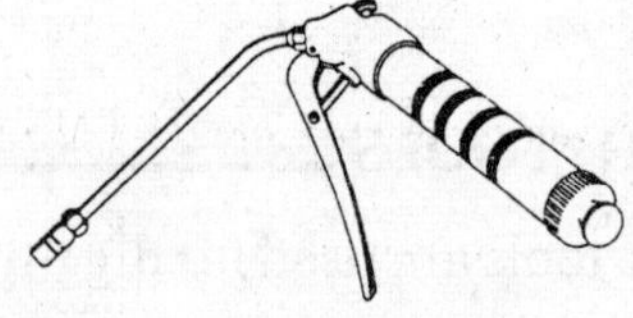

Grease Gun: used to force grease into a grease fitting, thereby lubricating the machine.

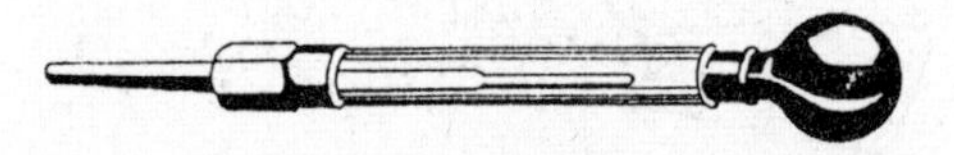

Hydrometer or Acidmeter: used to measure the density of the fluid in a wet cell battery.

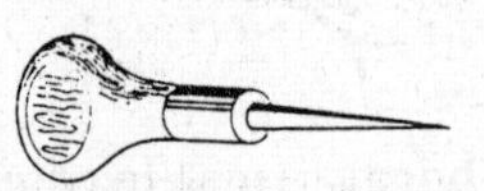

Awl: used to mark a starter hole for screws, to scribe a line, etc.

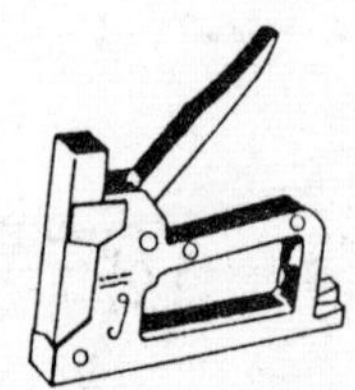

Staple Gun: used to adhere materials to surfaces rapidly but temporarily.

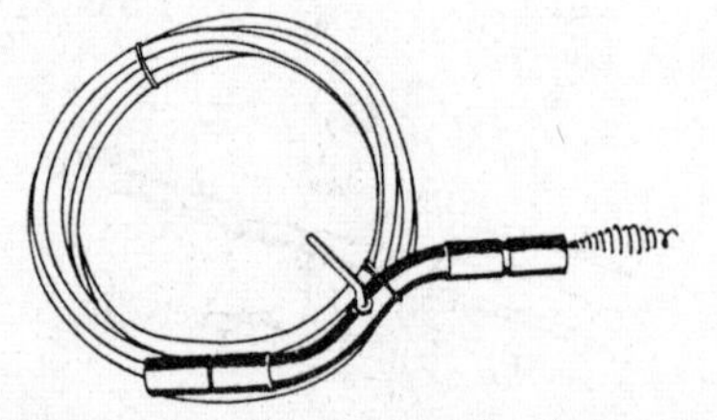

Hand Auger: used to feed through a clogged pipe to reach and free an obstruction.

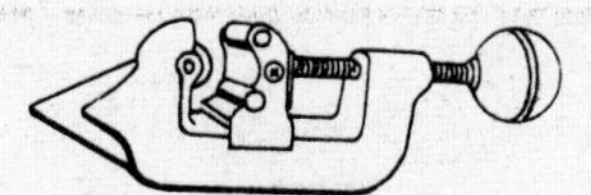

Pipe and Tube Cutter: used to cut light metal pipe.

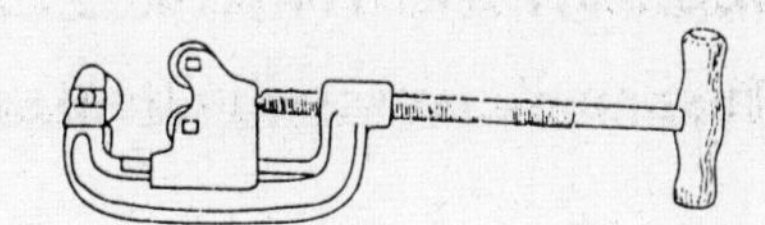

Pipe Cutter: used to cut heavy metal pipe

POWER TOOLS

CUTTING TOOLS

Power cutting tools are used to rapidly cut to size all types of materials.

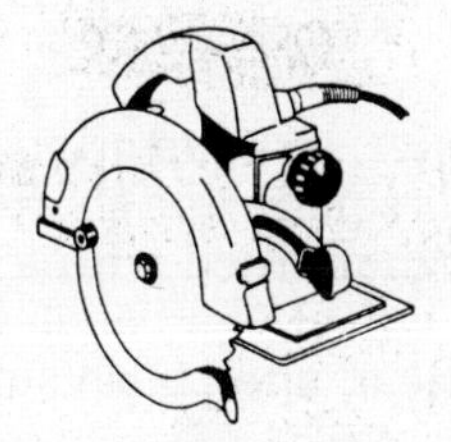

Circular Saw: used to cut lumber and plywood to size.

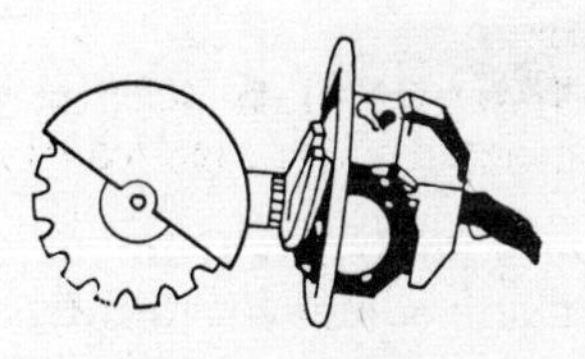

Gasoline-Powered Circular Saw: used for heavy-duty cutting of large areas.

Chain Saw: used for hevy-duty cutting of large beams or trees and for pruning trees.

Reciprocating Saw: used in the same manner as a hand saw or a keyhole saw.

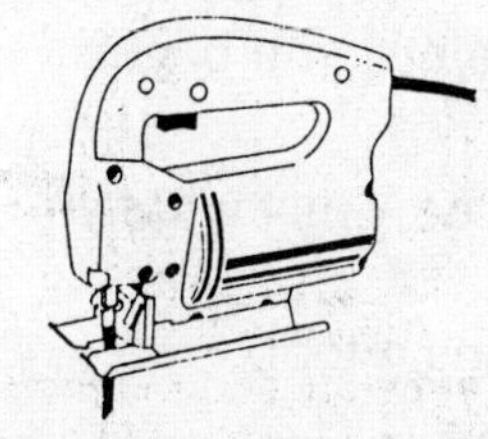

Saber Saw: acts as a powered jig or coping saw.

Acetylene Torch: used to cut steel and heavy metal structures.

OTHER POWER TOOLS

These tools perform specialized tasks.

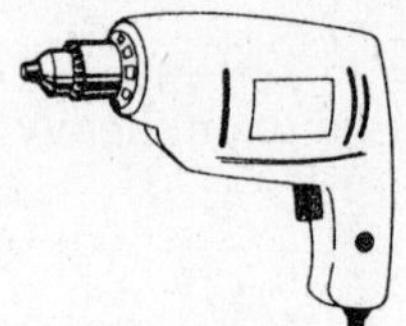

Drills (available in many sizes): can be adapted to perform a number of tasks, such as sanding, buffing, and cutting; however, the main use is to drive a drill bit, that is, to create a hole. To increase their versatility and effectiveness, drills can be reversible and have variable speeds.

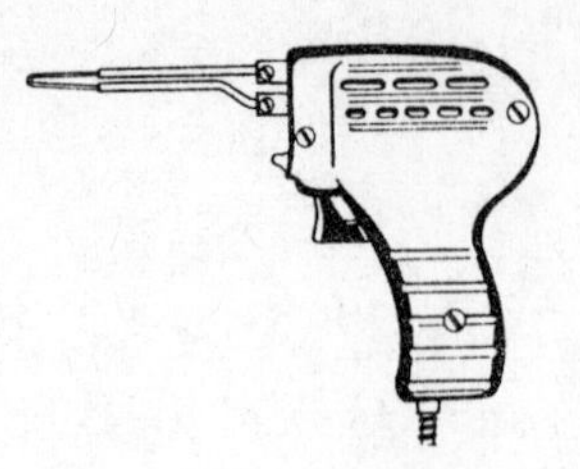

Soldering Gun: used for light soldering of small parts.

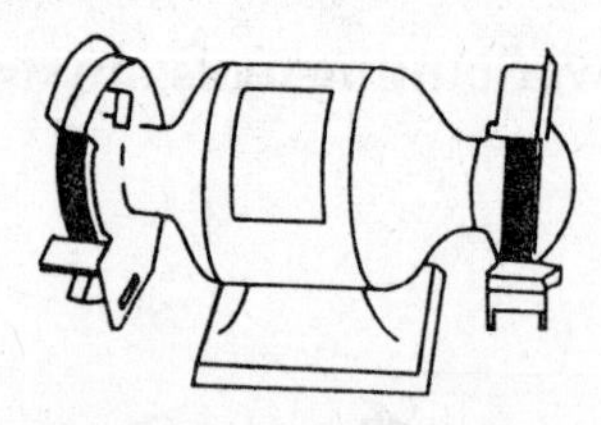

Bench Grinder: used to grind the cutting edges of tools square and sharp, to remove burrs, etc. When fitted with a wire brush and buffing wheel, the bench grinder can be used to clean and polish surfaces.

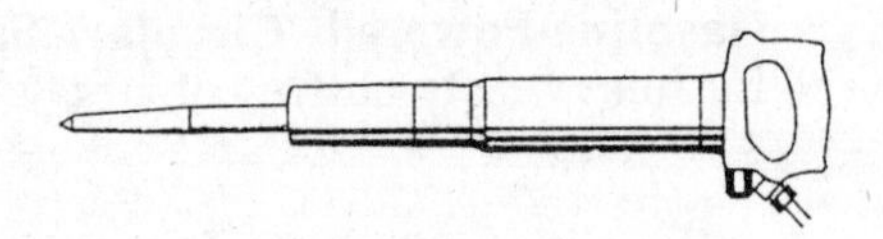

Heavy-Duty Soldering Iron: used to heat metal and soft solder for joining.

Pneumatic Hammer: used as a hammer, pick, concrete breaker, or digging tool. It can be operated by air pressure or electric current.

Proper Tool Safety

SAFETY RULES FOR HAND TOOLS

- Use sharp, clean tools and know the right way to use them before you start.
- Wear appropriate protective clothing and eye shields. (See the section entitled "Safety Equipment.")
- Clean the area where you will be working before you begin.
- Select the correct tool for the job.
- Plan all cutting operations before you begin. Know just what you will do and in what order you will do things.
- Look behind and around you before you swing a tool.
- Don't overreach.
- Put tools away in a safe place when you are through with them, and keep them there when not in use.

SAFETY RULES FOR POWER TOOLS

- Familiarize yourself with the stop switch before you start the tool.
- Keep tools sharp and clean.
- Study the operating and safety instructions before operating the tool.
- Clean the area where you will be working before you begin.
- Plan all cutting operations before you begin. Know just what you will do and in what order you will do things.
- If possible, clamp the work before you cut.
- Use protective guards, if provided.
- When electrical tools are used, make sure that they are properly grounded.
- Where possible, keep both hands on the tool.
- Avoid loose clothing.
- Pay attention to what you are doing; avoid distractions.
- Don't become overconfident and careless.
- Don't approach a person operating a tool from the back.
- Never clean scraps or debris away from the operating area of the tool with your hands or feet.

SAFETY EQUIPMENT

This equipment is designed to protect the tool user from injury.

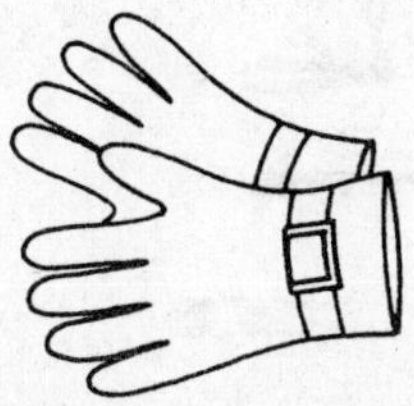

Gloves: there are various types of gloves made of cotton, leather, and rubberized materials. They are designed to protect the hands against burns and cuts.

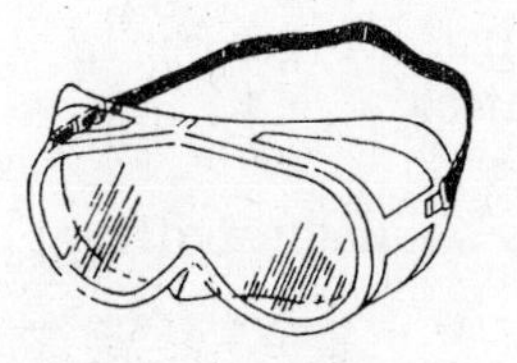

Eye Protectors: there are three types—goggles, spectacles, and the full face mask.

Shoes: steel-tipped and steel-sole-plated safety shoes come in high and low cut.

Head Protectors: a variety of head protectors is available. Examples are the fire fighter's helmet and the construction worker's hard hat.

Practice Exercises

There are three basic types of tool questions. The first measures your ability to recognize a tool and then identify it by proper or commonly used name. The second is a measure of your ability to relate a tool to a specific task; it asks you to demonstrate that you know what the tool is used for. The third type of question tests your knowledge of how to use the tool safely and properly.

The following exercises are representative of the types of questions that may be asked on your examination.

GROUP ONE

Directions: This first group of questions tests your ability to identify tools. It is divided into five sets, each with 10 questions. For each set, first write on your answer paper the numbers given in column A. Then, for each tool in column A, find the corresponding tool among the illustrations in column B and write the letter of the illustration next to the number of the tool.

Answer all questions; guess when you don't know the answer. Be sure to look up and check out every answer that you get wrong, as well as any answers that you guessed at. This will reinforce your learning.

SET 1

Match the tool named in column A with the correct illustration in column B.

Column A

1. Ball peen hammer
2. Backsaw
3. Coping saw
4. Utility knife
5. Half axe
6. Hand brace
7. Twist drill
8. Pipe cutter
9. Plunger
10. Phillips head screwdriver

Column B

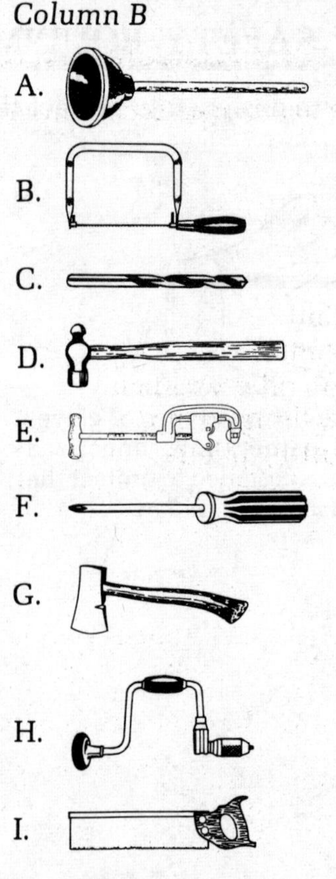

SET 2

Match the tool named in column A with the correct illustration in column B.

Column A

11. Hacksaw
12. "C" clamp
13. Offset rip
14. Tin snips
15. Double open-end wrench
16. Keyhole saw
17. Stillson wrench
18. Electrician's multipurpose tool
19. Strap wrench
20. Grease gun

Column B

A.

B.

C.

D. 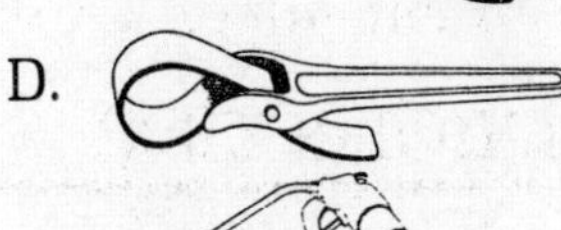

E. 

F.

G.

H.

I.

J.

SET 3

Match the tool named in column A with the correct illustration in column B.

Column A

21. Hand drill
22. End cutting pliers
23. Jeweler's screwdriver
24. Pinch Bar
25. Expansive bit
26. Spanner wrench
27. Combination plier wrench
28. Adjustable gluing clamp
29. Diagonal cutting pliers
30. Nail claw

Column B

A. 

B.

C.

D.

E.

F.

G.

H.

I.

J.

SET 4

Match the tool named in column A with the correct illustration in column B.

Column A

31. Offset box wrench
32. Digging fork
33. Needle-nose pliers
34. Tube cutter
35. Hydrometer
36. Nail hammer
37. Auger
38. Hand saw
39. Nut driver
40. Crescent wrench

Column B

A. 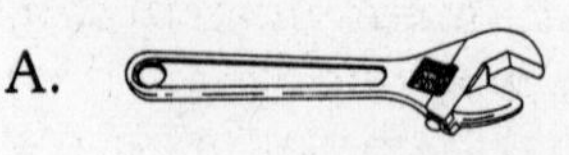

B. 

C. 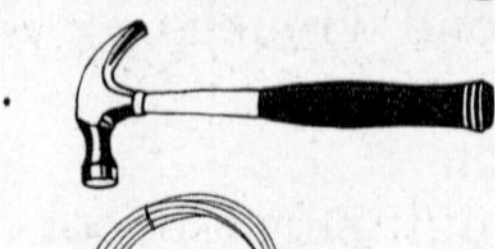

D. 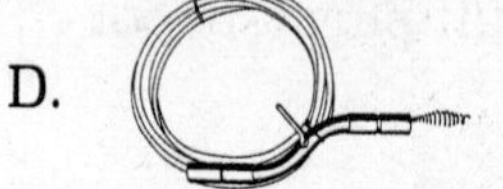

E. F. 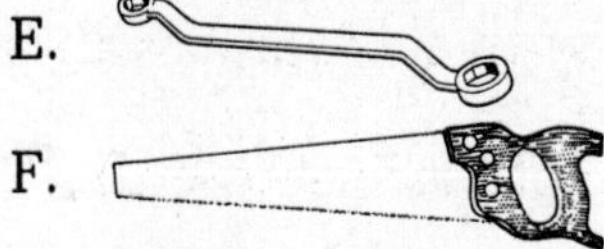

G. 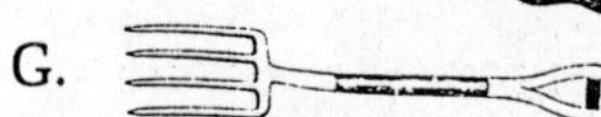

H. I. 

J. 

SET 5

Match the tool named in column A with the correct illustration in column B.

Column A

41. Reciprocating saw
42. Power drill
43. Gasoline-powered circular saw
44. Chain saw
45. Circular saw
46. Soldering gun
47. Bench grinder
48. Heavy-duty soldering iron
49. Saber saw
50. Acetylene torch

Column B

A. B. 

C. 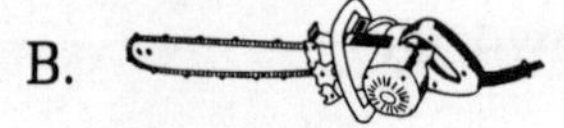

D. 

E. F. 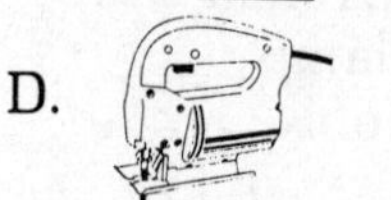

G. 

H. 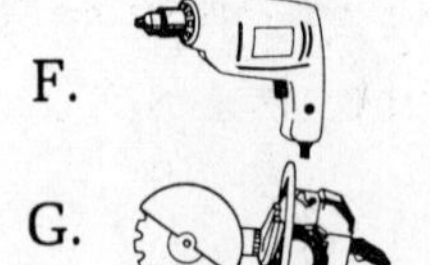

I. 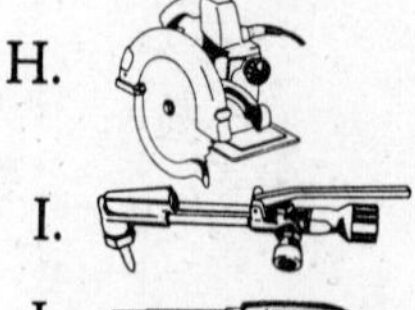

J. 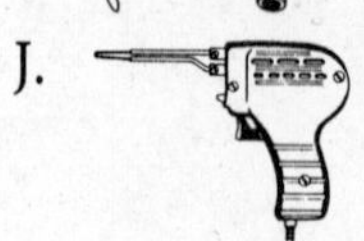

GROUP TWO

Directions: This second group measures your knowledge of how tools are used. It is divided into two sets, each with 10 questions. For each set, first write on your answer paper the numbers given in column A. Then, for each use in column A, find the corresponding tool among the illustrations in column B and write the letter of the illustration next to the number of the use.

Answer all questions; guess when you don't know the answer. Be sure to look up and check out every answer that you get wrong, as well as any answers that you guessed at.

SET 1

Match the use in column A with the correct illustration in column B.

Column A

51. Cut a hole in plaster board wall (sheet rock)
52. Chip a hole in poured concrete
53. Make a hole in a steel deck
54. Turn a pipe
55. Remove a bolt
56. Cut light rope
57. Clear a clogged drain
58. Drive a chisel
59. Cut a hole in a roof
60. Drive a nail

Column B

A. 

B. 

C. 

D. 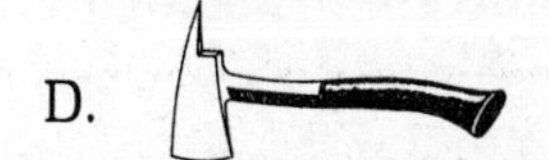

E. 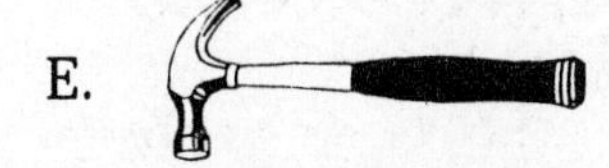

F.

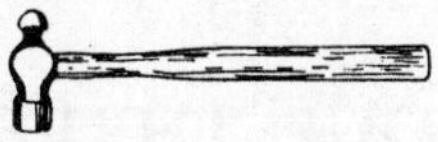

G.

H.

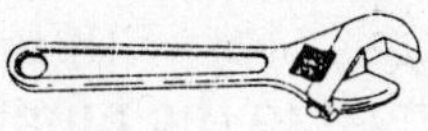

I.

J.

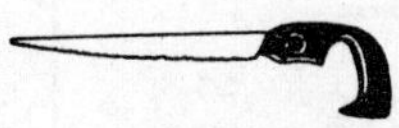

SET 2

Match the use in column A with the correct illustration in column B.

Column A

61. Loosen small bolts holding a cover in place
62. Drive a cold chisel
63. Remove a wood screw
64. Cut rivets on steel plate
65. Open a heavy steel trapdoor
66. Cut a board quickly
67. Couple two pipes of conduit
68. Cut light sheet metal
69. Sharpen and clean tools
70. Loosen a lag bolt holding a wood cornice

Column B

A. 

B. 

C. 

D. 

E. 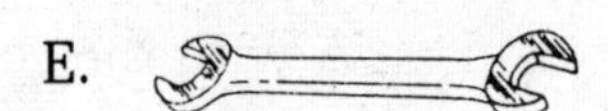

F. 

G. 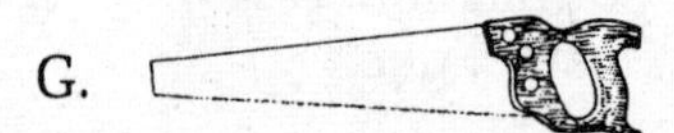

H. 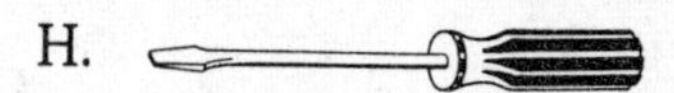

I. 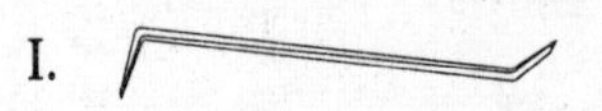

J. 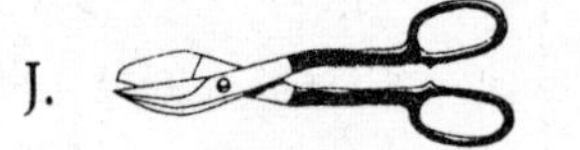

GROUP THREE

Directions: Questions 71 through 80 also test your knowledge of tools and their uses. For each question, first look carefully at the picture on the left and identify what is shown. Then choose from the boxes labeled (A), (B), (C), and (D) the item that is used with or goes best with what is shown in the first picture. Write the corresponding letter next to the number of the question on your answer paper.

71. 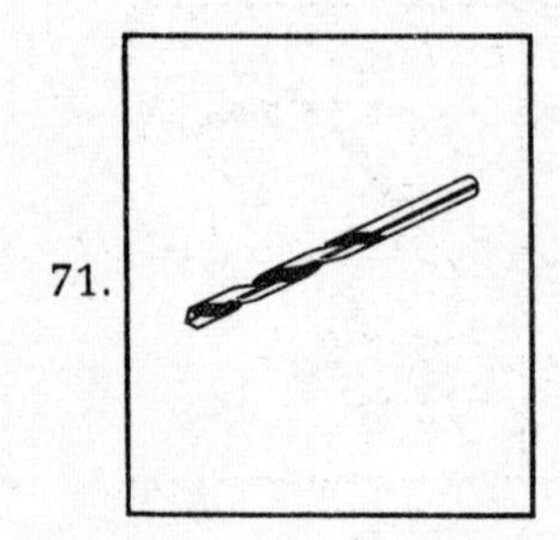

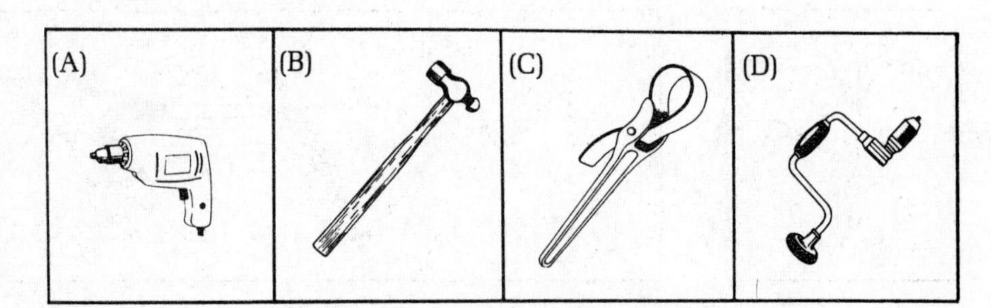

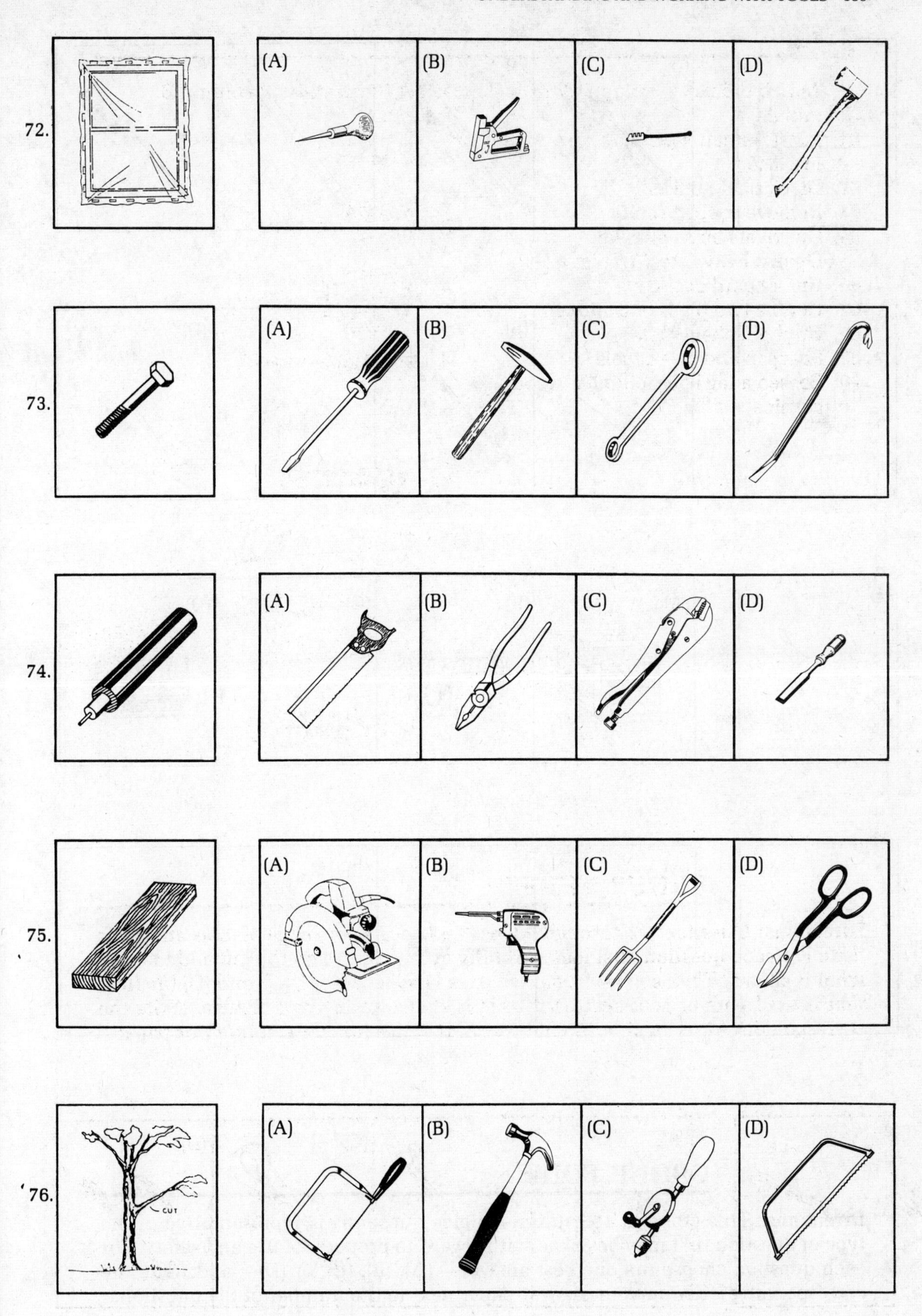
72.
(A)
(B)
(C)
(D)
73.
(A)
(B)
(C)
(D)
74.
(A)
(B)
(C)
(D)
75.
(A)
(B)
(C)
(D)
76.
CUT
(A)
(B)
(C)
(D)

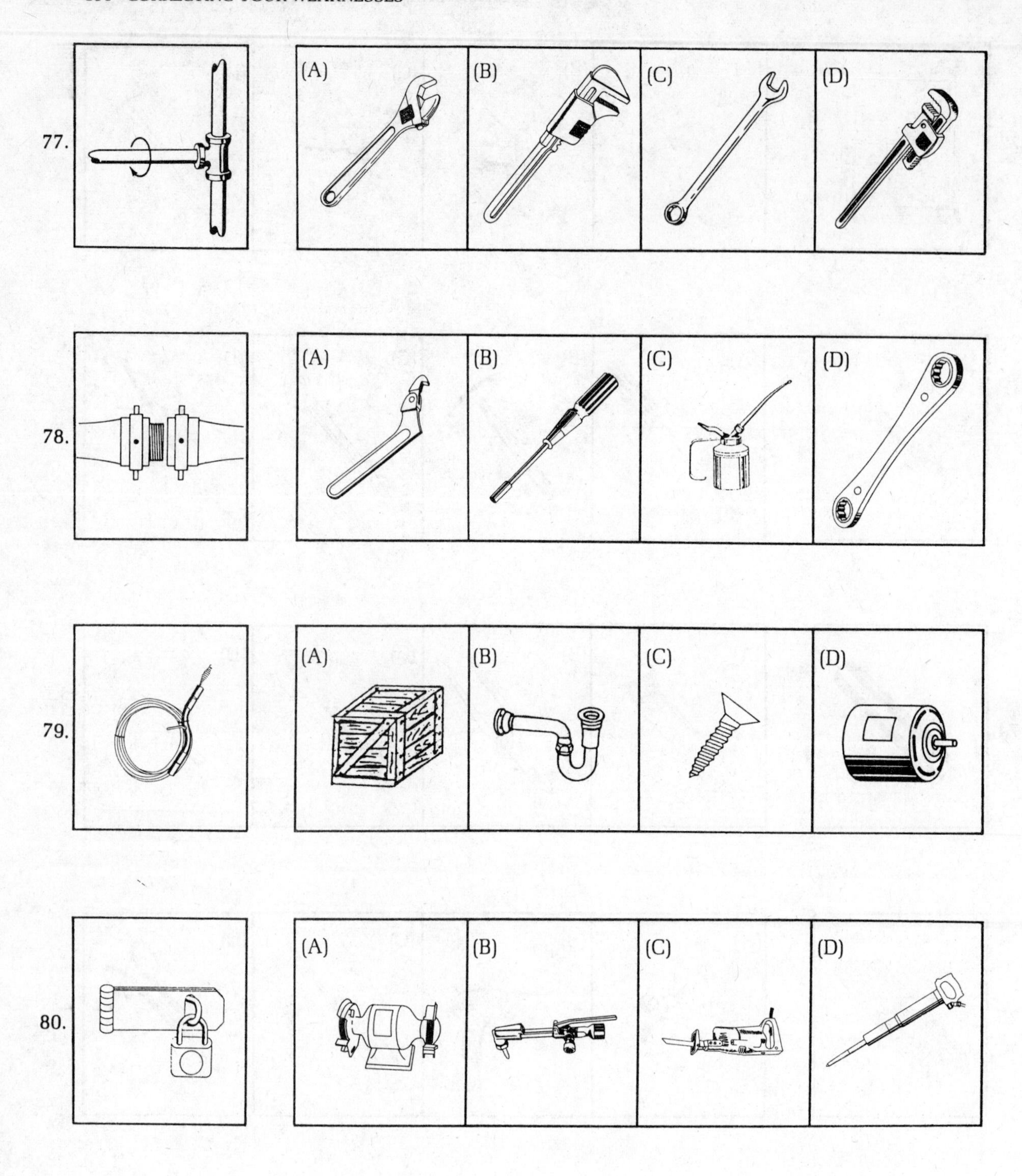

GROUP FOUR

Directions: This group of 15 multiple-choice questions is representative of the type of question that may be asked with regard to proper tool use and safety. For each question choose the one best answer—(A), (B), (C), or (D)—and write the corresponding letter on your answer paper next to the number of the question.

81. There are two major categories of tools, all-purpose and specialized. The most important reason for using specialized tools is that they are
    (A) much more impressive to the citizen.
    (B) cheaper.
    (C) easier to obtain.
    (D) more effective.

82. Which of the following describes the proper use of the ax to gain entrance to a building through a window with two panes, one above the other?
    (A) Place cutting edge of the ax at the center of the lower rail of the bottom window, and pry upward.
    (B) Place the cutting edge of the ax at the center where the two windows meet, and pry the windows apart.
    (C) Use the handle of the ax to break out both the top and the bottom window.
    (D) Use the flat side of the ax head to break out the center rails of both windows.

83. In the proper grip for holding a nail hammer, the thumb is
    (A) around the hammer handle.
    (B) on top of and in line with the claw.
    (C) on the same side of the hammer handle as the fingers.
    (D) in any of the above positions; all are correct.

84. A firefighter is discussing the proper use of hand tools. Which of the following statements would NOT be correct to make?
    (A) A soft-faced hammer should be used when there is danger of damaging the surface of the work.
    (B) The most common fault in using a hammer is to hold the handle too far from the head.
    (C) A wrench should be pulled, not pushed.
    (D) When using an adjustable wrench, the handle should be pulled toward the side having the fixed jaw.

85. A firefighter explaining what was learned at the fire academy that day would be INCORRECT in making which of the following statements?
    (A) Before a power tool is connected to a source of power, be sure that the switch of the tool is in the ON position.
    (B) Keep all safety guards in position, and use shields or goggles when necessary.
    (C) Never try to clear a jammed tool unless you disconnect the source of power first.
    (D) Have sufficient lighting available; use extension lights where necessary.

86. The primary use of an oxyacetylene torch is to
    (A) weld two pieces of steel together.
    (B) solder heavy cables.
    (C) mold metal for future use.
    (D) cut metal plates.

87. Lubricants such as grease and oil are used for all of the following purposes EXCEPT to
    (A) reduce friction.
    (B) cool.
    (C) prevent rust.
    (D) reduce speed.

88. A turnbuckle is a
    (A) general-purpose wrench.
    (B) valve on a water supply line.
    (C) coupling used to adjust the tension between two rods or wires.
    (D) device to show which circuit has been activated.

89. A "sweat joint"
    (A) joins two pieces of copper tubing.
    (B) allows small amounts of water to leak out.
    (C) expands under pressure.
    (D) uses "sweat" to make the joint tight.

90. A firefighter carrying an electric power tool by the cord would be
    (A) incorrect—this could lead to a short circuit.
    (B) correct—the tool is lighter when carried that way.
    (C) incorrect—the tool is heavier when carried that way.
    (D) correct—this helps to stretch the wire and keep it pliable.

91. The purpose of the third prong on electrical plugs is to
    (A) hold the plug securely in the outlet
    (B) insure that the prongs are always plugged into the correct holes
    (C) act as a guide and make the plug easier to insert
    (D) ground the appliance or tool

92. While cutting the roof of a fire building, the gasoline-powered saw becomes wedged in the cut. After closing the OFF switch the FIRST action the fire fighters should take is to
    (A) notify the officer in command that the roof cannot be opened
    (B) use a pry bar to pry the saw free from the cut
    (C) disconnect the spark-plug wire
    (D) disconnect the gasoline tank from the saw

93. The most appropriate tool to turn the wing cock on a gas supply line is
    (A) a screwdriver
    (B) needle-nose pliers
    (C) a vice grip
    (D) a ratchet wrench

94. Fire fighters investigating a fire are told that the fire was in a ballast. The fire fighters would be correct if they checked which of the following to determine whether this statement was true?
    (A) The fluorescent light at the ceiling.
    (B) The flushometer in the lavatory.
    (C) The oil burner in the basement.
    (D) The fuse box in the basement.

95. A couple returning from vacation find that while they were away the heat went off and the water pipes froze. They ask a fire fighter what the safest way is to defrost the pipes. Which of the following is the most appropriate advice?
    (A) Have a plumber remove the section of frozen pipe and replace it with new pipe.
    (B) Turn on the heat in the house and let the pipes thaw naturally.
    (C) Use a propane torch to heat the pipes and melt the ice.
    (D) Heat only the frozen areas with an electric hair dryer.

Answer Key and Explanations

ANSWER KEY

GROUP ONE

Set 1	Set 2	Set 3	Set 4	Set 5
1. D	11. I	21. H	31. E	41. E
2. I	12. F	22. D	32. G	42. F
3. B	13. A	23. B	33. I	43. G
4. J	14. H	24. A	34. J	44. B
5. G	15. B	25. E	35. B	45. H
6. H	16. C	26. I	36. C	46. J
7. C	17. J	27. C	37. D	47. C
8. E	18. G	28. G	38. F	48. A
9. A	19. D	29. J	39. H	49. D
10. F	20. E	30. F	40. A	50. I

GROUP TWO		GROUP THREE	GROUP FOUR
Set 1	**Set 2**		
51. J	61. D	71. A	81. D
52. I	62. F	72. B	82. A
53. G	63. H	73. C	83. A
54. B	64. A	74. B	84. B
55. H	65. I	75. A	85. A
56. A	66. G	76. D	86. D
57. C	67. B	77. D	87. D
58. F	68. J	78. A	88. C
59. D	69. C	79. B	89. A
60. E	70. E	80. B	90. A
			91. D
			92. C
			93. C
			94. A
			95. B

EXPLANATIONS

GROUP ONE

It is important that you become able to identify each tool. You must look up each wrong answer to determine what the correct tool looks like.

Set 1

1. **D** See Hammers.
2. **I** See Saws.
3. **B** See Saws.
4. **J** See Miscellaneous Hand Tools.
5. **G** See Axes.
6. **H** See Drilling Tools.
7. **C** See Drilling Tools.
8. **E** See Miscellaneous Hand Tools.
9. **A** See Miscellaneous Hand Tools.
10. **F** See Screwdrivers.

Set 2

11. **I** See Saws.
12. **F** See Clamps.
13. **A** See Tools for Prying—offset ripping chisel.
14. **H** See Pliers.
15. **B** See Wrenches.
16. **C** See Saws.
17. **J** See Wrenches.
18. **G** See Pliers.
19. **D** See Wrenches.
20. **E** See Miscellaneous Hand Tools.

Set 3

21. **H** See Drilling Tools.
22. **D** See Pliers.
23. **B** See Screwdrivers.
24. **A** See Tools for Prying.
25. **E** See Drilling Tools.
26. **I** See Wrenches.
27. **C** See Pliers.
28. **G** See Clamps.
29. **J** See Pliers.
30. **F** See Tools for Prying.

Set 4

31. **E** See Wrenches.
32. **G** See Digging Tools.
33. **I** See Pliers.
34. **J** See Miscellaneous Hand Tools.
35. **B** See Miscellaneous Hand Tools.
36. **C** See Hammers.
37. **D** See Drilling Tools.
38. **F** See Saws.
39. **H** See Wrenches.
40. **A** See Wrenches.

Set 5

41. **E** See Power Saws.
42. **F** See Other Power Tools.
43. **G** See Power Saws.
44. **B** See Power Saws.
45. **H** See Power Saws.
46. **J** See Other Power Tools.
47. **C** See Other Power Tools.
48. **A** See Other Power Tools.
49. **D** See Power Saws.
50. **I** See Power Saws.

GROUP TWO

Set 1

It is important that you become familiar with the use of each tool. You must look up each wrong answer to determine what the correct tool looks like.

51. **J** See Saws.
52. **I** See Drilling Tools.
53. **G** See Drilling Tools.
54. **B** See Wrenches.
55. **H** See Wrenches.
56. **A** See Miscellaneous Hand Tools.
57. **C** See Miscellaneous Hand Tools.
58. **F** See Hammers.
59. **D** See Axes.
60. **E** See Hammers.

Set 2

61. **D** See Wrenches.
62. **F** See Hammers.
63. **H** See Screwdrivers.
64. **A** See Miscellaneous Hand Tools.
65. **I** See Tools for Prying.
66. **G** See Saws.
67. **B** See Wrenches.
68. **J** See Pliers.
69. **C** See Other Power Tools.
70. **E** See Wrenches.

GROUP THREE

71. **A** The left-hand picture shows a high-speed twist drill. The appropriate tool to be used with this would be a power hand drill (A). Choice B is a ball peen hammer, choice C is a strap wrench, and choice D is a hand brace.
72. **B** The left-hand picture shows a window covered with plastic. A stapler (B) is the correct tool for adhering the plastic to the window frame. Choice A is an awl, choice C is a glass cutter, and choice D is an axe.

73. **C** The left-hand picture shows a bolt. A box wrench (C) is the correct tool to use with this. Choice A is a screwdriver, choice B is a tack hammer, and choice D is a pry bar.

74. **B** The left-hand picture shows a stripped electric wire. Electrician's pliers (B) are best for cutting this wire. Choice A is a back saw, choice C is a vice grip, and choice D is a wood chisel.

75. **A** The left-hand picture shows a board. The tool most appropriate for use with this is the electric saw (A). Choice B is an electric soldering gun, choice C is a digging fork, and choice D is a tin snips.

76. **D** The left-hand picture shows a tree limb. A tubular saw (D) is best for cutting this. Choice A is a coping saw, choice B is a hammer, and choice C is a hand drill.

77. **D** The left-hand picture shows a pipe. A Stillson wrench (D) or pipe wrench is required to turn this. Choice A is a crescent wrench, choice B is a monkey wrench, and choice C is a combination wrench.

78. **A** The left-hand picture shows a male and female hose coupling. A spanner wrench (A) is used to make them tight. Choice B is a nut driver, choice C is an oil can, and choice D is a ratchet wrench.

79. **B** The left-hand picture shows a hand auger. This tool would be used to release an obstruction in a P-trap (B). Choice A is a wood box, choice C is a screw, and choice D is a motor.

80. **B** The left-hand picture shows a hasp with a lock. An acetylene torch (B) would be the best tool for cutting this hardened steel lock. Choice A is a bench grinder, choice C is a reciprocating saw, and choice D is a pneumatic hammer.

GROUP FOUR

81. **D** A specialized tool is specifically designed for the task at hand. All-purpose tools are intended for a wide variety of applications and therefore must be adaptable. This need to be adaptable causes them to be less effective. Choice A is a trivial concern, and choices B and C are likely to be untrue.

82. **A** The object is to gain access to the building with the least amount of damage while using the safest method. Opening a window is always the first choice. Prying the window from the bottom will cause the window lock screws to come loose and release the window. Choice B will lead to breaking the windows and possibly to serious cuts. Choices C and D will result in serious damage and cuts.

83. **A** This provides for a firm grip with the least possibility of the hammer slipping out of the hand and flying away uncontrolled.

84. **B** This should read ". . . too close to the head." Choice D—this results in the power being applied to the solid side of the tool. Choices A and C are both correct statements.

85. **A** This should read "in the OFF position"; this is a very important safety rule. Choices B, C, and D are all correct statements.

86. **D** The oxyacetylene torch is used to cut metals by heating them to their melting points. Choices A, B, and C are for joining or shaping.

87. **D** Lubricants are used to protect (C) parts that rub against each other and increase their ease of operation. By reducing friction (A) and heat (B), speed is maintained or increased.

88. **C** A turnbuckle consists of a loop of metal with two thread ends. As the metal is turned, the threaded rods move either toward or away from each other, depending on the amount of tension desired.

89. **A** A "sweat joint" is the correct name for a solder joint in plumbing. It is created by joining two copper pipes, heating them, and then applying solder containing tin.

90. **A** The tool should be carried by the handle; carrying tools by the wires loosens the connections and leads to short circuits and severe shocks. This is another important safety rule.

91. **D** The third prong safely transmits any electrical discharge into the ground.

92. **C** Disconnecting the spark-plug wire will ensure that the saw does not accidentally start when it is freed. Choice A—this is not completely true; the saw can be freed rapidly or an axe can be used. Choice B—the spark plug must be removed first. Choice D—this would be an unsafe act; since the saw is designed to be self-contained, the tank should not be removed.

93. **C** A wing cock has a flat surface; a heavy-duty tool such as vice grip is needed to turn it. Note these valves are usually quarter-turn valves (see Pliers).

94. **A** The ballast is used to step up the electric power and charge the molecules of the gas in a fluorescent tube.

95. **B** This is the safest method, as the heat will thaw the pipes evenly and avoid a pressure build-up in them. Choice A—this would be necessary only if the frozen section of pipe is broken. Choice C—this could lead to a fire or possible steam explosion. Choice D—although using the hair dryer will work, it is dangerous if not done properly (from the faucet back to and just past the frozen area). Trapped, heated water can lead to an explosion.

CHAPTER 10

Mathematics, Machines, Science, and Information Ordering

Mathematics

WHAT IS MATHEMATICS?

Mathematics involves the manipulation of data represented by symbols to produce information in a form that is more useful than the original data. For example, a water tank is in the form of a cylinder. The upper and lower bases of the tank are parallel, congruent circles. The distance between the top and the bottom is the height of the tank.

The rule for finding the volume of a cylinder is as follows: The volume of a cylinder is equal to the area of the base multiplied by the height.

Since the base of a cylinder is a circle whose area is πr^2, the formula for the volume is

$$\text{Volume } (V) = \text{Base} \times \text{Height } (h)$$
$$V = \pi r^2 \times h$$

Example: The radius of a cylindrical tank is 3½ feet. Its height is 8 feet. Find the volume in cubic feet, letting $\pi = 22/7$.

$$\text{Volume } (V) = \pi r^2 h$$

$$V = \frac{22}{7} \times 3\frac{1}{2}\,ft. \times 3\frac{1}{2}\,ft. \times 8\,ft.$$

$$V = \frac{\overset{11}{\cancel{22}}}{\cancel{7}} \times \frac{\cancel{7}}{\cancel{2}} \times \frac{7}{\cancel{2}} \times \overset{4}{\cancel{8}}$$

$$V = 308 \text{ cu. ft. } (\text{ft.}^3)$$

The final answer is of greater use to anyone who wishes to fill the tank than knowledge of just the radius and the height. The beauty of mathematics is that it can be applied with equal success to different problems.

A second advantage is that mathematics lets us do efficiently what would be impossible without it. A simple example should make this point clear.

Any kind of bar resting on a fixed point or edge can be used as a lever; the point or edge is called the *fulcrum*. A lever will just balance when the numerical product of the effort (E) and its distance (d) from the fulcrum (F) is equal to the numerical product of the resistance (R) and its distance (D) from the fulcrum, that is, when

$$\text{Effort } (E) \times \text{Distance } (d) = \text{Resistance } (R) \times \text{Distance } (D)$$

Let's consider the following problem:

$E = 15$ lb. $R = ?$
↓ $d = 7$ △ $D = 3$ ↓

If we solve for R, we will find what resistance an effort of 15 pounds will support, by means of the lever shown, if $d = 7$ feet and $D = 3$ feet.

$$E \times d = R \times D$$
$$15 \text{ lb.} \times 7 \text{ ft.} = R \times 3 \text{ ft.}$$
$$\frac{15 \text{ lb.} \times 7 \text{ ft.}}{3 \text{ ft.}} = R$$
$$35 \text{ lb.} = R$$

WHY SHOULD A FIRE FIGHTER LEARN MATHEMATICS?

The most important reason for learning mathematics is that it is useful in our everyday life. We are constantly called upon to make decisions based on arithmetic, algebra, geometry, measurement estimations, maps, and scaled drawings.

What underlying skills should you possess to solve mathematical problems? You need to:

- Understand whole numbers and rational numbers (decimals and fractions), and their use in counting and measuring.
- Know the basic facts of the four common operations of arithmetic—addition, subtraction, multiplication, and division.
- Know how to measure and how to estimate measurements to determine height, weight, temperature, volume, etc. Estimating before actually taking measurements is good experience, and it will make you a better estimator. You should be able to make estimates in both the traditional (American) and the metric system, and you should be comfortable with both. The best way is to learn each system independently, and then, when they have become second nature, to learn to convert from one to the other.

  It is helpful to have a general idea of some approximate equivalents. A yard (36 inches) is a little shorter than a meter (39 inches). A quart (32 ounces) is a little less than a liter (34 ounces). A pound (16 ounces) is a little less than half a kilogram (17.6 ounces). For temperatures, if you multiply a Celsius or centigrade reading by 2 and add 30, you will get a fairly good estimate of the Fahrenheit reading; the exact formula is

  $$F = \tfrac{9}{5}C + 32$$

  where F = Fahrenheit degrees and C = Celsius degrees.
- Be familiar with and able to use both two- and three-dimensional geometric concepts, and to be aware of the relationship between them. Do you remember the Pythagorean theorem? It states that in a right triangle the

square of the hypotenuse equals the sum of the squares of the other sides: $a^2 + b^2 = c^2$

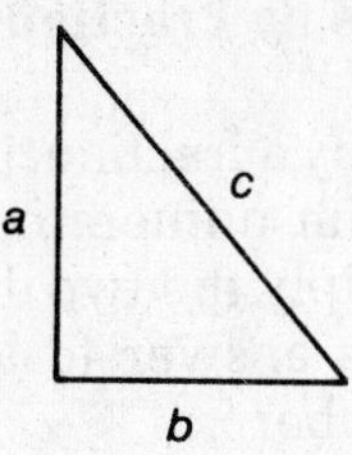

- Be able to use functions to describe and analyze the relationships between variables (time and height, size and temperature, etc.). You may have studied functions in algebra, geometry, or trigonometry.

BRIEF REVIEW OF FUNDAMENTAL TOPICS AND PRINCIPLES

ARITHMETIC

Mixed Numbers, Fractions, and Improper Fractions

1. A *mixed number*, for example, 3¼, consists of a whole number (3, in this case) and a *fraction* (¼). In an *improper fraction*, for example, $\frac{13}{4}$, the numerator (number above the bar) is greater than, or equal to, the denominator (number below the bar).
2. *To change a mixed number to an improper fraction*, first find the product of the whole number and the denominator; then add the numerator to this product, and place the result over the denominator.

   Example: Change 25⅔ to an improper fraction.

Multiply the whole number	(25):	25
by the denominator	(3):	× 3
to obtain the product	(75):	75
Add the numerator	(2):	+ 2
Place the result over the denominator	(77): (3):	$\frac{77}{3}$ = 25⅔

3. *To change an improper fraction to a mixed or whole number*, first divide the numerator by the denominator. If there is a remainder, write it as the numerator of a fraction that has the same denominator as the improper fraction. Reduce the fraction to lowest terms.

   Example: Change $\frac{57}{6}$ to a mixed number.

Divide the numerator	(57):	6)57 = 9 3/6 = 9½
		54
by the denominator	(6):	3
Place the remainder	(3):	3
over the denominator	(6):	6
reduce		$\frac{3}{6} = \frac{1}{2}$

Operations with Fractions

Multiplication of Mixed Numbers by Fractions and Multiplication of Fractions by Fractions

To multiply a mixed number by a fraction, (1) change the mixed number to an improper fraction, (2) multiply the numerators of the two fractions to get the numerator of the answer, (3) multiply the two denominators to get the denomina-tor of the answer, (4) reduce the answer to lowest terms, and (5) change the improper fraction to a mixed number.

Example: Find the product of $3\frac{1}{4}$ and $\frac{2}{3}$.

$$(1):\quad 3\tfrac{1}{4} = \frac{13}{4}$$

$$(2)\text{ and }(3):\quad \frac{13}{4} \times \frac{2}{3} = \frac{26}{12}$$

$$(4):\quad \frac{26}{12} = \frac{13}{6}$$

$$(5):\quad \frac{13}{6} = 2\tfrac{1}{6}$$

Therefore $3\frac{1}{4} \times \frac{2}{3} = 2\frac{1}{6}$.

Division of Mixed Numbers and Fractions

From the illustration below, you can see that 1/3 of a circle can be divided into two equal parts so that each of the parts is 1/6 of the circle; in other words, 1/3 divided by 2 is 1/6, or

$$\frac{1}{3} \div \frac{2}{1} = \frac{1}{6}$$

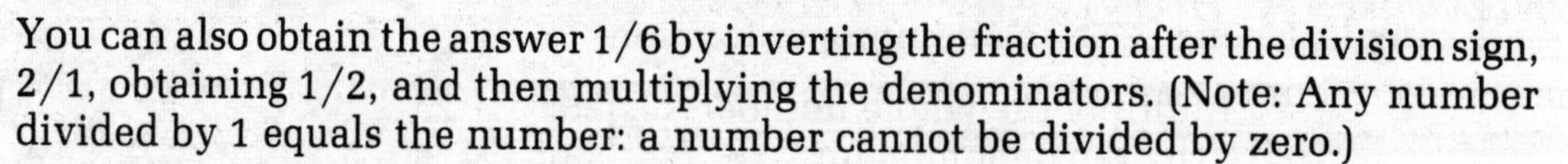

You can also obtain the answer 1/6 by inverting the fraction after the division sign, 2/1, obtaining 1/2, and then multiplying the denominators. (Note: Any number divided by 1 equals the number: a number cannot be divided by zero.)

$$\frac{1}{3} \div 2 = \frac{1}{3} \times \frac{1}{2} = \frac{1}{6}$$

To divide a number or a fraction by a fraction, (1) invert the divisor, (2) multiply the two numerators, and (3) divide by the product of the two denomina-tors.

Example: Divide $4\frac{1}{3}$ by $\frac{1}{4}$.
First change $4\frac{1}{3}$ to an improper fraction:

$$4\frac{1}{3} = \frac{13}{3}$$

Then:

$$\frac{13}{3} \div \frac{1}{4} = \frac{13}{3} \times \frac{4}{1} = \frac{52}{3} = 17\frac{1}{3}$$

Addition and Subtraction of Fractions

To add or subtract fractions whose denominators are different, first change the fractions to equivalent fractions with the lowest common denominator. Then add the numerators, and write the sum over the common denominator. Finally, reduce the answer.

Example: Add 1½, 4⅔, 5¼.

The lowest common denominator is 12.

$$1\frac{1}{2} = 1\frac{6}{12}$$
$$4\frac{2}{3} = 4\frac{8}{12}$$
$$5\frac{1}{4} = 5\frac{3}{12}$$
$$10\frac{17}{12} = 11\frac{5}{12}$$

Operations with Decimals

A decimal is a part of a whole number and therefore is a fraction whose denominator is 10 or a multiple of 10: 100, 1000, etc. To change a fraction to a decimal fraction, divide the numerator by the denominator to as many places as are necessary.

Example: Change 7/8 to a decimal.

$$\frac{7}{8} = 8\overline{)7.000}\quad 0.875$$

$$\begin{array}{r} 6\,4 \\ \hline 60 \\ 56 \\ \hline 40 \\ 40 \\ \hline \end{array}$$

Note: It will help you to remember that ⅞ means 7 divided by 8.

WORKING WITH UNKNOWNS (ALGEBRA)

Many of the problems with which you will be confronted will require you to write a formula and to solve for an unknown. A formula is a rule in which letters represent numbers. For example, to find the area (*A*) of a rectangle, use the formula $A = L \times W$. Let us assume that the length (*L*) is 4 feet and the width (*W*) is 3 feet. What is the area of the rectangle?

$$A = L \times W$$
$$A = 4 \text{ ft.} \times 3 \text{ ft.} = 12 \text{ sq. ft. (ft.}^2\text{)}$$

But what if you know the area and width and need to find the length? In that case, you write the formula ($A = L \times W$) and solve for the unknown. Let us assume that the area is 12 square feet and the width is 3 feet. What is the length of the rectangle?

$$A = L \times W$$
$$12 \text{ ft.}^2 = x \times 3 \text{ ft.}$$

Remember that both sides of a formula may be divided by the same number:

$$\frac{12 \text{ ft.}^2}{3 \text{ ft.}^2} = x \times \frac{\cancel{3 \text{ ft.}}}{\cancel{3 \text{ ft.}}}$$
$$x = \frac{12}{3}$$
$$x \text{ (Length)} = 4 \text{ ft.}$$

Many problems involving number relationships can be solved by remembering the four principles that may be applied to a formula:

1. Both sides of a formula may be divided by the same number.
2. Both sides of a formula may be multiplied by the same number.
3. The same number may be added to both sides of a formula.
4. The same number may be subtracted from both sides of a formula.

Example: Let us use one of the principles to solve the following problem: Find the time needed for a fire truck traveling at the rate of 55 miles per hour to go 165 miles.

First write the formula that expresses the relationships stated in the problem. The correct formula is

$$\text{Distance} = \text{Rate} \times \text{Time}$$

or, using symbols,

$$D = R \times T$$

Inserting the numbers given in the problem, you have

$$165 \text{ miles} = 55 \times T$$

To solve for T, divide both sides of the formula by R (55 miles per hour)

$$\frac{165 \text{ miles}}{55 \text{ mph}} = \frac{55}{55} \times T$$

$$T = \frac{165}{55}$$

$$T \text{ (Time)} = 3 \text{ hr.}$$

Finally, check to see whether your answer satisfies the relationships given in the problem.

Ratio and Proportion

When you compare two quantities by division, you are finding their *ratio.* To find the ratio of two quantities, for example, 5 fire lieutenants and 30 fire fighters, divide the first quantity into the second:

$$\frac{5}{30} = 5:30$$

You may express the ratio either as a fraction (5/30), with a colon (5:30), or with the word *per* (5 fire lieutenants per 30 fire fighters). *Note:* The symbol "/" is read as "per" and indicates division.)

You know that the ratio 5/30 is equal to the ratio 1/6. The equation 5/30 = 1/6 is called a *proportion.* A proportion is an equation that tells you that two ratios are equal. Another way of writing a proportion is to replace the equal sign with a double colon (::).

To solve a proportion problem you must first insert the knowns and unknowns into their proper places, and then solve for the unknown.

Example: If a fire truck can travel 2 miles in 3 minutes, how far can it travel in 5 minutes at the same rate of speed?

Let x represent the number of miles traveled in 5 minutes. Then

$$x \text{ miles} : 2 \text{ miles} :: 5 \text{ min.} : 3 \text{ min.}$$

or

$$\frac{x \text{ miles}}{2 \text{ miles}} = \frac{5 \text{ min.}}{3 \text{ min.}}$$

Multiply the inner terms (2 and 5) together and the outer terms (x and 3) together. The product of the inner terms (10) equals the product of the outer terms (3x).

(x) miles $\times$ 3 minutes $=$ 2 miles $\times$ 5 minutes

$$3x = 10$$

$$x = \frac{10}{3}$$

$$x = 3\tfrac{1}{3} \text{ miles}$$

GEOMETRY

To measure the distance around a room, the volume of a container, or any other quantity pertaining to a particular geometrical form, you need to know the formulas for determining the perimeter or circumference, area, and volume of a triangle, rectangle, or circle.

Shape:

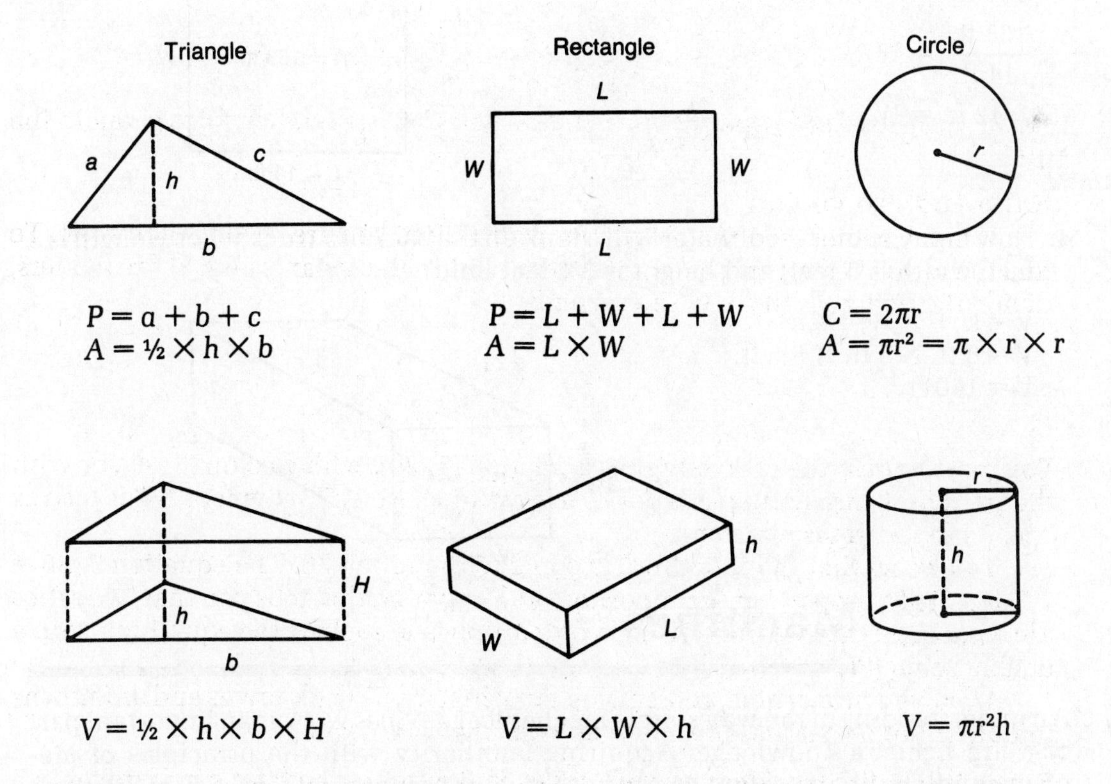

Geometry Examples

1. How many feet of rope are necessary to tie off a protective area around a hazardous building that is 25 wide by 30 long?

$P = L + W + L + W$
$P = 30 \text{ ft.} + 25 \text{ ft.} + 30 \text{ ft.} + 25 \text{ ft.}$
$P = 110 \text{ ft.}$

30 ft.
25 ft.

2. The diameter of a wheel is 14 feet. How many times must the wheel turn to travel 440 feet? (Use $\pi = 22/7$.)

$C = 2\pi r$

$C = 2 \times \frac{22}{\not 7} \times \not 7 \text{ ft.}$

$C = 44 \text{ ft. (for one turn)}$
$= 440 \text{ ft} : 44 \text{ ft.} :: x \text{ turns} : 1 \text{ turn}$

$\frac{440 \text{ ft.}}{44 \text{ ft.}} = \frac{x}{1} = 10 \text{ times}$

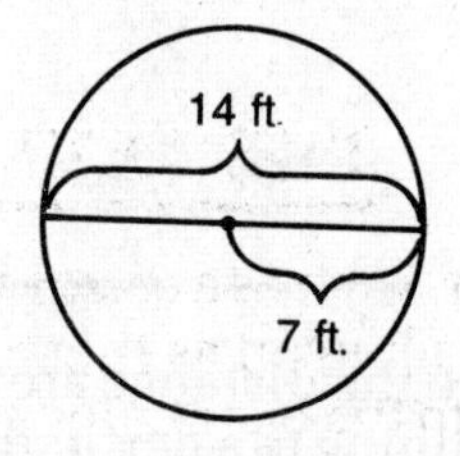

3. A room has an area of 168 square feet. If its length is 14 feet, what size salvage cover will be needed to protect the floor?

$A = L \times W$
$168 \text{ ft.}^2 = 14 \text{ ft.} \times W$
$\frac{168 \text{ ft.}^2}{14 \text{ ft.}} = W$
$12 \text{ ft.} = W$

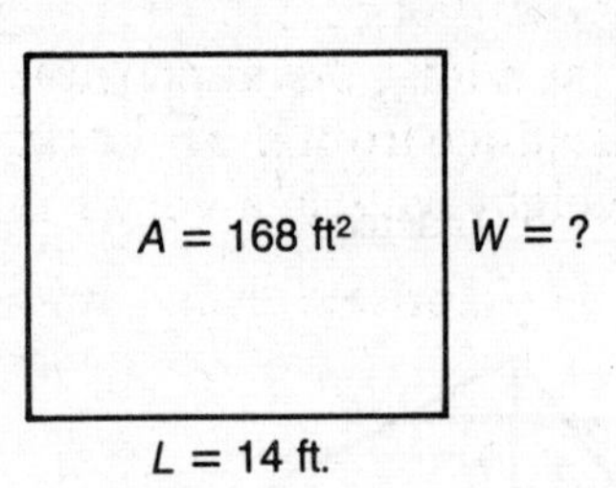

4. How many cubic feet of water will a tank on the back of a truck whose length is 8 feet, width is 5 feet, and height is 3½ feet hold?

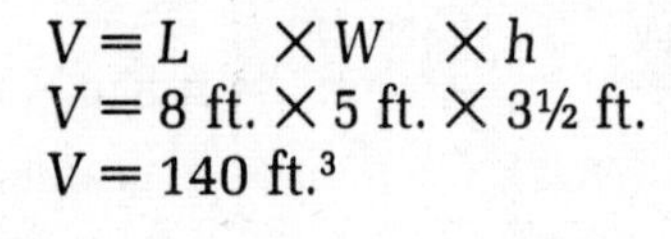

$V = L \times W \times h$
$V = 8 \text{ ft.} \times 5 \text{ ft.} \times 3\tfrac{1}{2} \text{ ft.}$
$V = 140 \text{ ft.}^3$

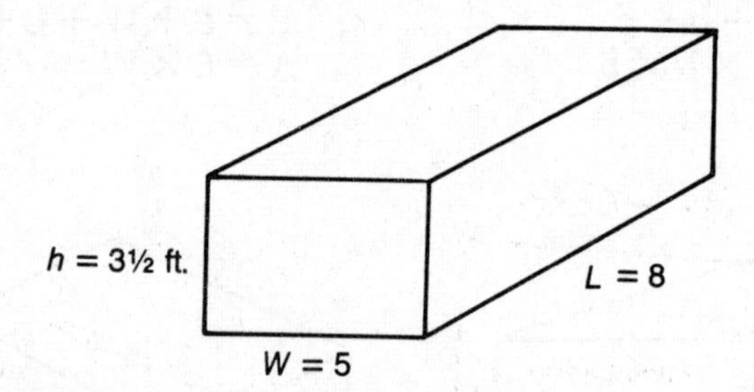

Machines

An understanding of the way simple mechanical devices work is an important part of a fire fighter's knowledge. Acquiring familiarity with the principles of mechanics will be an important component of your preparation for the Fire Fighter's Examination.

Machines can be grouped into two categories, *simple machines* and *compound machines*. The simple machines are the lever, inclined plane, wheel and axle, pulley, and screw. Compound machines are composed of any combination of simple machines.

Machines permit human beings to overcome physical weakness and accomplish heavy work. In other words, machines give human beings a *mechanical advantage*; this is the number of times a machine multiplies an applied force. The mechanical advantage of a machine is the ratio of the resistance to the effort. For example, if you want to lift 200 pounds using a machine and you exert an effort of 50 pounds, what is the mechanical advantage? mechanical advantage =

$$\frac{\text{resistance}}{\text{effort}} = \frac{200 \text{ pounds}}{50 \text{ pounds}} = 4$$

In answering this type of question on your examination, you need not consider such things as friction loss and efficiency unless you are specifically told to in the directions.

THE LEVER

A lever is a bar, such as a crowbar, that rotates about a point. The point at which the lever rotates is known as the *fulcrum*. The significance of the lever is that a small force applied to the bar at a long distance from the fulcrum can move or lift a heavy weight situated a short distance from the fulcrum. In a lever illustration, the fulcrum is usually designated by a small triangle (Δ).

A look at illustration A will help to make the principle of the lever clear. The fulcrum is designated by the triangle (Δ), the fire fighter is applying the force (effort) at a long distance from the fulcrum, and the heavy box (the resistance) a short distance from the fulcrum is being lifted. To determine the ability of a lever to lift a weight we use a very simple relationship:

$$\text{Effort} \times \text{Effort Distance} = \text{Resistance} \times \text{Resistance Distance}^{*}$$

Illustration A

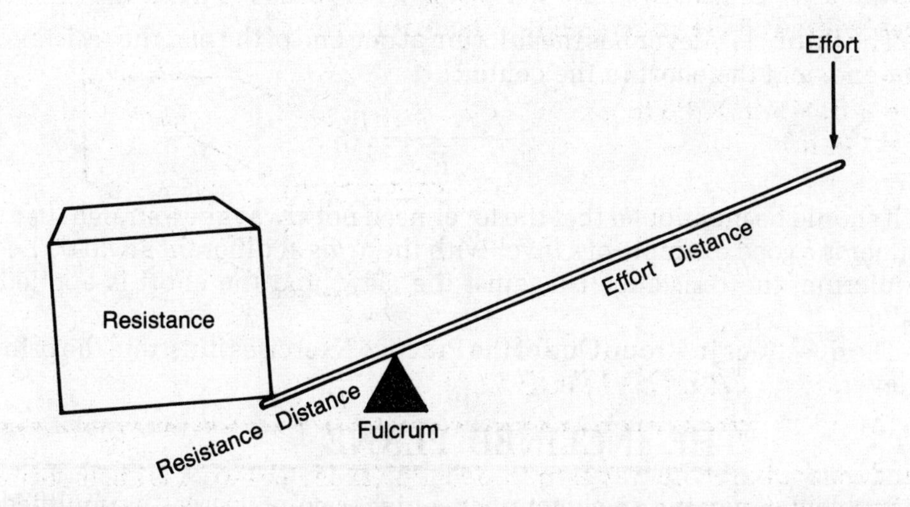

* Effort Distance is the distance from the fulcrum to the point where the effort is applied. Resistance Distance is the distance from the fulcrum to the point where the resistance is applied.

Let's look at illustration A again, but this time we will insert numbers. If the box weighs 1000 pounds and is 1 foot from the fulcrum, how much force (effort) will the fire fighter have to exert at a distance of 5 feet from the fulcrum?

$$\text{Effort} = E \qquad \text{Resistance} = 1000 \text{ pounds}$$
$$\text{Effort Distance} = 5 \text{ feet} \qquad \text{Resistance Distance} = 1 \text{ foot}$$
$$\text{Effort} \times \text{Effort Distance} = \text{Resistance} \times \text{Resistance Distance}$$
$$E \times 5 \text{ feet} = 1000 \text{ pounds} \times 1 \text{ foot}$$
$$E = \frac{1000 \text{ pounds} \times 1 \text{ foot}}{5 \text{ feet}}$$
$$E = 200 \text{ pounds}$$

The force (effort) required is 200 pounds.

The mechanical advantage of a lever system is determined by dividing the Effort Distance by the Resistance Distance. For example, if a 12-foot lever has an Effort Distance of 8 feet and a Resistance Distance of 4 feet, the mechanical advantage is

$$\frac{\text{Effort Distance}}{\text{Resistance Distance}} = \text{mechanical advantage (MA)}$$
$$\frac{8 \text{ feet}}{4 \text{ feet}} = \text{MA}$$
$$2 = \text{MA}$$

Because the force, resistance, and fulcrum can be located in different positions, there are three classes of levers. In the first-class lever, which is the most common, the fulcrum is between the effort (*E*) and the resistance (*R*):

E R
Δ

The second-class lever has the fulcrum at one end of the bar and the effort at the other end, with the resistance in between:

E
Δ
R

The third-class lever has the fulcrum at one end of the bar, the resistance at the other end, and the effort in the center:

∇ R
E

It should be mentioned that the lever need not always be a straight bar. A claw hammer is a good example of a lever with the arms at different angles. The head is the fulcrum, the resistance is against the claw, and the effort is applied to the handle.

The questions in Group One of the Practice Exercises illustrate the principle of the lever.

THE INCLINED PLANE

The problem of moving an object up to a higher point is greatly simplified by the use of an inclined plane. The inclined plane is a slanted surface connecting a lower point to an upper point. By pulling or pushing an object up the inclined plane, a significant mechanical advantage can be gained over lifting the object straight up.

The mechanical advantage of the inclined plane is equal to the length of the inclined plane divided by the height the object is to be lifted. For example, if the length of the plane is 10 feet and the height is 2 feet, the mechanical advantage is

$$\frac{\text{length of inclined plane}}{\text{height of inclined plane}} = \text{mechanical advantage (MA)}$$

$$\frac{10 \text{ feet}}{2 \text{ feet}} = \text{MA}$$

$$5 = \text{MA}$$

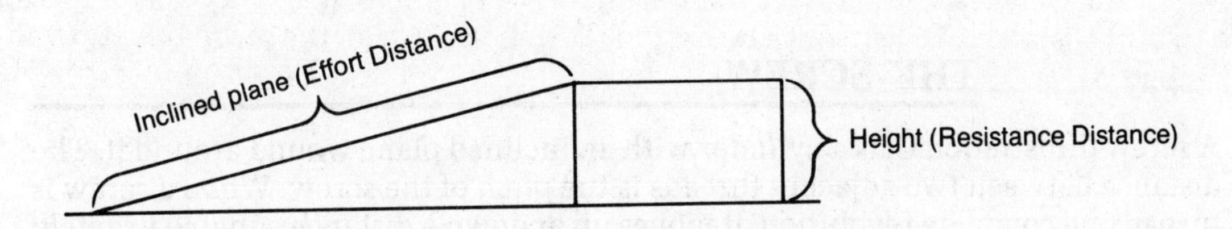

$$\text{Mechanical advantage} = \frac{\text{Effort Distance}}{\text{Resistance Distance}}$$

The following equation is used to solve an inclined plane problem:

$$\text{Effort} \times \text{Length of plane} = \text{Resistance} \times \text{Height to be raised}$$

The known values are inserted in their proper place.

The questions in Group Two of the Practice Exercises illustrate the principle of the inclined plane.

THE PULLEY

A pulley is a wheel turning on an axle over which a rope, belt, or chain is passed for the purpose of transmitting energy and doing work. Pulleys can be used in two ways: (1) fixed position, where the block is not movable; and (2) movable position, where the block is movable. For the simplest type of pulley, single fixed block, there is no mechanical advantage, only a change of direction.

The mechanical advantage of a pulley is determined by counting the number of ropes available to pull *upward*. The rope used to pull downward is *not* included. In illustration B the mechanical advantages (M.A.) are shown for several situations.

Three basic types of problems may be asked about pulleys: (1) to determine the load that can be lifted; (2) to determine how high the load will be lifted; and (3) to make an observation about how the systems work.

The questions in Group Three of the Practice Exercises illustrate the principles of the pulley.

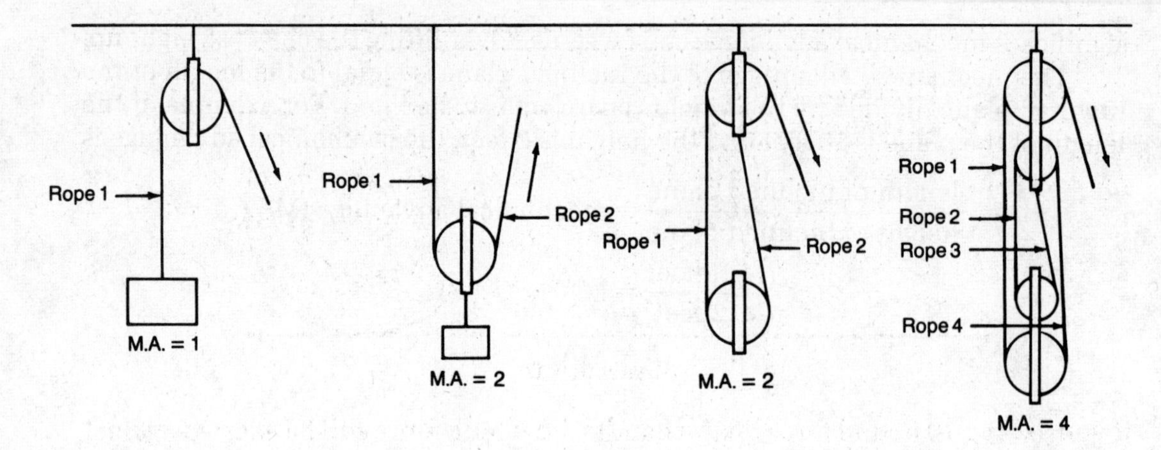

Illustration B

THE SCREW

A screw (illustration C) is a cylinder with an inclined plane wound around it. The distance between two adjacent threads is the *pitch* of the screw. When a screw is turned one complete revolution, it moves up or down a distance equal to its pitch; the force applied will travel a distance equal to the circumference made by the handle.

To calculate the effort in a screw problem the following equation is used:

$$\text{Effort} \times \text{Effort Distance (Circumference)} = \text{Resistance} \times \text{Resistance distance (Pitch)}$$

A common example of the principle of the screw used in the fire service is the common jackscrew.

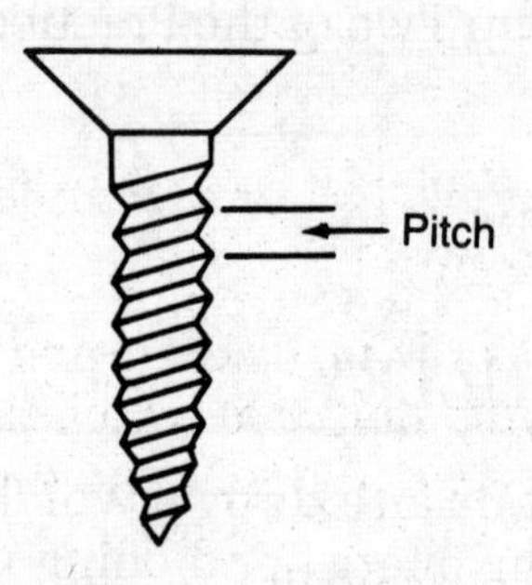

Illustration C

To understand how the screw works, look at illustration D. If you could pull the lever handle around one turn, it would move a distance of $2\pi \times E$ (the circumference of a circle whose radius equals E). At the same time, the screw has made one revolution, and moved a distance equal to its pitch (p). The mechanical advantage is equal to the effort arm distance ($2\pi E$) divided by the resistance arm distance (P). For example, what is the mechanical advantage if the length of the lever is 14 inches and the pitch for the screw thread is 1/8 inch? ($\pi = 22/7$)

$$\text{Mechanical advantage} = \frac{2 \times 22/7 \times 14 \text{ in.}}{1/8} = \frac{88}{1/8} = 704$$

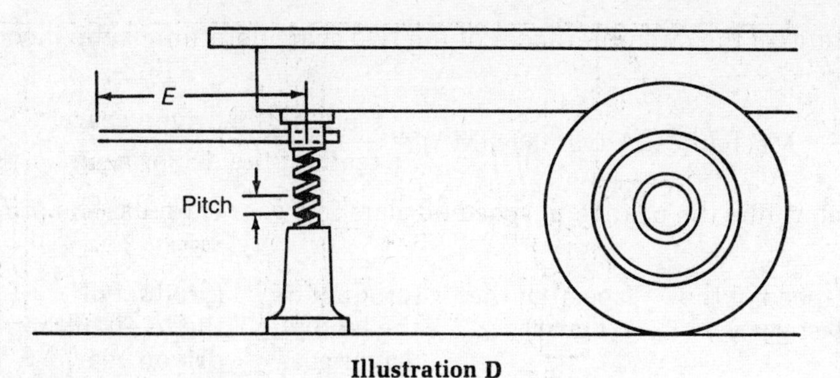

Illustration D

If you exert a 10 pound force on the handle, how big a force will be exerted against the resistance?

$$\text{Mechanical Advantage} = \frac{\text{Resistance}}{\text{Effort}}$$

$$704 = \frac{\text{Resistance}}{10 \text{ pounds}}$$

$$\text{Resistance} = 7040 \text{ pounds}$$

The questions in Group Four of the Practice Exercises illustrate the principle of the screw.

THE WHEEL AND AXLE

The wheel and axle, as the term implies, is composed of two parts: (1) a *large wheel*, which is attached to (2) an *axle*, which is actually a smaller wheel. The wheel and the axle turn together.

When force is applied to the large wheel, the resistance attached to the small wheel can be overcome. By turning the large wheel one full turn, it is possible to lift a weight equal to the distance of the circumference of the smaller wheel (circumference $= 2\pi r$).

To determine how much weight can be lifted by means of the wheel and axle, the following equation is used:

Example: The screwdriver is a wheel handle and axle tip. If you exert 10 pounds of force to the wheel handle, the radius of which is 2 inches, how big a force will be exerted against the resistance at the axle tip, the radius of which is 1/4 inch?

Effort × Circumference of large wheel
= Resistance × Circumference of small wheel

$$10 \text{ lbs} \times (2)(\pi)(2 \text{ in.}) = \text{Resistance} \times (2)(\pi)(\tfrac{1}{4} \text{ in.})$$

$$\tfrac{4}{1} \times 10 \times (2)(\pi)(2 \text{ in.}) = \text{Resistance} \times (2)(\pi) \times \tfrac{1}{4} \times \tfrac{4}{1}$$

$$80 \text{ lbs.} = \text{Resistance}$$

The questions in Group Five of the Practice Exercises illustrate the principle of the wheel and axle.

GEARS

Gears are used to change direction, increase or reduce speed, and increase or reduce force.

The ratio of the circumferences of the two gears determines the mechanical advantage.

$$\text{Mechanical Advantage (MA)} = \frac{\text{teeth of the driven gear}}{\text{teeth of the driver gear}}$$

To determine the change in speed of gears in a gear train use the proportion

s_2	:	s_1	::	t_1	:	t_2
(speed of the last gear)		(speed of the 1st gear)		(product of the teeth of driver gear		(product of the teeth of the driven gear)

The questions in Group Six of the Practice Exercises illustrate the principle of gears.

BELT DRIVES

There are two types of belt drives, positive and nonpositive. A positive drive consists of two wheels with sprockets that mesh with a chain or belt. A nonpositive drive makes use of a smooth wheel and smooth belt; an example is the V-belt found in the front of an automobile engine.

Belt drives are used to increase or reduce speed or force and to change the direction of a turning wheel. The belt drive transmits power and changes speed in the same manner as gears, but permits the two wheels to be separated and connected only by the belt. A change in direction is accomplished by twisting the belt.

The questions in Group Six of the Practice Exercises illustrate the principle of the belt drive.

Science

This short review is not intended to teach you science; its sole purpose is to jog your memory and bring back facts you have already learned but may have forgotten. If the materials do not seem familiar, you should spend some time reviewing a general science textbook. Keep in mind that most of the science questions you will encounter on the Fire Fighter Examination are based on principles of physics and chemistry, and these are the areas of science on which you should concentrate.

MATTER

Matter is anything that takes up space and has weight. It exists in three states: gas, liquid, and solid.

A *gas* is a substance without shape or definite volume. It expands to fill the space available.

A *liquid* has no specific shape but has a definite volume. A liquid takes the shape of the container in which it is stored. Examples: a cola bottle, a storage tank, a lake.

A *solid* has a definite shape and a definite volume. Examples: a car, a piece of wood, a book.

ELECTRICITY

The preferred kinetic molecule state of an object is to be neutral, that is, to have equal numbers of protons and electrons.

When something is not neutral, there is a chance that an electron flow can develop. The flow will always be the same; electrons flow from the point where there is an excess through a conductor (metal wire—silver, gold, aluminum, or copper) to the place where there is a deficiency. The flow is prevented by using a resistor or insulator.

A charge can accumulate on a substance. When the excess accumulation leaves the surface, we have a "static (electric) spark." This spark may have sufficient heat potential to ignite some substance, particularly a flammable liquid.

An electric circuit is made up of three parts:

1. The electric source: a battery or generator.
2. The conductor: wires.
3. The appliance: a motor, light bulb, or household appliance.

A circuit is *open* when the flow of electrons is able to make a complete loop back to the source, and is *closed* or *broken* when the electrons cannot return to the source. Since electricity can flow in only one direction at a time, two wires are necessary to make a current—one from the source to the appliance, the other from the appliance to the source.

There are two types of circuits. In a *series* system the flow of current must pass through all appliances before returning to the source; any break in the flow stops the flow of the whole system. In a *parallel* system the current can flow through each subcircuit independently and then back to the source; a break in a subsystem flow affects only the appliances on the subsystem, all others remaining operative.

Resistance, limiting the flow of electrons, will produce heat. As the resistance increases, so does the heat and so does the chance of fire. The resistance in a wire depends on three factors:

1. The *type* of wire. Aluminum, copper, and silver are good conductors; tungsten and nichrome are good resistors.
2. The *thickness* of the wire. As the thickness increases, the resistance is reduced.
3. The *length* of the wire. The longer the wire, the greater is the resistance.

Since resistance results in an increase in heat, the possibility of overloading or short circuiting is increased. Overloading can lead to burning of the insulation around the wire. A short circuit can lead to a powerful short electrical discharge. To prevent overloading and short circuits, a fuse or circuit breaker is placed in the system. The fuse contains a small metal strip that melts at a predetermined temperature lower than the temperature the wire is designed to handle. A household wiring system would normally have its fuses in parallel to prevent an overload in any subsystem from shutting down the whole system. Bypassing the fuse or using an improper fuse has often led to a serious fire.

HEAT

The temperature of a substance determines whether it will give off or take on heat. Heat always flows from the warmer to the colder body.

The two most common scales for measuring heat are the Fahrenheit scale and the Celsius or centigrade scale. Both scales use the freezing and boiling points of water as reference points. On the Fahrenheit scale the freezing point of water is 32 degrees; on the Celsius scale it is 0 degrees. The boiling point of water on the Fahrenheit scale is 212 degrees; on the Celsius scale it is 100 degrees.

As water is heated, it absorbs heat. For each British thermal unit (Btu) of heat, one pound of water rises 1 degree Fahrenheit. This statement is true for all conditions except at the points of freezing and boiling. At these two points, additional heat is necessary to allow the water to change state from a solid (ice) to a liquid (water) or from a liquid (water) to a gas (steam). To go from ice to water requires 143.4 additional Btu's, and to go from water to steam requires 970.3 Btu's.

Heat is transferred from one object to another in three ways: by *conduction*, the direct touching of objects; by *convection*, the movement or circulation of a heated gas (air); and by *radiation*, energy traveling as an electromagnetic wave through space.

Sources of heat energy include the combustion of solids, liquids, and gases; electrical heat energy; mechanical heat energy (such as friction); the heat of compression (known as the diesel effect); and nuclear heat energy.

FIRE

Fire is rapid combustion resulting in heat and light. For a fire to take place, four conditions must be met:

1. There must be a fuel (wood, gasoline, paper, etc.).
2. There must be oxygen (air).
3. There must be heat.
4. There must be sufficient energy to get the chemical reaction started.

To understand this concept better, think about the conditions around you at the present moment. This book is a fuel, the air around you contains oxygen, and you are probably in an area where it is warm and comfortable (heat)—yet there is no fire. To create a fire you would need sufficient additional energy to cause the substances to react and hence burn.

There are four basic types of fires:

1. Fires that burn and leave an ash such as wood are called CLASS A FIRES.
2. Fires in liquids are called CLASS B FIRES.
3. Fires in electric circuits or appliances in which the current is ON are called CLASS C FIRES. (*Note:* When the current is OFF or disconnected, the same fire can be classed as A or B, depending on what is burning—isulation or oil).
4. Fires in reactive metals such as magnesium, lithium, titanium, and zirconium are called CLASS D FIRES.

There are four ways to extinguish a fire:

1. Remove the fuel. Turn off the gas, shut off the oil flow, take the combustible materials away.
2. Remove the oxygen. Put a cover on the combustible materials.
3. Remove the heat. Put water on the fire; cool it.
4. Interfere with the chemical reaction. Put a special extinguishing agent, a dry chemical, on the fire.

Note: Fires most often occur from carelessness, ignorance, and/or failure to maintain equipment properly.

WATER

Water, the universal solvent and primary fire-extinguishing agent, exists as a solid, a liquid, and a gas. As water is cooled, it contracts in volume until just before freezing, at which point (34 degrees Fahrenheit or 4 degrees Celsius) it expands.

For this reason, in cold climates pipes, if not protected, often break. The freezing point of water can be lowered by adding a freezing point depressant such as ethylene glycol (antifreeze).

BASIC CHARACTERISTICS OF WATER

- Water boils at 212 degrees Fahrenheit.
- Water solidfies at 32 degrees Fahrenheit.
- Water changing state from solid to liquid requires an additional 143.4 Btu's per pound of water.
- Water changing state from liquid to steam requires an additional 970.3 Btu's per pound of water.
- At all times except when changing states, 1 Btu per pound of water is required to raise the water temperature by 1°F.
- Water is a solvent that can wash away and/or dilute many combustible products.
- Water in the liquid state has a very stable viscosity and can easily be moved by gravity or pump through pipe or hose.
- The high surface tension of water allows it to be thrown in droplet form and allows many extinguishing agents to work effectively with it.
- One gallon of water weighs 8.35 (approximately 8½) pounds.
- Fresh water has a density of 62.4 pounds per cubic foot.
- Salt water has a density of 64 pounds per cubic foot.
- A 1-cubic-foot container can hold 7.48 (approximately 7½) gallons of water.
- A 13.54-inch column of water exerts the same force as a 1-inch column of mercury.
- In theory, water can be drafted up to 33 feet high; however, 28 feet is the practical limit.
- A water droplet, when converted to steam, becomes 1700 times larger than its original size.
- A 1-inch-square, 1-foot-high column of water exerts a pressure of 0.434 pound.

The questions in Group Eight of the Practice Exercises deal with science.

SIX PRINCIPLES OF FLUID PRESSURE

- Liquid pressure is exerted in a perpendicular direction to any surface on which it acts.
- At any given point beneath the surface of a liquid, the pressure is the same in all directions.
- Pressure applied to a confined liquid from outside is transmitted in all directions without any loss of force.
- The pressure of a liquid in an open vessel is proportional to the depth of the liquid.
- The pressure of a liquid in an open vessel is proportional to the density of the liquid.
- Liquid pressure on the bottom of a vessel is unaffected by the size of the vessel.

Information Ordering

Information ordering is the process of putting several related items into a meaningful order or sequence. Fire fighting procedures, fire inspection processes, and emergency medical practices are generally prepared and carried out in some predetermined order. When presenting a question that requires the candidate to order items, the examiner will give instructions on the rules of ordering. It is very important for you to understand the ordering instructions and then to follow them precisely.

The candidate may be asked to read a paragraph and then select from a list of possible choices the activity that comes first, second, or third, or sometimes all three. The question may ask to arrange the events in the order they happen (timed events), or it may present a series of events and ask what activity should come next.

When reading a passage that requires you to order the information, number each item as you encounter it. When you have completed the passage, you can go back quickly and determine where each item begins and its relationship to the other items. The items may not always be presented in the correct order, but there should be some indication of their positions in the sequence of events.

Practice Exercises

For each question select the one *best* answer—(A), (B), (C), or (D)—and write the corresponding letter on your answer paper next to the number of the question.

GROUP ONE: THE LEVER

1. How many pounds of force would be required to lift the 300-pound box in illustration 1?
   (A) 100
   (B) 200
   (C) 600
   (D) 1200

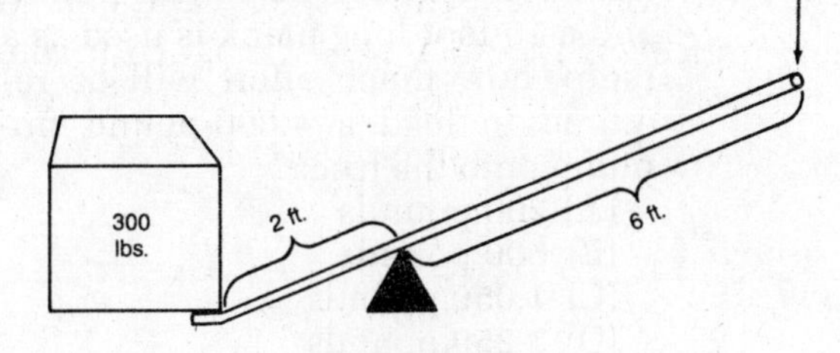

Illustration 1

2. The force that must be applied to the handle of the hammer in illustration 2 is
   (A) 10 pounds.
   (B) 15 pounds.
   (C) 75 pounds.
   (D) 175 pounds.

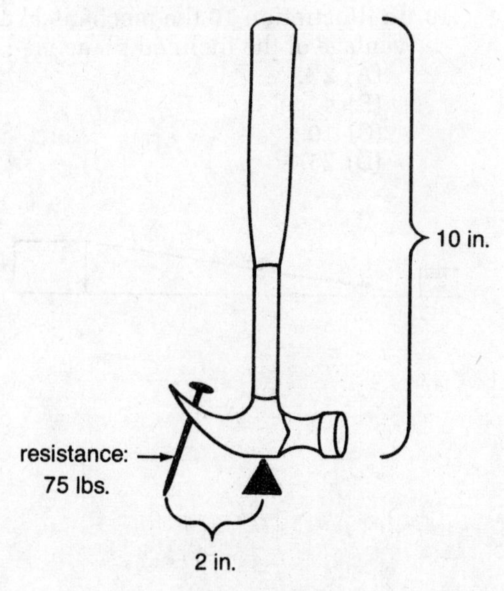

Illustration 2

3. The best way to make it easier to lift the weight with the board as indicated in illustration 3 would be to
   (A) move the box closer to the fire fighter
   (B) use a shorter board
   (C) move the weight closer to the fulcrum
   (D) turn the box to an upright position

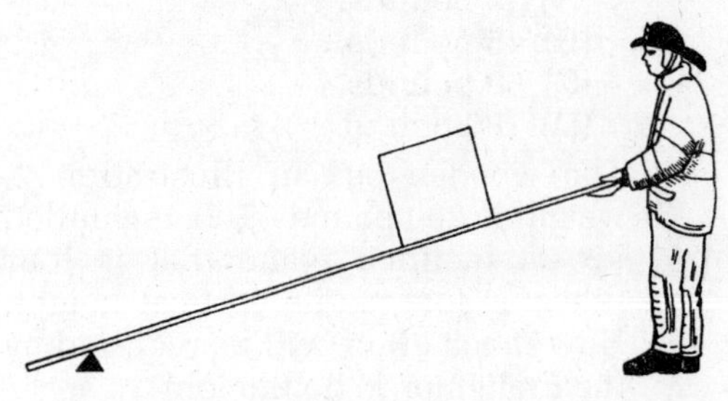

Illustration 3

4. In illustration 4, a fire fighter is using a tool to pry up a floor board to search for hidden fire. If the resistance is 600 pounds, the adz is 6 inches, and the fire fighter applies a force at a point 3 feet from the fulcrum, how much force will be needed?
   (A) 50 pounds.
   (B) 100 pounds.
   (C) 300 pounds.
   (D) 600 pounds.

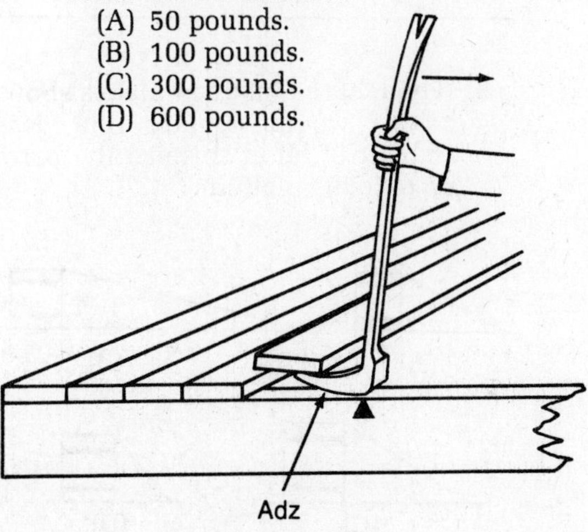

Illustration 4

5. In illustration 5, the firefighter is applying a force to a 25-foot ladder at a point 10 feet from the fulcrum. If the resistance of the ladder is 64 pounds, how much force must the firefighter use?
   (A) 64 pounds.
   (B) 80 pounds.
   (C) 160 pounds.
   (D) 640 pounds.

GROUP TWO: THE INCLINED PLANE

6. If a 300-pound bale of paper is to be raised to a platform 6 feet above, rolling the bale up a 30-foot ramp would require an effort of
   (A) 30 pounds.
   (B) 45 pounds.
   (C) 60 pounds.
   (D) 180 pounds.

7. The wooden box in illustration 7, weighing 200 pounds, is being pulled up an inclined plane that is four times as long as its vertical height. How much effort will be required by the firefighter to do the job?
   (A) 50 pounds
   (B) 100 pounds
   (C) 200 pounds
   (D) 400 pounds

8. Which of the inclined planes shown below would require the least amount of effort to raise the barrel up onto the platform?

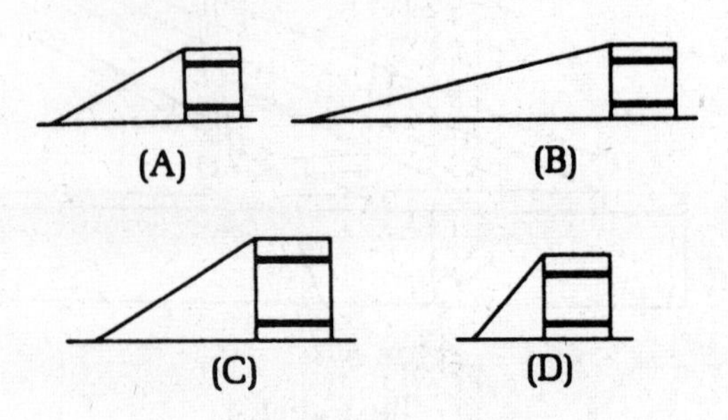

9. If the tailboard of the truck in illustration 9 is 5 feet above the ground and a 16-foot-long plank is used as a ramp, how much effort will be required to load a 4,000-pound fire pump onto the truck?
   (A) 200 pounds
   (B) 600 pounds
   (C) 1,050 pounds
   (D) 1,250 pounds

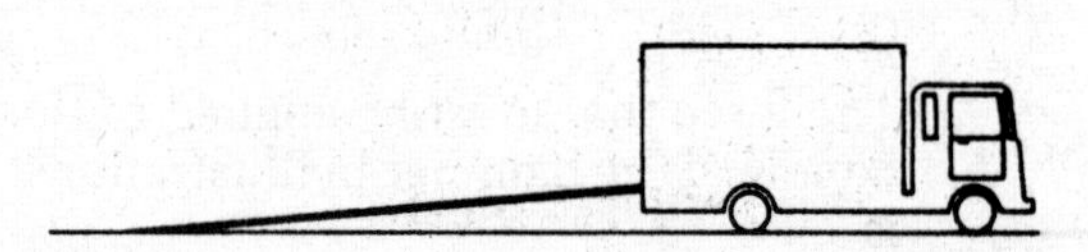

10. In illustration 10 the mechanical advantage of the inclined plane is
    (A) 2½.
    (B) 4.
    (C) 10.
    (D) 25.

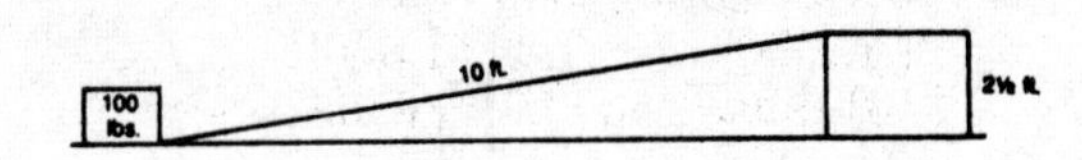

GROUP THREE: THE PULLEY

11. In illustration 11 a weight of 360 pounds is being raised with the help of a single and movable pulley system. How much effort is required to lift this weight?
    (A) 360 pounds
    (B) 180 pounds
    (C) 120 pounds
    (D) 85 pounds

360 lbs.

12. If the weight in illustration 12 must be lifted a height of 2 feet, how much rope will a firefighter have to pull through the pulley?
    (A) 1 foot (C) 3 feet
    (B) 2 feet (D) 4 feet

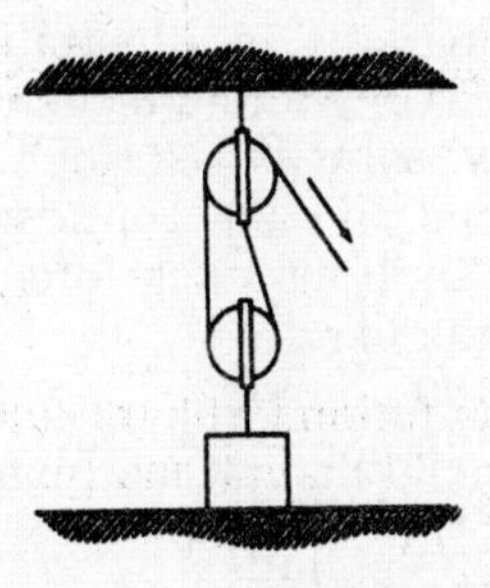

13. If the weight being lifted by the firefighters in illustration 13 is 2,000 pounds, how many pounds of effort must the firefighters exert?
    (A) 250 (C) 1,000
    (B) 500 (D) 2,000

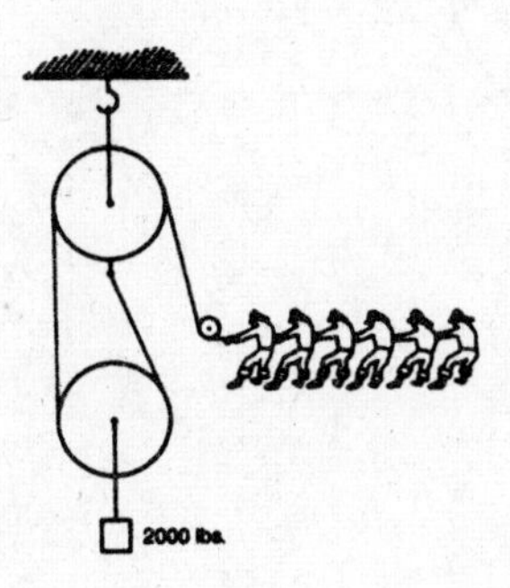

14. Which of the pulley arrangements in illustration 14-15 would require the greatest effort to lift the can?
    (A) 1 (C) 3
    (B) 2 (D) 4

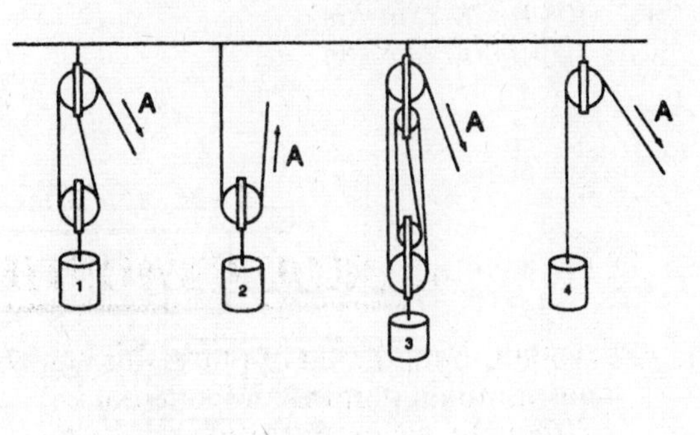

15. What is the maximum weight that a pulling force of 100 pounds applied to rope *A* in arrangement 4 of illustration 14-15 would be able to lift?
    (A) 100 pounds
    (B) 600 pounds
    (C) 1,200 pounds
    (D) 1,800 pounds

16. If a pull equal to 300 pounds is applied to line *A* in illustration 16, how much weight can be lifted?
    (A) 600 pounds
    (B) 900 pounds
    (C) 1,500 pounds
    (D) 1,800 pounds

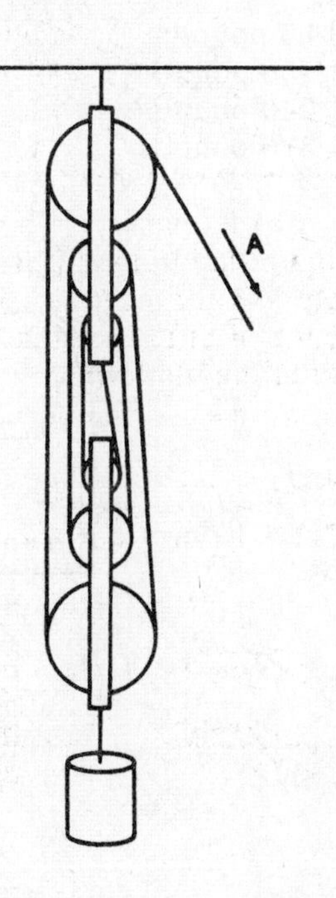

GROUP FOUR: THE SCREW

17. How much weight will a jackscrew with a pitch of ⅕ inch and a handle 28 inches long lift when a force of 5 pounds is applied to the handle?
    (A) 2,200 pounds
    (B) 3,500 pounds
    (C) 4,400 pounds
    (D) 6,000 pounds

18. If the pitch in the jackscrew in illustration 18 is ½ inch and the jack handle measures 14 inches, approximately how much force will the firefighter have to apply to lift a 2,000-pound truck?
    (A) 1 pound
    (B) 4½ pounds
    (C) 5½ pounds
    (D) 11½ pounds

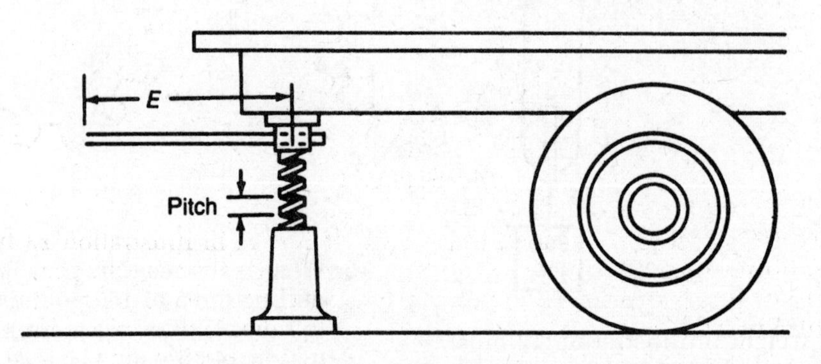

Illustration 18

GROUP FIVE: THE WHEEL AND AXLE

19. The axle in illustration 19-20 is 6 inches in diameter, and the handle when turned makes a circle with a diameter of 24 inches. If a firefighter uses a force of 60 pounds to turn the handle, how much weight can be lifted?
    (A) 90 pounds
    (B) 150 pounds
    (C) 240 pounds
    (D) 370 pounds

20. Assume that the booster reel on the fire apparatus in illustration 19-20 has an axle with an 8-inch diameter and a wheel with a 20-inch diameter. If a load of 300 pounds must be pulled in, how much force need the firefighter exert?
    (A) 100 pounds
    (B) 120 pounds
    (C) 160 pounds
    (D) 200 pounds

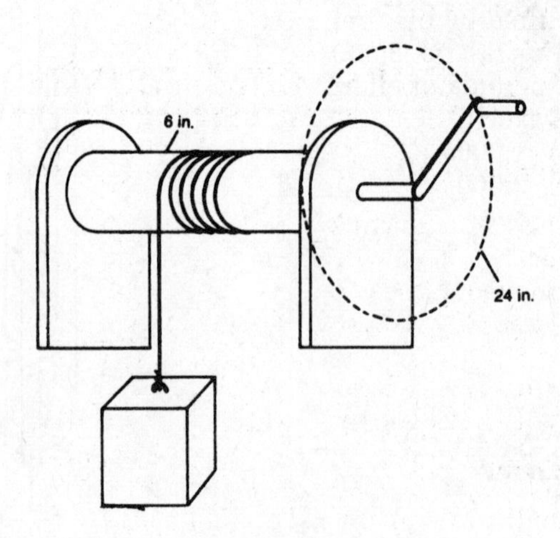

Illustration 19–20

GROUP SIX: GEARS

21. If gear A in illustration 21 turns clockwise, gear D will
    (A) also turn clockwise
    (B) be unable to turn
    (C) turn counterclockwise
    (D) turn in the same direction as gear C

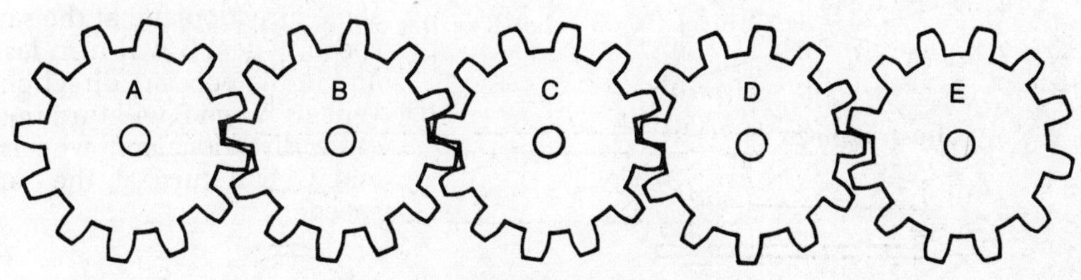

Illustration 21

22. It would be correct to state about the gears in illustration 22 that:
    (A) If gear B turns once, gear A will turn twice.
    (B) If gear A turns clockwise, gear B will turn clockwise.
    (C) If gear B turns twice, gear A will turn once.
    (D) If gear B turns counterclockwise, gear A will turn clockwise.

Illustration 22

23. If gear *A* in illustration 23 turns one revolution per second, gear *D* will turn
    (A) one revolution per second
    (B) six revolutions per second
    (C) one-sixth of a revolution per second
    (D) sixty revolutions per second

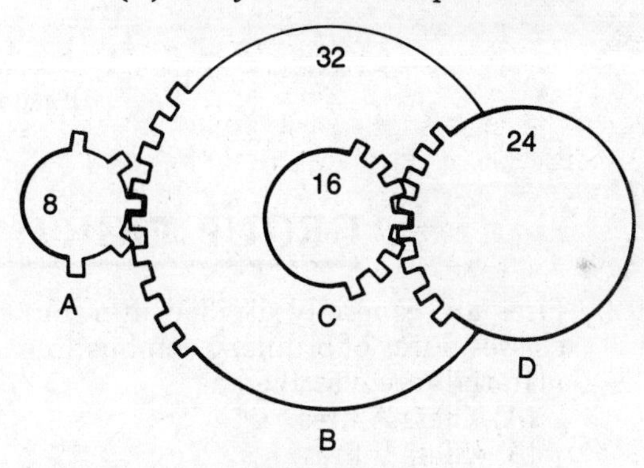

Illustration 23

GROUP SEVEN: BELT DRIVES

24. It would be INCORRECT to say about illustration 24 that:
    (A) Wheel X will turn in the same direction as wheel Y.
    (B) If wheel Y is moved closer to wheel X, then wheel X will turn faster.
    (C) If wheel X is the drive wheel, then wheel Y will have an increase in speed.
    (D) If wheel Y is the drive wheel, then wheel X will have an increase in force.

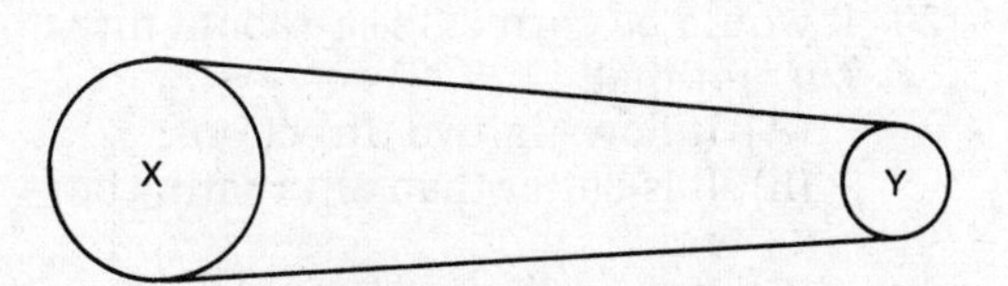

Illustration 24

25. The most appropriate reason for having the belt cross over as shown in illustration 25 is to
    (A) increase the speed of pulley Y
    (B) decrease the force of pulley Y
    (C) change the direction of the force
    (D) increase the safety of the drive system

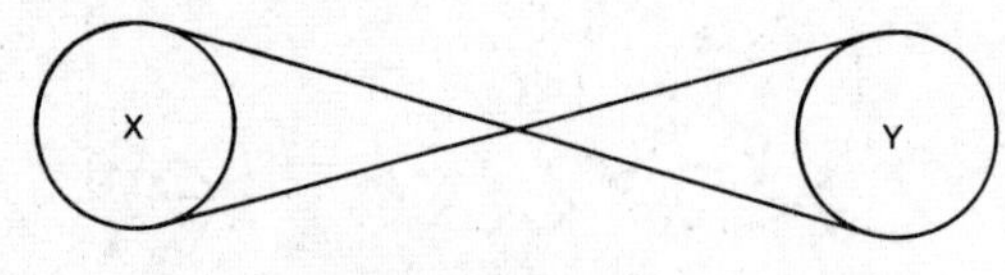

Illustration 25

26. A fire fighter would be correct in stating about illustration 26 that:
    (A) All wheels will turn at the same speed, but wheel C will turn in the opposite direction from A and B.
    (B) All wheels will turn in the same direction, but wheel C will turn more slowly than A or B.
    (C) Wheels A and C will turn in the same direction and at the same speed; wheel B will turn faster and in the opposite direction.
    (D) Wheels A and B will turn in the same direction, and wheels B and C will turn at the same speed.

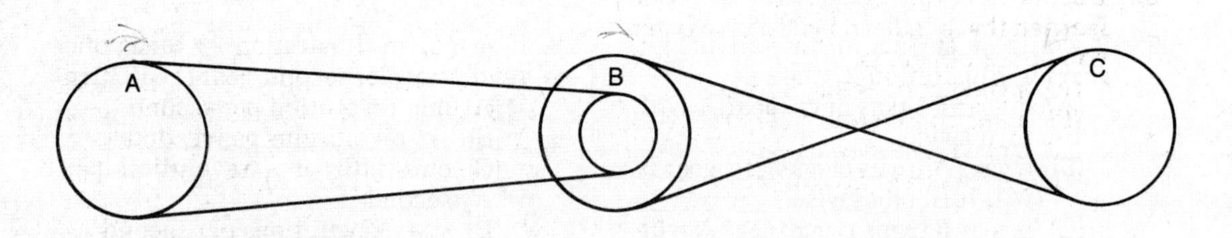

Illustration 26

GROUP EIGHT: SCIENCE

27. Fires are generally divided into four classes. Fires of ordinary combustible materials are classified as
    (A) Class A fires
    (B) Class B fires
    (C) Class C fires
    (D) Class D fires

28. There are two types of current, direct and
    (A) indirect
    (B) modular
    (C) alternating
    (D) circular

29. It would be correct to say about direct current that:
    (A) It flows in two directions.
    (B) It is better than alternating current.
    (C) It is produced by a battery.
    (D) It always takes the shortest route.

30. When heated, the bar in illustration 30 bends because
    (A) heat bends metals
    (B) the bar is defective
    (C) the two metals expand at different rates
    (D) the chemical reactions of the metals cause one to expand and the other to contract

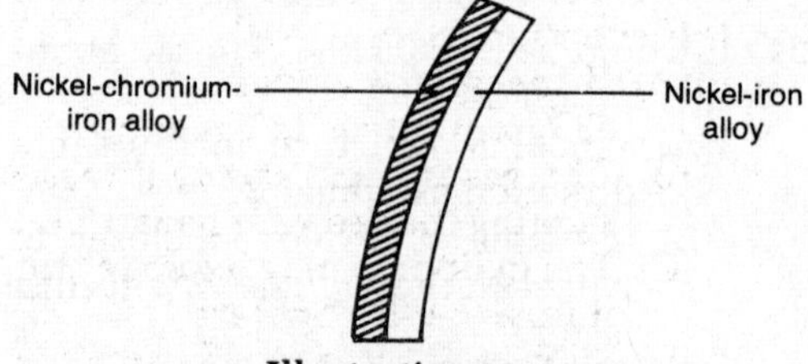

Illustration 30

31. An instructor at the fire academy told a group of school children, "It takes more energy to push water through a long hose line than through a short one

of the same diameter." The need for the extra energy is due to
(A) inertia (C) friction
(B) gravity (D) ebullition

32. If the fire fighter operating the pump sets the pressure at 200 pounds per square inch, the pressure at the unopened nozzle of a 300-foot length of hose will most likely be
(A) less than the pressure at the pump
(B) equal to the pressure at the pump
(C) greater than the pressure at the pump
(D) undeterminable without additional information

33. During a training session it was learned that a material which is a poor conductor of heat is
(A) a good conductor of sound
(B) a fair conductor of electricity
(C) a good insulator of heat
(D) an excellant fire retardant

34. Water, when changing from ice to steam, absorbs
(A) air (C) heat
(B) volume (D) density

35. During the winter months in northern climates many water pipes break. The most likely reason is that:
(A) Water expands when it freezes and produces great pressure.
(B) Ice is much heavier than water.
(C) Cold pipes are very brittle.
(D) Cold makes the pipes contract and separate.

36. At a major fire in a factory a number of compressed gas cylinders containing carbon dioxide (CO_2) are threatened by the flames. It is correct to say about this situation that:
(A) If the safety valves let go, the CO_2 will intensify the fire.
(B) The gas is toxic and will make fighting the fire very hazardous.
(C) The cylinders may explode and injure the fire fighters.
(D) Since CO_2 is a nonflammable gas, there is no hazard and no problem.

37. According to the National Fire Codes, every building with a standpipe system must be equipped with sufficient stations so that all portions of each story of the building are within 30 feet of a nozzle attached to not more than 100 feet of hose. Which of the following illustrations best shows this concept?

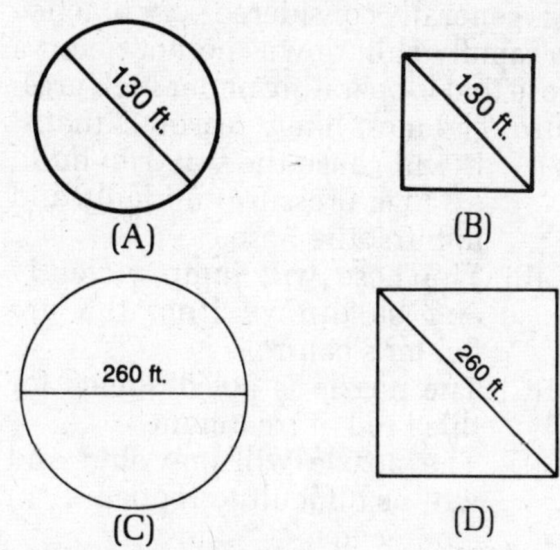

38. In the winter months, cars are more difficult to start. Once started, racing the engine is
(A) good—this warms the car up fast
(B) poor—this causes an excessive drain of the electrical system
(C) good—the battery will be rapidly charged
(D) poor—the oil is thick and will not properly lubricate the motor parts

39. Sprinkler systems have consistently proved their value. The one major exception to their outstanding record has occurred in instances of explosion. The most likely reason for their failure in such cases is that:
(A) The sprinkler heads are clogged by the explosion.
(B) These fires burn out too fast.
(C) The piping to the system is blown apart.
(D) An arsonist has shut the system down.

40. Fire fighters should not use water on live electrical equipment because
(A) it will cause short circuits and damage the equipment
(B) it is ineffective on electrical equipment
(C) there is a danger of conducting electricity and endangering the fire fighter
(D) this expensive, sophisticated equipment may be ruined

41. The most common method fire fighters use for extinguishing fires is
    (A) fuel removal
    (B) flame inhibition
    (C) oxygen exclusion
    (D) cooling

42. It is generally considered poor practice to rapidly shut down the nozzle on a hose that is operating under high pressure. The most likely reason is that:
    (A) It will cause the water to build up high pressure suddenly and to burst the hose.
    (B) The hose will jump violently and be thrown from the fire fighter's control.
    (C) The nozzle is not designed for this kind of treatment.
    (D) The nozzle will jam shut and will be difficult to reopen in an emergency.

43. At a serious fire the officer ordered two fire fighters to cut a hole in the roof over the fire and another fire fighter to break open a fixed window. The officer's orders were
    (A) good—these actions will allow fresh air to come in through the roof and will keep all his fire fighters busy
    (B) good—these actions will allow fresh air to come in through the window and the smoke to escape through the roof
    (C) poor—only a hole in the roof was necessary
    (D) poor—there was no need for a hole in the roof if the window could be broken

44. Which of the following would be considered a good conductor of heat?
    (A) Air (C) Iron
    (B) Wood (D) Stone

45. The practice of using booster cables to jump a car whose battery is worn down is
    (A) dangerous, provided that the cables are connected correctly rectly
    (B) dangerous, if not done correctly, because of the possibility of escaping hydrogen vapors
    (C) not dangerous, provided that proper ground is established
    (D) dangerous, if done in cold weather

46. While visiting a friend, you find that he is repairing his car and is using a large open pan filled with gasoline to bathe and wash the parts. This action is
    (A) correct—gasoline is an excellent solvent that rapidly cleans the parts
    (B) incorrect—gasoline emits a flammable vapor at temperatures of 40° below zero and can ignite easily
    (C) correct—gasoline is less expensive than other solvents and works almost as well
    (D) incorrect—gasoline is a toxic substance and requires the use of special respiratory equipment when used as a cleaning solvent

47. A fire that starts by "spontaneous combustion"
    (A) is impossible
    (B) is quickly extinguished
    (C) is common in cases of arson
    (D) has started without any outside source of ignition

48. The proper method for extinguishing a grease fire in a frying pan is to cover it. This method of extinguishment is
    (A) removal of fuel
    (B) removal of oxygen
    (C) cooling
    (D) flame inhibition

49. The reason for the compressed air in the portable extinguisher in illustration 49 is to
    (A) prevent the water from freezing
    (B) push the water out
    (C) reduce the weight of the extinguisher
    (D) allow the extinguisher to float if dropped in the water

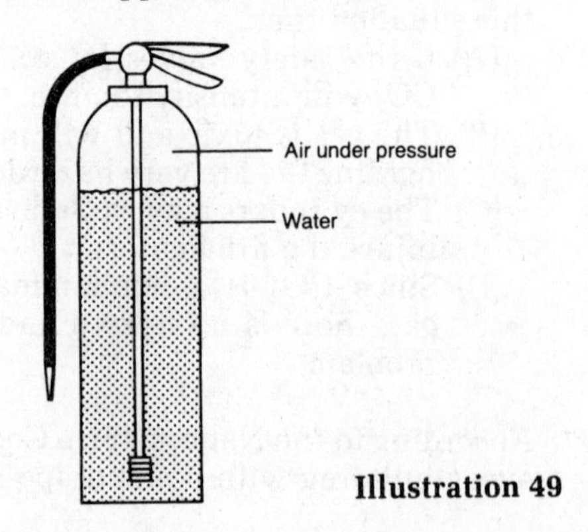

Illustration 49

GROUP NINE: MATHEMATICS

Answer questions 50 through 55 based on the information given in the following chart:

*One gallon of water weighs about 8.5 pounds (specifically, 8.35 pounds).
*A container measuring 1 foot by 1 foot (1 cubic foot) can hold approximately 7.5 gallons of water.
*A 1-cubic-foot container filled with water weighs approximately 62.5 pounds.

50. If the water supply tank on a fire apparatus holds 400 gallons of water when full, how much weight does the water add to the apparatus when the tank is full?
    (A) 3400 pounds.
    (B) 2800 pounds.
    (C) 2400 pounds.
    (D) 1650 pounds.

51. A 5-gallon container of water would most nearly weigh
    (A) 5 pounds.
    (B) 37 pounds.
    (C) 42 pounds.
    (D) 62.5 pounds.

52. If the freshwater tank on the roof of a fire station measures 10 feet × 10 feet × 10 feet, how much weight will a half-full tank exert on the roof supports?
    (A) 31,200 pounds.
    (B) 62,400 pounds.
    (C) 80,000 pounds.
    (D) 97,500 pounds.

53. If a tank measuring 5 feet × 7 feet × 3 feet is full, how many gallons of water will it hold?
    (A) 455 gallons.
    (B) 587 ½ gallons.
    (C) 787 ½ gallons.
    (D) 949 gallons.

54. At a major fire three hose lines are being used. Each line delivers 250 gallons of water per minute. If no water runs out of the building, how long will it take for a floor measuring 20 feet × 20 feet to be filled with water to a height of 1 foot?
    (A) 4 minutes.
    (B) 8 minutes.
    (C) 10 minutes.
    (D) 16 minutes.

55. A commonly used formula in the fire service for the placement of ladders against a building is to divide the length of the ladder by 5 and then add 2 to the result. This gives the number of feet from which to place the base of the ladder from the wall. If this formula is applied, how many feet from the wall should the base of the ladder in illustration 55 be?
    (A) 3 (C) 7
    (B) 5 (D) 9

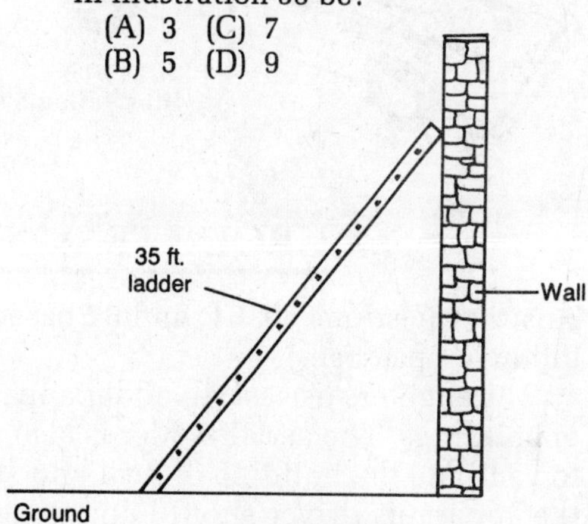

Illustration 55

56. The table in illustration 56 measures 6 feet in diameter. The smallest salvage cover that can be put over the table to protect the top is
    (A) a rectangle 6 feet × 8 feet
    (B) a square 7 feet × 7 feet
    (C) a rectangle 6 feet × 5 feet
    (D) a square 8 feet × 8 feet

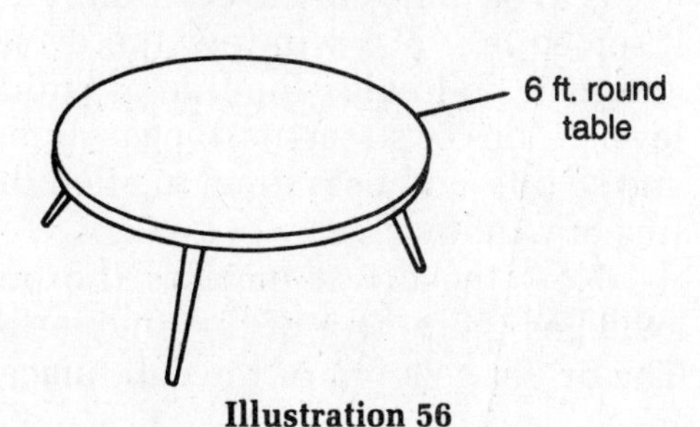

Illustration 56

57. If you compare the amount of fire hose that can be stored in compartment A of illustration 57 with the amount of fire hose that can be stored in compartment B, you would be correct in stating that:
    (A) Compartment A will hold more hose.
    (B) Compartment B will hold more hose.
    (C) The two compartments will hold the same amount.
    (D) There is no way to determine the capacities of the compartments without more information.

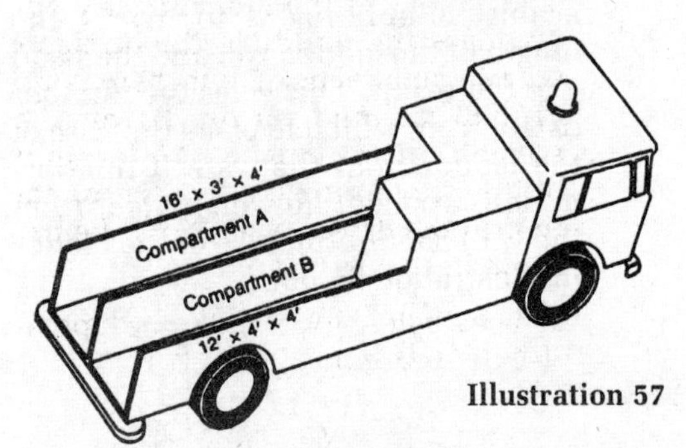

Illustration 57

58. "Pressure tanks shall be acceptable as primary water supply systems provided that an additional volume equivalent to one half of the required water storage space is provided for the required air." The most likely meaning of this statement is that:
    (A) The tank can be two-thirds full of water and one-third full of air.
    (B) The tank must not be more than one-half full of water and one-half full of air.
    (C) The tank must be one-third full of water and two-thirds full of air.
    (D) The tank must be full of water, and a supply of air must be near by.
59. If a water tank measures 8 feet × 4 feet × 6 feet, how many cubic feet of water can it hold?
    (A) 24
    (B) 32
    (C) 156
    (D) 192

GROUP TEN: INFORMATION ORDERING

Answer questions 60, 61, and 62 based solely on the information provided in the following passage.

Firefighters use aerial ladder apparatus to rescue people from the upper floors of buildings. The aerial ladder is also used to gain entry to search for victims and to ventilate the building of heat and smoke. When approaching the fire building, the apparatus driver should anticipate the use of the ladder. Deciding where to set up the ladder is based on an information-gathering process called "size-up." The driver looks at the building as the apparatus slowly approaches the scene and then determines if a person is in urgent need of rescue, if someone is exposed to heat and smoke and in need of assistance, or if the fire area needs venting. If there is no obvious need for a rescue, or if the fire is not readily visible, then the driver will position the apparatus for future use and set up in the center of the building so the ladder when raised can reach a maximum number of windows or the roof.

The size-up process leads to the decision about the best place to locate the ladder truck. The driver slows down to view the situation, to determine what needs to be done, and to accurately line-up the turntable with the target objective. If someone is at a window, the driver notes the window the person is sitting in and then, following the line of windows down to the street level, locates street-level guides or street markings—lines in the sidewalk, telephone poles, hydrants, and so on—and uses them to align the center of the apparatus turn-table with the line of windows.

Now the driver engages the brakes and activates the power take-off; this switches the power of the engine from driving mode to aerial operation mode. The driver gets out of the cab, places chocks under the wheels and engages the tormentor stabilizer system. When the base of the truck is set in a firm position, the bed ladder can be lifted from its cradle, elevated to the proper angle, and rotated into position below the window. Then the fly ladder can be raised to rescue the person.

60. When arriving at a building fire, Ladder 21 finds a person in extreme danger at the fifth floor window of a building fire. It would be most correct if the driver
    (A) after lining up and properly placing the turn-table, engaged the tormentors, put the chocks under the wheels, and proceeded to elevate the bed ladder.
    (B) engaged the power take-off, engaged the tormentor stabilizer system, elevated the bed ladder, raised the fly, and rotated the aerial ladder.
    (C) used the fifth floor window as the guide to line up the turntable, engaged the brakes, put the power take-off to the on position, and raised the ladder.
    (D) after applying the break, engaged the power take-off, got out of the cab, put the chocks under the wheels, and elevated the ladder.
61. At a fire on John Street, a person must be removed from the upper floors of a building by use of an aerial ladder. The following steps would be used to make this rescue (The steps listed below are not in correct order.):
    1. Elevate bed ladder from the cradle.
    2. Raise fly ladder.
    3. Engage power take-off.
    4. Engage the tormentors.

    The above steps would best be performed in the following order:
    (A) 1, 3, 4, 2
    (B) 3, 4, 1, 2
    (C) 4, 3, 1, 2
    (D) 3, 1, 4, 2
62. Ladder Company 55 has been dispatched to 122 East Broadway for a reported smoke condition on the sixth floor. As the fire company approaches the building, the firefighters can smell smoke, but there is no visible fire, and no one is on the scene to direct them to the location of the fire. Following is a list of activities the driver might do. (The list is not in the correct order.)
    1. Locate the ladder in the best position for future use.
    2. Line up the turn-table with a crack in the sidewalk.
    3. Follow the line of windows to the street.
    4. Determine if the fire floor needs venting.

    The above procedures should be done in the following order
    (A) 4, 3, 2, 1
    (B) 1, 3, 2, 4
    (C) 4, 2, 3, 1
    (D) 3, 2, 4, 1

Answer questions 63 and 64 based solely on the information provided in the following passage.

An effectively placed and operated hose line saves lives by quickly extinguishing the fire. When people are trapped in a burning building, the first hose line should be placed so the hose stream can be directed between them and the fire. When no one is exposed to danger from the fire, the hose line should be put at a point that protects the property that is most severely exposed. If a second hose line is needed, the rules for proper hose placement dictate the second hose line be used to back up or supplement the first line. If the second hose line is not needed as a backup, it should be brought to the adjoining apartment or to the floor above the fire, whichever is in more danger. A third hose line could be used to protect the secondary means of egress, persons trapped at a window on the floors above the fire, or to stop the fire from extending to an adjoining building.

63. Engine 321 responds to a fire in a three-story building and finds fire on the first floor with some extension to the upper floors. The following steps would be taken. (The items listed are not in priority order.)

    1. Direct a line to be taken to the floor above the fire floor.
    2. Direct a line to be used to protect the adjoining building.
    3. Direct a line to protect the secondary means of egress.
    4. Direct a line to the first floor to protect the stairway exit and extinguish the fire.

    The most appropriate sequence for the hose line placement is:
    (A) 4, 3, 2, 1
    (B) 2, 4, 1, 3
    (C) 1, 2, 3, 4
    (D) 4, 1, 3, 2

64. Consider the following scenario:

    "A fire on the first floor is growing rapidly beyond the capabilities of the first hose line and is extending to the floor above."

    The proper action for the firefighter on the first hose line is to request
    (A) a second hose line be brought in to go to the floor above.
    (B) a third hose line be put into operation to protect the adjoining building, which may become exposed.
    (C) a second line be brought in to back up the first line.
    (D) a third line be brought in to back up the first line and then to advance up to the floor above.

Answer Key and Explanations

ANSWER KEY

GROUP ONE	GROUP TWO	GROUP THREE	GROUP FOUR
1. A	6. C	11. B	17. C
2. B	7. A	12. D	18. D
3. C	8. B	13. C	
4. B	9. D	14. D	
5. C	10. B	15. B	
		16. D	

GROUP FIVE	GROUP SIX	GROUP SEVEN	GROUP EIGHT	
19. C	21. C	24. B	27. A	39. C
20. B	22. D	25. C	28. C	40. C
	23. C	26. D	29. C	41. D
			30. C	42. A
			31. C	43. B
			32. B	44. C
			33. C	45. B
			34. C	46. B
			35. A	47. D
			36. C	48. B
			37. C	49. B
			38. D	

GROUP NINE		GROUP TEN
50. A	55. D	60. D
51. C	56. A	61. B
52. A	57. C	62. B
53. C	58. A	63. D
54. A	59. D	64. C

ANSWER EXPLANATIONS

GROUP ONE

1. **A** This is a class 1 lever. Using this formula:

   Effort × Effort Distance = Resistance × Resistance Distance

   and inserting the known values, we can solve for the effort (x).

   $$\text{Effort} \times \text{Effort Distance} = \text{Resistance} \times \text{Resistance Distance}$$

   $$x \times 6\text{ ft.} = 300\text{ lb.} \times 2\text{ ft.}$$

   $$x = \frac{300\text{ lb.} \times 2\text{ ft.}}{6\text{ ft.}}$$

   $$x = \frac{600\text{ foot} - \text{pounds (ft.} - \text{lb.)}}{6\text{ ft.}}$$

   $$x = 100\text{ lb.}$$

2. **B** This is a class 1 lever with the arms at different angles.

$$\text{Effort} \times \text{Effort Distance} = \text{Resistance} \times \text{Resistance Distance}$$

$$x \times 10 \text{ in.} = 75 \text{ lb.} \times 2 \text{ in.}$$

$$x = \frac{75 \text{ lb.} \times 2 \text{ in.}}{10 \text{ in.}}$$

$$x = \frac{150 \text{ inch-pounds (in.-lb.)}}{10 \text{ in.}}$$

$$x = 15 \text{ lb.}$$

3. **C** This is a class 2 lever system. Increasing the length of the effort distance while decreasing the length of the resistance distance will reduce the effort required to lift the box. To see that this is correct, try putting weights at different distances from the fulcrum and then working out the solutions.

4. **B** This is a class 1 lever.

$$\text{Effort} \times \text{Effort Distance} = \text{Resistance} \times \text{Resistance Distance}$$

$$x \times 3 \text{ ft.} = 600 \text{ lb.} \times 6 \text{ in.}$$

$$x = \frac{600 \text{ lb.} \times 6 \text{ in.}}{3 \text{ ft.}}$$

Note: Before you continue, you must change inches to feet so that you are working with the same units. Since 6 inches = ½ foot,

$$x = \frac{\frac{\overset{300}{\cancel{600}} \text{ lb.} \times \cancel{1} \text{ ft.}}{\cancel{1} \quad \cancel{2}}}{3 \text{ ft.}}$$

$$x = \frac{300 \text{ ft.-lb.}}{3 \text{ ft.}}$$

$$x = 100 \text{ lb.}$$

5. **C** This is a class 3 lever. The resistance is the weight of the ladder and the resistance distance is the full length of the ladder.

$$\text{Effort} \times \text{Effort Distance} = \text{Resistance} \times \text{Resistance Distance}$$

$$x = \frac{64 \text{ lb.} \times 25 \text{ ft.}}{10 \text{ ft.}}$$

$$x = \frac{1600 \text{ ft.-lb.}}{10 \text{ ft.}}$$

$$x = 160 \text{ lb.}$$

GROUP TWO

6. **C**

$$\text{Effort} \times \text{Length of plane} = \text{Resistance} \times \text{Height to be raised}$$

$$x \times 30 \text{ ft.} = 300 \text{ lb.} \times 6 \text{ ft.}$$

$$x = \frac{300 \text{ lb.} \times 6 \text{ ft.}}{30 \text{ ft.}}$$

$$x = \frac{1800 \text{ ft.-lb.}}{30 \text{ ft.}}$$

$$x = 60 \text{ lb.}$$

7. **A** The problem requires an understanding of the mechanical advantage of the inclined plane. A person might think that an item of information (the vertical height) had been left out of the question, but this is not the case. The mechanical advantage of the inclined plane is determined by dividing the length of the plane by the height to be raised. Since the plane is four times as long as its vertical height, the mechanical advantage is 4 (4/1 = 4). Using the standard format gives

$$\text{Effort} \times \text{Length of plane} = \text{Resistance} \times \text{Height to be raised}$$

$$x \times 4 = 200 \text{ lb.} \times 1$$

$$x = \frac{200 \text{ lb.} \times 1}{4}$$

$$x = 50 \text{ lb.}$$

Note: There are no units of measure after the numbers 1 and 4.

8. **B** This question requires an understanding of the relationship involved in the mechanical advantage of the inclined plane and the ability to visualize this relationship, keeping in mind that

$$\text{M.A.} = \frac{\text{Length of plane}}{\text{Height to be raised}}$$

On a small piece of paper, measure off the distance to the ground from the top of the box—the height to be raised. Now use this as a guide to measure how many times this height can fit into the inclined plane. Choice B = 4, choice A = 2, choice C = 2, choice D = 1. The choice with the greatest mechanical advantage requires the least effort.

9. **D**

$$\text{Effort} \times \text{Length of plane} = \text{Resistance} \times \text{Height to be raised}$$

$$x \times 16 \text{ ft.} = 4000 \text{ lb.} \times 5 \text{ ft.}$$

$$x = \frac{4000 \text{ lb.} \times 5 \text{ ft.}}{16 \text{ ft.}}$$

$$x = \frac{20000 \text{ ft.-lb.}}{16 \text{ ft.}}$$

$$x = 1250 \text{ lb.}$$

10. **B** The mechanical advantage of the inclined plane is equal to the length of the plane divided by the height to be raised:

$$\text{M.A.} = \frac{10 \text{ ft.}}{2\frac{1}{2} \text{ ft.}} = 4$$

GROUP THREE

11. **B** A single movable pulley has two ropes, which share the load equally. To determine the mechanical advantage, count the number of ropes that carry the load, in this case two, and then divide the load by the mechanical advantage:

$$\text{Effort} = \frac{360 \text{ lb.}}{2} = 180 \text{ lb.}$$

12. **D** The amount of rope that must be pulled is related to the mechanical advantage. Multiply the height of the weight to be raised by the mechanical advantage to determine how much rope must be pulled:

$$\text{Height to be raised} \times \text{Mechanical advantage} = \text{Length of rope}$$

$$2 \text{ ft.} \times 2 = 4 \text{ ft.}$$

13. **C** The mechanical advantage of this pulley system is 2. Dividing the weight to be lifted (2000 lb.) by the mechanical advantage (2) gives 1000 lb. The small block near the first fire fighter is known as a snatch block and has no mechanical advantage; it serves to change the direction of the pull. The number of fire fighters on the line does not affect the mechanical advantage of the pulley system; it distributes the load that each fire fighter must exert but does not change the total force required.
14. **D** This pulley serves only to change direction and does not provide any mechanical advantage. The mechanical advantage in choice A = 4; in choice B = 2; in choice C = 2.
15. **A** As stated above, the mechanical advantage of this system is 1; in other words, there is no mechanical advantage, only a change of direction of the pull.

$$1 \times 100 \text{ lb.} = 100 \text{ lb.}$$

16. **D** The mechanical advantage is 6. Multiplying the mechanical advantage by the force to be exerted (300 lb. × 6) gives the total weight that can be lifted.

GROUP FOUR

17. C

Effort × E.D. (circumference) = Resistance × R.D. (pitch)

$$5 \text{ lb.} \times 2\pi r = x \times \frac{1}{5}$$

$$5 \text{ lb.} \times 2 \times \frac{22}{\not{7}} \times \overset{4}{\not{28}} = \frac{1}{5} \times x$$

$$\frac{1}{5} x = 880 \text{ lb.}$$

$$x = 5 \times 880 \text{ lb.}$$

$$x = 4400 \text{ lb.}$$

18. D

Effort × E.D. (circumference) = Resistance × R.D. (pitch)

$$x \times 2\pi r = 2000 \text{ lb.} \times \frac{1}{2}$$

$$x \times 2 \times \frac{22}{\not{7}} \times \overset{2}{\not{14}} = \overset{1000}{\not{2000}} \times \frac{1}{\not{2}}$$

$$x = \frac{1000}{88}$$

$x = 11.36$ (when rounded to the nearest ½, the answer is approximately 11½ lbs.)

GROUP FIVE

19. C

Effort × Circumference of large wheel = Resistance × Circumference of small wheel

$$60 \text{ lb.} \times 2 \times \pi \times r = x \times 2 \times \pi \times r$$

$$60 \text{ lb.} \times 2 \times \pi \times 12 = x \times 2 \times \pi \times 3$$

(*Note:* Radius = one-half the diameter.)

$$x = \frac{60 \text{ lb.} \times \not{2} \not{\times} \not{\pi} \times 12 \text{ in.}}{\not{2} \not{\times} \not{\pi} \times 3 \text{ in.}}$$

$$x = \frac{60 \text{ lb.} \times 12 \text{ in.}}{3 \text{ in.}}$$

$$x = \frac{720}{3}$$

$$x = 240 \text{ lb.}$$

20. B

Effort × Circumference of large wheel = Resistance × Circumference of small wheel

$$x \times 2\pi r = 300 \text{ lb.} \times 2\pi r$$

$$x = \frac{300 \text{ lb.} \times \not{2} \not{\times} \not{\pi} \times 4 \text{ in.}}{\not{2} \not{\times} \not{\pi} \times 10 \text{ in.}}$$

$$x = \frac{300 \text{ lb.} \times 4}{10}$$

$$x = 120$$

GROUP SIX

21. **C** When two gears mesh together, the second gear turns in the opposite direction. Each successive gear turns the next adjacent gear in the opposite direction.

22. **D** Two gears that mesh turn in opposite directions. Choices A and C—the number of times a gear will turn is related to the number of sprockets. In illustration 22 the drive gear (A) has 8 sprockets; the second gear (B) has 12 sprockets. If gear B turns once (choice A), gear A will turn 1½ times. If gear B turns twice (choice C), gear A will turn 3 times. Choice B—the gears will turn in opposite directions.

23. **C**

$$S_2 : S_1 :: T_1 : T_2$$

$$\frac{S_2}{S_1} = \frac{T_1}{T_2}$$

$$S_2 = S_1 \times \frac{T_1}{T_2}$$

$$S_2 = 1 \times \frac{8 \times 16}{32 \times 24}$$

$$S_2 = \frac{128}{768} = \frac{1}{6}$$

$$S_2 = \frac{1}{6} \text{ of a revolution per second}$$

GROUP SEVEN

24. **B** As the wheels move closer to each other, the belt will become loose and will not drive the other wheel. The other statements are correct. *Note:* with belt drive system, the distance apart has nothing to do with the relative speeds of the wheels.

25. **C** By having the belt take a single twist, the direction of the turn can be reversed. A second twist will return it to the same direction.

26. **D** The belt from drive wheel A is connected to a smaller wheel (B), which will therefore turn faster. The belt connecting the outer part of wheel B to wheel C is twisted, causing a change in direction.

GROUP EIGHT

27. **A** Class A fires burn in ordinary combustible materials such as wood, paper, and cloth. The distinguishing characteristic of this class of fire is that it leaves an ash.

28. **C** This is the type of current available at the ordinary electric outlet; it differs from direct current in that the flow of electricity reverses direction at regular intervals.

29. **C** Direct current is produced by a flow of electrons from one plate in a battery to another. It flows in one direction and can be obstructed by a resistor.

30. **C** All metals expand when heated; however, they do not expand at the same rate. The piece of metal shown in illustration 30 is known as a bimetallic strip and is used in thermostats.

31. **C** Friction is the resistance created between two surfaces that are in contact with each other. Choice A—inertia is the tendency of a body at rest to remain at rest or of a body in motion to remain in motion. In either case the body will remain as is until acted upon by an outside force. Choice B—gravity is the force of mutual attraction between bodies, such as the earth and the moon. Choice D—ebullition is the boiling or bubbling up of a substance.

32. **B** The pressure in a closed system (no fluid flowing) is undiminished throughout the system.

33. **C** An insulator is a material that serves as a nonconductor. Choice D—you might reason that a poor conductor should act as a fire retardant. However, consider a piece of wood; it is a poor conductor of heat but burns readily.

34. **C** As a substance absorbs heat, its molecules are activated; this activation of the molecules results in ice (solid) changing state to water (liquid) and then expanding still further to become steam (gas). Choice A—it does not absorb air, but does displace it. Choice B—volume is space in three dimensions, and usually is expressed as cubic feet, cubic inches, etc. Choice D—density is a measure of the compactness of the parts of a substance.

35. **A** Water is a unique substance in that it expands as it approaches the freezing point; other materials contract.

36. **C** The heat will cause the CO_2 to expand, thereby leading to cylinder failure. When this happens, parts of the cylinder may fly in all directions, and the cylinder may fly off as a rocket.

37. **C** From the standpipe outlet it is possible to go 130 feet in any direction. Think of the standpipe as the center of a circle, and the hose (100 feet) plus 30 feet as the radius. The diameter is 260 feet.

38. **D** The cold causes the oil to thicken; the thick oil will take longer to circulate throughout the engine. Until this is accomplished, there is a good chance that metal will rub against metal, causing damage to the engine.

39. **C** In most cases the system is blown apart and the water cannot reach the fire.

40. **C** Water can conduct electricity back to the fire fighter through the stream. For this reason special care must be exercised.

41. **D** Most fires are extinguished by putting water on them (cooling). Choices A, B, and C are the other three methods that can be used given the right conditions.

42. **A** The pressure in a hose stream is undiminished when no flow occurs; whatever pressure the pump is putting out, will be transmitted throughout the hose line. Rapid shutdown will result in a high-pressure shock wave, which can burst the hose.

43. **B** Heat rises. By allowing the cold air to enter from the bottom (through the window) and warm air to escape from the top (through the roof), an effective air movement system will be developed. Choice A—only limited amounts of fresh air will be able to enter through the roof; this also implies that fire fighters cut holes in roofs just to keep busy—a ridiculous idea. Choice C—no; there must also be a way to allow the fresh air to enter. Choice D—a broken window would allow for only limited smoke ventilation and would lead to increased fire damage and danger to the fire fighters.

44. **C** Iron is a good *conductor*; choices A, B, and D are good *insulators.*

45. **B** When the acid in the battery is heated, it breaks down and gives off hydrogen gas. Hydrogen gas is very flammable and has led to many accidents and fires.

46. **B** Gasoline is a volatile flammable liquid that gives off vapors at very low temperatures. Its vapors are heavier than air; they travel long distances and are easily ignited by a remote ignition source. There are appropriate solvents for cleaning automotive parts, and only these solvents should be used. Gasoline should not be used.

47. **D** Spontaneous combustion occurs from the decay of materials by chemical or biological reaction. Choice B—nothing in the question relates to extinguishment; this is purely a distractor. Choice C—spontaneous combustion is an act of nature; arson is a deliberately set fire.

48. **B** The cover excludes air, which contains 21 percent oxygen, and allows the smoke to fill up the area under the cover.

49. **B** When the valve is opened, the lower atmospheric pressure outside the container allows the compressed air in the extinguisher to expand and push out the water.

GROUP NINE

50. A The chart on page 180 indicates that water weighs approximately 8½ pounds per gallon.

$$x = 400 \text{ gal.} \times 8\frac{1}{2} \text{ lb.}$$

$$x = \overset{200}{\cancel{400}} \text{ gal.} \times \frac{17}{\cancel{2}} \text{ lb.}$$

$$x = 200 \text{ gal.} \times 17 \text{ lb.}$$

$$x = 3400 \text{ lb.}$$

51. C

$$x = 8½ \text{ lb./gal.} \times 5 \text{ gal.}$$

$$x = \frac{17}{2}\text{lb.} \times 5 \text{ gal.}$$

$$x = \frac{85 \text{ lb.}}{2}$$

$x = 42.5$ lb., which is most nearly 42 (choice C)

52. A

Volume = Length × Width × Height

$V = 10 \text{ ft.} \times 10 \text{ ft.} \times 10 \text{ ft.}$

$= 1000 \text{ ft.}^3$

If the water weighs 62.5 lb/ft.^{3}, and we have 1000 ft.^{3}, a full tank would weigh

$$\frac{1000 \text{ ft.}^3 \times 62.4 \text{ lb.}}{1 \text{ ft.}^3} = 62{,}400 \text{ lb.}$$

Since the tank in this problem is only one-half full, we must divide 62,400 lb. by 2; this gives 31,200 lb.

53. C

Volume = Length × Width × Height

$V = 5 \text{ ft.} \times 7 \text{ ft.} \times 3 \text{ ft.}$

$V = 105 \text{ ft.}^3$

$105 \text{ ft.}^3 \times 7½ \text{ gal.}/1 \text{ ft.}^3$ = number of gallons.

787.5 = number of total gallons

54. A This is a multiple-step problem.

(1) Find the volume.

Volume = Length × Width × Height

$V = 20 \text{ ft.} \times 20 \text{ ft.} \times 1 \text{ ft.}$

$V = 400 \text{ ft.}^3$

(2) Compute the number of gallons in 400 ft.^{3}

$400 \text{ ft.}^3 \times 7.5 \text{ gal/ft.}^3 = 3000$ gallons

(3) Add the three hose streams.

Flow rate = 250 gal./min. + 250 gal./min. + 250 gal./min. = 750 gal/min.

(4) Divide the total water (3000 gal.) by the flow rate (750 gal./min.).

$$\frac{3000 \text{ gal.}}{750 \text{ gal./min.}} = 4 \text{ min.}$$

55. D The length of the ladder is 35 feet. Using the formula given:

$$\frac{\text{Length}}{5} + 2 = \text{distance from the wall}$$

we obtain

$$\frac{35}{5} + 2 = 9$$

Note: multiplication and division should be done before addition and subtraction.

56. A We can immediately eliminate choice C; it cannot cover the table since it has a side of only 5 ft. We can also eliminate choice D, because choices A and B are both smaller than D but large enough to cover the table. To determine which is smaller, A or B, we must calculate their areas. It is not necessary to compute the area of the table.

Area = Length × Width

Area of A = 6 ft. × 8 ft. = 48 ft.^{2}

Area of B = 7 ft. × 7 ft. = 49 ft.^{2}

The 6 ft. × 8 ft. salvage cover is the best choice.

Note: In this problem we are concerned with only the top of the table, not the overhang.

57. C The volume of the compartment will determine the amount of hose that can be stored.

Volume = Length × Width × Height

Volume of A = 16 × 3 × 4 = 192 ft.^{2}

Volume of B = 12 × 4 × 4 = 192 ft.^{2}

58. **A** If the tank is $\frac{2}{3}$ full of water, then $\frac{1}{2}$ of $\frac{2}{3} = \frac{1}{3}$, and $\frac{2}{3}$ water + $\frac{1}{3}$ air = a full tank. Choice (B) is incorrect because, if the tank is only $\frac{1}{2}$ filled with water, then $\frac{1}{2}$ of $\frac{1}{2} = \frac{1}{4}$; adding the air and water we get: $\frac{1}{2}$ water + $\frac{1}{4}$ air = $\frac{3}{4}$ of a tank.

59. **D** To determine cubic feet, multiply all three dimensions: length (8 ft.), width (4 ft.), and height (6 ft.) = 192 ft.^{3}

GROUP TEN

60. **D** This follows the pattern established in the reading; it omits steps but keeps the correct order. Choice A is incorrect because the chocks should be placed before the tormentors. Choice B is incorrect because the ladder should be rotated before the fly ladder is raised. Choice C is incorrect because the driver would use the street guide below the window, not the fifth floor window.

61. **B** This would be the most correct sequence as outlined in the passage. Choice A is incorrect because before raising the ladder or engaging the tormentors the driver would need to engage the power take off. Choice D is incorrect because the driver would stabilize the apparatus before elevating the bed ladder. Choice C is incorrect because the driver must engage the power take-off before engaging the tormentors.

62. **B** This follows the pattern established in the passage. Choices A and C are incorrect; the passage implies the location of the fire is unknown and not easily determined from the exterior of the building. The building may need venting, but before this occurs, the firefighters will have to locate the fire, so this will happen at some future point after their arrival. Choice C is incorrect because before the driver can determine if venting is needed, the location of the fire must be found. The fire will be found at some time in the future.

63. **D** This follows the intent of the passage: Protect the property most severely exposed and extinguish the fire. Choices A, B, and C are incorrect because the sequences would result in an incorrect hose line placement. Choices A and B are positions for the third hose line, and choice C is a position for the second hose line.

64. **C** The priority role of the second hose line is to back up the first hose line. Choice A is incorrect because the second line would be used on the floor above only after it was determined it was not needed as a back-up for the first line. Choices B and D are incorrect because the firefighter would call for a second line before requesting a third line.

CHAPTER 11

Interacting With People

As a firefighter you will have to interact with many people: other firefighters, superiors, school children, business owners, and the public in general. This chapter will expose you to different types of situations that could occur and that would require you to handle them properly. The questions in the Practice Exercises require no special knowledge of fire department policies, regulations, or operations. Rather, they are designed to evaluate your common sense, judgment, and general attitudes toward people and teamwork.

There are several important concepts to keep in mind when answering questions of this type.

- Firefighters are expected to do their job with a high degree of professionalism.
- Firefighters are expected to have and to show concern for the public, their peers, and their superiors.
- Firefighters are expected to treat all people with courtesy and respect.

At a fire, the firefighter's priorities, listed in order of importance, are as follows:

- Life safety.
- Fire extinguishment.
- Prevention of property damage.

One effective technique to arrive at the correct answer choice for these questions is to empathize, that is, to put yourself in the "shoes" of other people. How would they expect to be treated? What would they know about the fire department's actions? What do they expect from the firefighter? What are their fears? From what point of view are they viewing their problems? From what point of view are they viewing the firefighter's problem?

Interacting With Other Firefighters and With Fire Officers

The occupation of fire fighting requires a very close working relationship with other firefighters and immediate supervisors. It calls for teamwork: an ability to count on the group for support and help, and a commitment to give support to the group and to be part of a team.

When a superior gives an order, you will be expected to carry it out. When another firefighter needs help, you will be expected to give it. When you encounter a problem, you will be expected to act responsibly and seek help through the chain of command.

Interacting With the Public

When interacting with the public, a firefighter must bear in mind that most people have little knowledge of:

- how and why fires start
- why and how a fire spreads
- how fires are extinguished
- why it takes time to control and extinguish a fire
- why firefighters sometimes cause property damage

The firefighter must also keep in mind the heightened emotions caused by civilians' personal involvement and the psychological pressure of the fire. People may be excited, defensive, or experiencing shock. For this reason they may not fully listen to what the firefighter is saying, or comprehend it.

Practice Exercises

For each question choose the one *best* answer—(A), (B), (C), or (D)—and write the corresponding letter on your answer paper next to the number of the question.

GROUP ONE

1. While on the midnight to 3 A.M. house watch, a firefighter is told by a civilian telephone caller that she thinks she saw smoke about a block from the firehouse but was late for work and couldn't check it out. The firefighter should
   (A) notify his officer of what was reported.
   (B) go up the block and make a quick investigation.
   (C) send the local police officer to check out the reported smoke condition.
   (D) disregard the call; if there is a fire, someone else will report it.
2. While performing a routine inspection in an office building, you, a firefighter, are asked a question about a matter of which you have little knowledge and which is under the jurisdiction of the building depart-

ment. The best course of action for you is to
(A) suggest that the person contact the building department.
(B) tell the person that you don't know the answer and then leave.
(C) answer the question with a positive tone and within your ability.
(D) tell the person that you will report the question to your superior who will contact the building department.

3. A firefighter complained to the captain about drafts from the loosely fitting windows in the firehouse. After several weeks the condition has not been corrected. The most appropriate action for the firefighter to take at this time is to
(A) ask the chief, the captain's boss, if the condition was reported.
(B) file a grievance through the union representative.
(C) ask the captain how the matter is coming along.
(D) write to the Fire Commissioner about the matter.

4. As a firefighter, you have been given an order by your immediate superior that you think is obviously wrong. You will be acting correctly if you
(A) carry out the order because it is your duty to do so.
(B) do what you think is correct.
(C) repeat the order to the officer and ask whether this is the correct action for you to take.
(D) don't do anything—that is, stall until the officer realizes the order is incorrect.

5. Firefighters sometimes indulge in fooling around or horseplay. This is
(A) good—it builds morale.
(B) poor—it leads to accidents.
(C) good—it builds physical coordination.
(D) poor—it wastes time.

GROUP TWO

6. While instructing a group of school children, a firefighter is asked, "What number should I dial to report a fire?" The appropriate answer for the firefighter to give is
(A) the telephone number of the local fire station.
(B) the number of fire headquarters.
(C) the Operator (O) or the emergency number (911).
(D) the international fire number (999-FIRE).

7. After attending a stimulating talk on public relations, a firefighter decides to take an active role in this area. The appropriate FIRST action for this firefighter is to
(A) contact the local newspaper's public relations department and volunteer to help.
(B) send a letter to local organizations, offering to speak at their meetings.
(C) ensure that his own firefighter duties are conducted professionally and courteously.
(D) inform the other firefighters of this decision and urge them to do the same.

8. While reading a firefighting magazine, a firefighter learns of what appears to be a better way to store the fire company's hose nozzles. The appropriate action for this firefighter is to
(A) set up the nozzles on the apparatus in this new way and then tell the group how it is done.
(B) forget about the idea because it didn't come from the firefighter's fire department.
(C) bring the idea to the attention of the immediate superior and discuss it with the group.
(D) put the suggestion in writing and send it to the union so that organization can make sure it is implemented.

9. You are a firefighter who has just been told of a work unit assignment change, and you find that operations are done somewhat differently from what you were taught. The correct action for you to take is to

(A) tell the officer that things can't be done that way.
(B) criticize the methods of the group and explain why the way it was taught at your previous assignment is correct.
(C) do things your way and not worry about the others.
(D) discuss the differences with the new officer and be guided by what the officer says.

10. About 10 minutes after an oil-burner fire in a private house has been extinguished, the owner returns and starts to complain to you about the broken cellar windows. What is the best action for you to take?
   (A) Explain the conditions found on your arrival and the reasons for this type of ventilation.
   (B) Listen patiently until the person has finished, and then say, "I'm sorry" and walk away.
   (C) Tell the person to put her complaint in writing to the fire commissioner.
   (D) Recommend that the owner install an automatic fire venting system.

GROUP THREE

11. While you are on housewatch in midafternoon, a visitor from the next county stops in and asks whether she and her family may look at the fire equipment. The appropriate response is to
   (A) explain that an appointment is necessary and ask whether she would like to make one.
   (B) let her and her family walk around the station and look at the equipment without supervision.
   (C) explain that you are on housewatch duty but that when you have finished you will be glad to show them around.
   (D) call the officer to get someone else to show them around.

12. It is recommended that firefighters on duty answer the telephone by stating their unit, rank, and name—for example, "Engine Co. 159, Firefighter Jones." This should be done
   (A) after finding out who is calling.
   (B) only if asked for.
   (C) at the beginning of the conversation.
   (D) only on the fire department's private telephone.

13. During a school-class visit, a firefighter is asked, "Why do firefighters wash the apparatus and equipment every day?" What is the most appropriate response?
   (A) It is department policy.
   (B) It looks good for the public.
   (C) It looks good for the chief.
   (D) It helps to reveal any defects or dangerous conditions.

14. While on fire-prevention inspection, you discover a serious fire hazard. When you confront the owner of the premises, you discover that he does not speak English. The proper action for you to take is to
   (A) find someone who can act as an interpreter.
   (B) make yourself understood by speaking slowly and loudly.
   (C) correct the condition yourself and show the owner how you did it.
   (D) give the owner a written order and, using sign language, direct him to comply.

15. While working at a fire, a fire chief directs you to go to your apparatus and get a special tool. While you are on your way, your company officer tells you to help remove some debris from in front of the building. What should you do?
   (A) Follow the last order given; officers always know what they are doing.
   (B) Continue on with what the chief ordered; he is the superior officer.
   (C) Explain the chief's directions to the company officer and then be guided by his decision.
   (D) Go back to the chief, explain your dilemma, and be guided by his decision.

16. While on fire-prevention inspection, you are stopped by a woman who tells you that near her home, on the other side of town, there is a supermarket that always has its rear exit doors locked and that keeps cardboard boxes piled in front of it. What is the most appropriate action for you to take?
    (A) Get the woman's name and address, and then refer her to the fire prevention section.
    (B) Get the address of the supermarket and the woman's name and address, and then go to the supermarket to inspect it.
    (C) Get the address of the supermarket and the woman's name and address, and then notify the fire unit responsible for that area to inspect the store.
    (D) Assure the woman that the local unit will be inspecting that store in the near future as part of its regular activities.

Answer Key and Explanations

ANSWER KEY

1. **A**	5. **B**	9. **D**	13. **D**
2. **A**	6. **C**	10. **A**	14. **A**
3. **C**	7. **C**	11. **D**	15. **C**
4. **C**	8. **C**	12. **C**	16. **C**

ANSWER EXPLANATIONS

GROUP ONE

1. **A** Notifying the officer will ensure that proper action is taken and will put the other firefighters on alert for a fast response if needed. Choice B—if a confirmed alarm of fire comes in while you are away, the officer will not know where you are, and there will be a delay in the normal response while the others look for you. Choice C—sending the police may be done in addition to choice A, but not in place of it. Choice D—the call would never be disregarded; all reports of fire must be investigated.

2. **A** This avoids giving any incorrect or misleading information and informs the person making the inquiry about how to get a proper answer.

3. **C** Checking back with the person to whom the firefighter reports is the first step. The firefighter should find out what, if anything, has been done and only then, if still dissatisfied, consider going to a higher level of management.

4. **C** There are two possibilities in this situation: (1) the order is wrong and should not be carried out, and (2) the order appears wrong to you, but the officer has information that you lack and that makes the order correct. *In either case your first objective is to inform the officer that you are unsure of what he wants done.* This is best accomplished by choice C.

5. **B** Horseplay or fooling around often starts out as harmless fun but has a consistent track record of leading to injury and property destruction.

GROUP TWO

6. **C** This telephone number is a universal emergency number throughout the United States. Teaching children this number helps them to report a fire from almost any location. Choices A and B are incorrect be

cause there will be times when the units are out and no one will be there to answer the call. Choice D is a fictitious number.

7. **C** This action can be implemented immediately and will have direct impact on the people who interact with the fire fighter, thereby leading to good public relations. Choices A, B, and D may be appropriate, depending on many conditions. The fire fighter should look into them *after* executing choice C—the appropriate *first* action.

8. **C** This will give the officer time to evaluate your suggestion and to consider its impact on fire operations and the other fire fighters. It will also allow the fire officer to explain why it may not be appropriate. Choice A—fire fighters depend on having tools and equipment in specific locations and setups. Making changes without first notifying the others can result in hard feelings. Choice B—no; if the idea is good, the group should know about it. Choice D—the suggestion should be sent, not to the union, but to the suggestion department or to the bureau of operations—but not until after you have discussed it with your fire officer.

9. **D** Such discussion will ensure that the fire fighter is aware of what is required of him and what the proper actions are. Choice A—this would be a form of direct disobedience. Choice B—this would create hard feelings; also, the old methods may not be best—there may be new and better ways. Choice C—this could create confusion and reduce the group's effectiveness.

10. **A** Ten minutes after fire extinguishment the conditions will have improved greatly and the owner will not be able to visualize what they were like when you arrived. By explaining the conditions and the reasons for breaking the windows, the actions will be more acceptable to the owner.

GROUP THREE

11. **D** The officer will assign another fire fighter to this task. Choice A—an appointment is generally not needed, although it may be if a large group such as a school class wants to come. Choice B—this could be dangerous and is not recommended or generally permitted. Choice C—this would result in making the people wait unnecessarily; choice D is a better solution.

12. **C** This allows the caller to know that he has reached the right party and to refer to you again if needed. Unit, rank, and name should be the very first words said routinely on all telephones while on duty.

13. **D** By explaining the maintenance procedures, including washing, the class will better understand the degree of preparation and the state of readiness of the fire unit.

14. **A** Using an interpreter is the only way you can be sure that the person understands what needs to be corrected. Choices B and D will not be of any help and may well lead to confusion. Choice C is unacceptable. You may not be able to make the appropriate corrections, you would be liable for any damage that might result, and the owner would still not understand the problem.

15. **C** You should inform the officer that you have already been assigned to a task; your company officer should then be able to redirect your actions. If you have done this and your officer then tells you to do something else, you must follow his directions. It is now his responsibility to notify the chief and to have someone else get the tool.

16. **C** The fire unit responsible for the supermarket should be the one to conduct this inspection. Getting the store's address will direct them to the proper place. Getting the woman's name and address will ensure that the unit can get back to her if they are unable to locate the store because the address is incorrect.

PART FOUR

Practice Examinations

CHAPTER 12

Practice Examination One

Official Examination City of New York

This chapter includes the first of four practice examinations that you will take.

Be sure that when you take this examination you allow yourself a 3½-hour uninterrupted time period. Do not try to take the examination in parts; if you do, you will defeat its purpose. By taking the examination at one sitting, you will become familiar with the test conditions and with your personal idiosyncrasies about sitting and working for this period of time.

Before Taking the Examination

First, go back and review quickly the test-taking strategies outlined in Chapter 3. Then, when you begin the examination, be sure to read and follow all instructions. Time for reading the instructions has been figured into the examination. Read each question carefully, and answer only what is asked of you. Select the answer that is the one best choice of those provided, and then record your selection on the answer sheet.

The answer sheet precedes the examination. The Diagnostic Procedure and Answer Explanations are given at the end of the chapter.

ANSWER SHEET
PRACTICE EXAMINATION ONE

Follow the instructions given in the test. Mark only your answers in the ovals below.
WARNING: Be sure that the oval you fill is in the same row as the question you are answering. Use a No. 2 pencil (soft pencil).
BE SURE YOUR PENCIL MARKS ARE HEAVY AND BLACK. ERASE COMPLETELY ANY ANSWER YOU WISH TO CHANGE.

START HERE　　DO NOT make stray pencil dots, dashes or marks.

1 A B C D	2 A B C D	3 A B C D
4 A B C D	5 A B C D	6 A B C D
7 A B C D	8 A B C D	9 A B C D
10 A B C D	11 A B C D	12 A B C D
13 A B C D	14 A B C D	15 A B C D
16 A B C D	17 A B C D	18 A B C D
19 A B C D	20 A B C D	21 A B C D
22 A B C D	23 A B C D	24 A B C D
25 A B C D	26 A B C D	27 A B C D
28 A B C D	29 A B C D	30 A B C D
31 A B C D	32 A B C D	33 A B C D
34 A B C D	35 A B C D	36 A B C D
37 A B C D	38 A B C D	39 A B C D
40 A B C D	41 A B C D	42 A B C D
43 A B C D	44 A B C D	45 A B C D
46 A B C D	47 A B C D	48 A B C D
49 A B C D	50 A B C D	51 A B C D
52 A B C D	53 A B C D	54 A B C D
55 A B C D	56 A B C D	57 A B C D
58 A B C D	59 A B C D	60 A B C D
61 A B C D	62 A B C D	63 A B C D
64 A B C D	65 A B C D	66 A B C D
67 A B C D	68 A B C D	69 A B C D
70 A B C D	71 A B C D	72 A B C D
73 A B C D	74 A B C D	75 A B C D
76 A B C D	77 A B C D	78 A B C D
79 A B C D	80 A B C D	81 A B C D
82 A B C D	83 A B C D	84 A B C D
85 A B C D	86 A B C D	87 A B C D
88 A B C D	89 A B C D	90 A B C D
91 A B C D	92 A B C D	93 A B C D
94 A B C D	95 A B C D	96 A B C D
97 A B C D	98 A B C D	99 A B C D
100 A B C D		

City of New York Department of Personnel Examination No #3040 Firefighter

Directions: Firefighters must be able to find their way in and out of buildings that are filled with smoke. They must learn the floor plan quickly for their own safety and to help fight the fire and remove victims.

Look at this floor plan of an apartment. There is an apartment on each side of this one. It is on the *fifth floor* of the building.

Doors are shown as

Doorways are shown as

Windows are shown as

You will have 5 minutes to memorize this floor plan. Then you will be asked to answer some questions about it without looking at it.

BEGIN NOW.

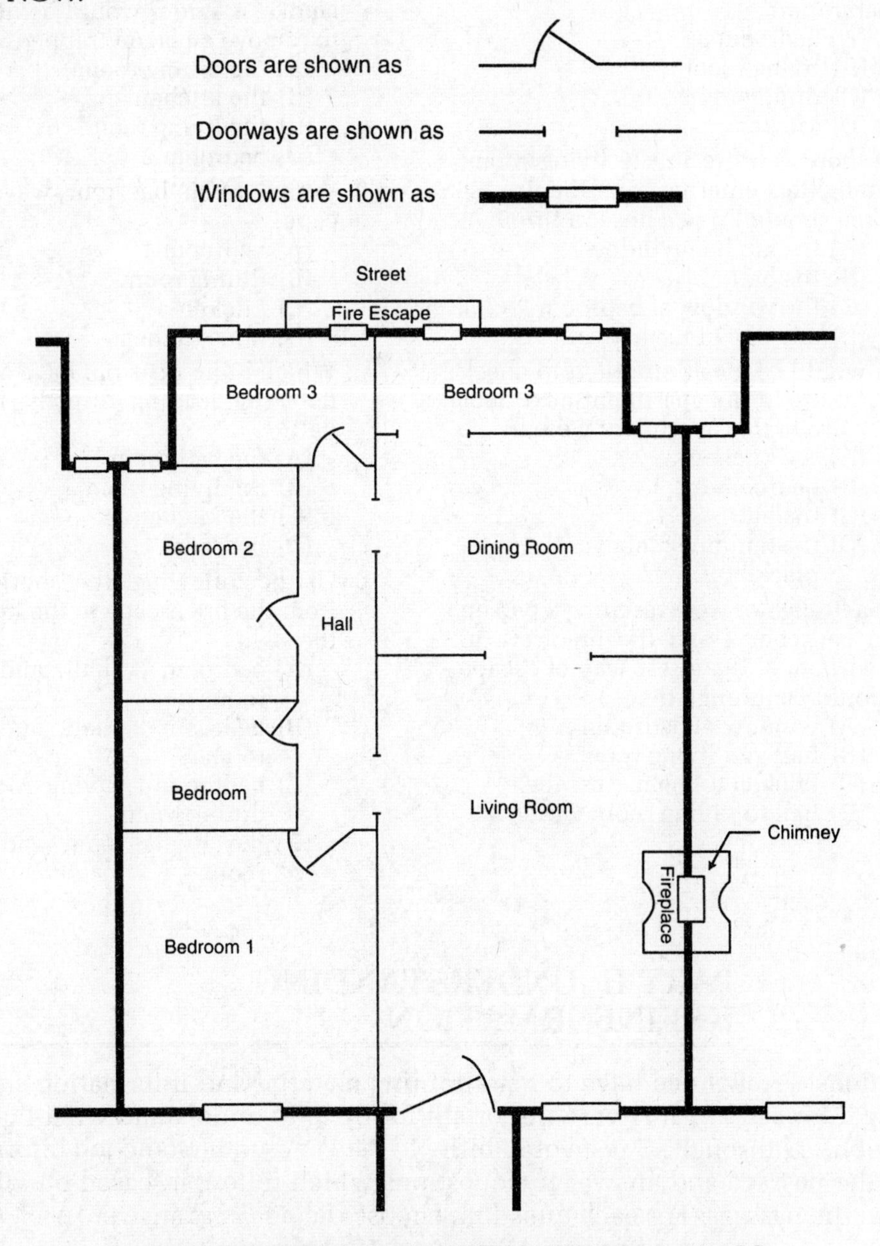

CLOSE THIS BOOKLET WHEN THE BELL RINGS

*The New York City Department of Personnel makes no commitment, and no implication is to be drawn, as to the content, style, or format of any future examination for the position of Firefighter.

PART I: RECALL OF DETAILS

Directions: Questions 1 through 10 test your ability to recall the details of the floor plan you just studied. Each question or statement is followed by four choices. For each question, choose the one best answer (A, B, C or D). Then, on your SEPARATE ANSWER SHEET, in the row with the same number as the question, blacken the oval containing the same letter as your answer.

1. Which room has no doors that can be closed?
   (A) bedroom 1
   (B) living room
   (C) dining room
   (D) none of these
2. Which room is farthest from the bathroom?
   (A) bedroom 3
   (B) living room
   (C) dining room
   (D) kitchen
3. If there is a fire in the living room, firefighters entering from the fire escape should bring a hose in through
   (A) the kitchen window.
   (B) the hall.
   (C) the window of bedroom 2.
   (D) any one of the above.
4. It would be most important to check for a fire in the apartment next door if a fire in this apartment were in
   (A) the kitchen.
   (B) bedroom 3.
   (C) the hall.
   (D) the chimney above the fireplace.
5. If a firefighter were rescuing a person in bedroom 2 and the fire were in bedroom 3, the safest way of escape would be through the
   (A) window of bedroom 2.
   (B) hall and living room.
   (C) kitchen to the fire escape.
   (D) hall to dining room window.
6. Which room has only one way of escaping from it?
   (A) the bathroom
   (B) the living room
   (C) bedroom 2
   (D) none of the above
7. If the hall were full of fire and heavy smoke, a ladder would be necessary to remove a person trapped in
   (A) the dining room.
   (B) the kitchen.
   (C) the living room.
   (D) bedroom 2.
8. Which room has four ways of escape?
   (A) bedroom 1
   (B) dining room
   (C) kitchen
   (D) none of them
9. Which room does *not* have a door or doorway leading directly into the hall?
   (A) the bathroom
   (B) the living room
   (C) the kitchen
   (D) bedroom 1
10. Of the following, the shortest way from the fire escape to the kitchen is through
    (A) bedroom 3, hall, and dining room.
    (B) bedroom 2, hall, and dining room.
    (C) bedroom 1, living room, and dining room.
    (D) the living room and dining room.

PART II: UNDERSTANDING JOB INFORMATION

Directions: Firefighters have to read training material and information sheets all during their careers. It is very important for them to understand what they read. Questions *11* through *45* test your ability to read and understand job information. Read the passage and answer the questions which follow it, based on what you read in the passage. For each question, choose the one *best* answer (A, B, C or D).

Then, on your SEPARATE ANSWER SHEET, in the row with the same number as the question, blacken the oval containing the same letter as your answer. Your answers are to be based *only* on information that is given or that can be assumed from information given in the reading passages. Start marking your answers with question 11.

Answer questions *11* through *15* on the basis of the passage below.

Arsonists are people who set fires deliberately. They don't look like criminals, but they cost the nation millions of dollars in property loss, and sometimes loss of life. Arsonists set fires for many different reasons. Sometimes a shopkeeper sees no way out of losing his business, and sets fire to it so he can collect the insurance. Another type of arsonist wants revenge, and sets fire to the home or shop of someone he feels has treated him unfairly. Some arsonists just like the excitement of seeing the fire burn and watching the firefighters at work; arsonists of this type have even been known to help fight the fire.

11. The writer of the passage feels that arsonists
    (A) usually return to the scene of the crime.
    (B) work at night.
    (C) don't look like criminals.
    (D) never leave their fingerprints.
12. An arsonist is a person who
    (A) intentionally sets a fire.
    (B) enjoys watching fires.
    (C) wants revenge.
    (D) needs money.
13. Arsonists have been known to help fight fires because they
    (A) felt guilty.
    (B) enjoyed the excitement.
    (C) wanted to earn money.
    (D) didn't want anyone hurt.
14. Shopkeepers sometimes become arsonists in order to
    (A) commit suicide.
    (B) collect insurance money.
    (C) hide a crime.
    (D) raise their prices.
15. The point of this passage is that arsonists
    (A) would make good firefighters.
    (B) are not criminals.
    (C) are mentally ill.
    (D) are not all alike.

Answer questions *16* through *21* on the basis of the passage below.

Water and ventilation are the keys to fire fighting. Firefighters put out most fires by hosing water on the burning material, and by letting the smoke and gases out. When burning material is soaked with cooling water it can no longer produce gases that burn. In a closed room, hot gases can raise the temperature enough for the room to burst into flame. This can happen even though the room is far away from the fire itself. Therefore, firefighters chop holes in roofs and smash windows in order to empty the house of gases quickly. This is called ventilation.

16. Burning material will stop giving off hot gases when it is
    (A) allowed to burn freely.
    (B) exposed to fresh air.
    (C) cooled with water.
    (D) sprayed with chemicals.
17. Hot gases cause a room to burn by
    (A) creating a draft.
    (B) exploding.
    (C) giving off sparks.
    (D) raising the room temperature.
18. A room can burst into flames even though it is
    (A) far from the fire.
    (B) soaked with water.
    (C) well ventilated.
    (D) cold and damp.
19. Firefighters sometimes smash windows and chop holes in roofs in order to
    (A) reach trapped victims.
    (B) remove burning materials.
    (C) ventilate a building.
    (D) escape from a fire.

20. Ventilation is important in firefighting because it
    (A) releases trapped smoke and gases.
    (B) puts out flames by cooling them.
    (C) makes it easier for firefighters to breathe.
    (D) makes the flames easier to see and reach with a hose.

21. Hot gases are most dangerous when they are in a room that is
    (A) large.
    (B) closed.
    (C) damp.
    (D) cool.

Answer questions *22* through *25* on the basis of the passage below.

When there is a large fire in an occupied apartment or tenement, the fire escapes often become overcrowded. To relieve this overcrowding, a portable ladder is often raised to the first level of the fire escape and put opposite to the drop ladder. For added help, an additional ladder can be raised from the ground to the second level. If the fire escape is located in the rear of the building, a "gooseneck" ladder that hooks over the roof can also be used. Then firefighters can help some occupants from the fire escape to the roof instead of to the ground.

22. Portable ladders are raised to fire escapes so that
    (A) firefighters can reach the roof from outside.
    (B) occupants can reach a higher level of the fire escape.
    (C) firefighters can enter windows more easily.
    (D) occupants can leave fire escapes more rapidly.

23. If all the ladders described in the passage are used, how many ways can the occupants reach the ground directly by ladder from the fire escape at the first level?
    (A) 1
    (B) 2
    (C) 3
    (D) 4

24. A "gooseneck" ladder is sometimes used
    (A) opposite the drop ladder.
    (B) from the top level.
    (C) from the first level.
    (D) from the second level.

25. The main topic of the paragraph is
    (A) relieving overcrowding on fire escapes.
    (B) setting up and using portable ladders.
    (C) rescuing occupants from apartments.
    (D) using the roof to escape from fires.

Answer questions *26* through *31* on the basis of the passage below.

During search operations the first step is usually to rescue victims who can be seen and heard, or those whose exact locations are known. Disorganized or careless search must be avoided since victims may be underneath rubble. Disorganized movement could cause injury or death. The best method is to start from the outer edge and work toward the center of an area. Sometimes a trapped or buried victim may be located by calling out or by tapping on pipes. Rescue workers should first call out, then have a period of silence to listen for sounds from a victim.

26. The main point of the paragraph is that
    (A) firefighters should call and listen often.
    (B) trapped victims can usually be heard.
    (C) searching should be an organized procedure.
    (D) many victims are buried in fires.

27. Normally the first victims to be rescued during a search are those who are
    (A) unconscious.
    (B) trapped under rubble.
    (C) easy to see and hear.
    (D) injured.

28. When searching for buried victims it is very important for firefighters to
    (A) have periods of silence.
    (B) keep moving constantly.
    (C) search rubble piles quickly.
    (D) stay away from rubble piles.

29. The best way to search an area is
    (A) around the edges.
    (B) from center to edge.
    (C) from corner to corner.
    (D) from edge to center.

30. Disorganized movement by a rescue worker can cause
    (A) panic and confusion.
    (B) property destruction.
    (C) wasted time.
    (D) death or injury.

31. Tapping on pipes is a good way to locate victims because
    (A) firefighters can use Morse code.
    (B) sound travels through a pipe.
    (C) firefighters can signal each other this way.
    (D) pipes usually aren't covered. by rubble.

Answer questions *32* through *36* on the basis of the passage below.

Fire often travels inside the partitions of a burning building. Many partitions contain wooden studs that support the partitions. The studs leave a space for the fire to travel along. Flames may spread from the bottom to the upper floors through the partitions. Sparks from a fire in the upper part of a partition may fall and start a fire at the bottom. Some signs that a fire is spreading inside a partition are: (1) blistering paint, (2) discolored paint or wallpaper, or (3) partitions that feel hot to the touch. If any of these signs is present the partition must be opened up to look for the fire. Finding cobwebs inside the partition is one sign that fire has not spread through the partition.

32. Fires can spread inside partitions because
    (A) there are spaces between studs inside of partitions.
    (B) fires can burn anywhere.
    (C) partitions are made out of materials that burn easily.
    (D) partitions are usually painted or wallpapered.

33. Cobwebs inside a partition are a sign that the fire has not spread inside the partition because
    (A) cobwebs are fire resistant.
    (B) fire destroys cobwebs easily.
    (C) spiders don't build cobwebs near fires.
    (D) cobwebs fill up the spaces between studs.

34. If a firefighter sees the paint on a partition beginning to blister, he should first
    (A) wet down the partition.
    (B) check the partitions in other rooms.
    (C) chop a hole in the partition.
    (D) close windows and doors and leave the room.

35. One way to tell if fire is spreading within a partition is the
    (A) temperature of the partition.
    (B) color of the smoke.
    (C) age of the plaster.
    (D) spacing of the studs.

36. The main point of the passage is
    (A) how fire spreads inside partitions.
    (B) how cobwebs help firefighters.
    (C) how partitions are built.
    (D) how to keep fires from spreading.

Answer questions *37* through *40* on the basis of the passage below.

When backing a fire truck into the firehouse, all firefighters should remain outside the building. Firefighters assigned to stop traffic should face traffic so they can alert the driver in case of an emergency. Additional firefighters should stand on the sidewalk in front of the firehouse to guide the driver. The truck should be slowly backed into the firehouse and immediately stopped upon orders of any firefighter. When the truck is completely in the firehouse, then and only then should the officer contact central headquarters for the placement of the company in service. Following this, the officer orders the entrance doors closed.

Use this diagram to help answer questions *37* through *39*. The letters indicate where firefighters are standing.

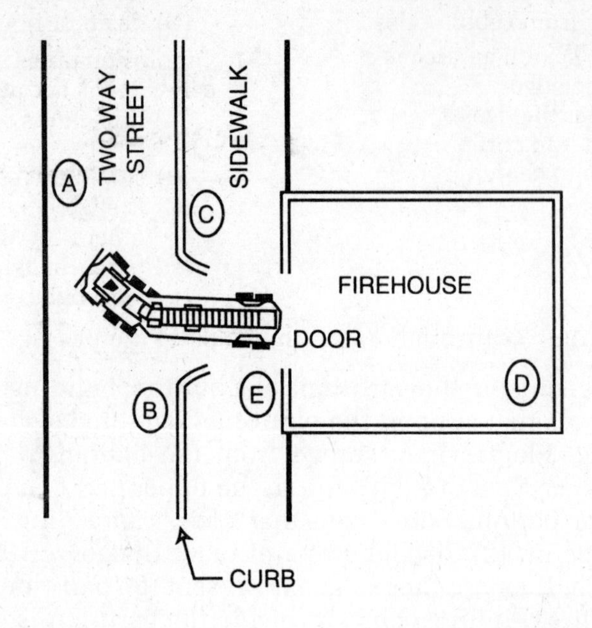

37. The truck is backing into the firehouse. Which firefighter is *not* needed according to the regulations?
    (A) firefighter A
    (B) firefighter B
    (C) firefighter C
    (D) firefighter D
38. Which firefighters are responsible for stopping cars?
    (A) firefighters A and C
    (B) firefighters C and E
    (C) firefighters A and B
    (D) firefighters B and E
39. Which firefighters can order the truck to stop?
    (A) firefighters A and B only
    (B) firefighters C and E only
    (C) firefighter D only
    (D) any of them
40. When is central headquarters notified that the company is ready to be put in service?
    (A) When the truck is returning from a fire.
    (B) After the truck is parked in the firehouse.
    (C) After the firehouse doors are closed.
    (D) When all of the firefighters have entered the firehouse.

Answer questions *41* through *45* on the basis of the passage below.

Unless they have had a fire, most people are not aware of the things firefighters do. Too often the public thinks of firefighters as lounging around a firehouse between fires. Firefighters can help change this image in small ways by their appearance, by greeting visitors who come to the firehouse, by their behavior on the street at a fire, and by treating the public in a courteous manner. For example, 90 percent of the rescues made by the average fire department take place at relatively small fires, not at spectacular extra-alarm fires. The public rarely hears about many rescues because the fire departments seldom let the press know about firefighters who have performed acts of bravery at routine fires.

41. What are firefighters doing when not fighting fires?
    (A) Lounging around the firehouse.
    (B) Working on public relations projects.
    (C) Making repairs to the equipment.
    (D) The passage doesn't say.

42. The passage places responsibility for improving the fire department's image on the
    (A) fire department itself.
    (B) press.
    (C) public.
    (D) people rescued by firefighters.

43. Most of the rescues made by firefighters take place at
    (A) extra-alarm fires.
    (B) special emergencies where no fire is involved.
    (C) relatively small fires.
    (D) spectacularly large fires.

44. The public rarely hears about rescues made by firefighters at routine fires because
    (A) information about fires must be kept confidential.
    (B) fire departments seldom report these rescues to the press.
    (C) most of these rescues take place late at night.
    (D) reporters aren't interested in covering routine fires.

45. What would be the best title for this passage?
    (A) An Inside Look at the Fire Department.
    (B) Making the Most of Fire Prevention Week.
    (C) Improving the Fire Department's Public Image.
    (D) Brave Acts Performed by Firefighters.

PART III: JUDGMENT AND REASONING

Directions: Questions *46* through *70* test judgment in situations that a firefighter might meet on the job. Read the question or statement. Each question or statement is followed by four choices. For each question, choose the one *best* answer (A, B, C or D). Then, on your SEPARATE ANSWER SHEET, in the row with the same number as the question, blacken the oval containing the same letter as your answer. Start marking your answers with question 46.

46. Which of these makes the best exit door from a large public building?
    (A) a door that opens out
    (B) a sliding door
    (C) a revolving door
    (D) a door that opens in

47. The best way to delay the spread of a fire from one room to the next is to
    (A) put a hole through the wall between the two rooms.
    (B) close all windows in both rooms.
    (C) remove all furniture from both rooms.
    (D) close the door between the rooms.

48. In which of the following one-story buildings would there most likely be a need for rescue work during a fire?
    (A) a high school with 70 students
    (B) a store with 70 customers
    (C) a nursing home with 70 residents
    (D) an office building with 70 workers

49. Fires in industrial plants are likely to cause more damage at night when they are closed than during the day when they are open because
    (A) it takes firefighters longer to travel to the fire at night.
    (B) fires are noticed sooner during work hours.
    (C) more arsonists work at night.
    (D) fire smolders longer when it is cooler.

50. Firefighters are getting ready to leave the scene after a fire when a reporter stops one of them and starts asking for details about the fire. The firefighter should tell the reporter
    (A) that firefighters are not allowed to talk to reporters.
    (B) to speak to the firefighter's supervisor.
    (C) to call for an appointment at the firehouse.
    (D) to check with other reporters. at the fire who might know the details.

51. A firefighter becomes trapped in a third-floor apartment when the stairs leading to the apartment catch fire. There is no fire escape. The best thing for the firefighter to do is
    (A) call for a portable ladder and escape through a window.
    (B) do nothing until help arrives.
    (C) jump from a window as quickly as possible.
    (D) move up to the next floor and wait to be rescued.

52. When firefighters give first aid to an unconscious person, there are several things to check for. Which of the following is *not* one of them?
    (A) Are the person's legs twisted in a way which might show that bones may be broken?
    (B) Is the person breathing?
    (C) Is there an open container of poison nearby?
    (D) Did the person try to commit suicide?

53. Mirrors can sometimes be a problem when firefighters are fighting a fire. Most likely this is because
    (A) firefighters may see the fire in the mirror and aim a hose at the mirror instead of at the fire itself.
    (B) mirrors are expensive and the owner will be angry if one is broken.
    (C) mirrors are usually heavy so firefighters have to be careful that a mirror doesn't fall on them.
    (D) mirrors shining on fires make them hotter.

54. A firefighter went to work the day after spraining an ankle while playing football. Even though in pain, the firefighter did not want to stay home because the firefighter felt needed on the job. The main reason the firefighter should have stayed home is that the firefighter
    (A) should know that no one is that necessary.
    (B) should get medical attention.
    (C) might not be able to do an equal share of the work at a bad fire.
    (D) was injured off duty and his medical expenses should not be paid by the department.

55. Pet dogs sometimes save their masters' lives by waking them up when there is a fire in the house. The most probable reason that dogs detect fire before their masters is that
    (A) dogs are more used to fires than their masters are.
    (B) dogs are more sensitive to smoke than their masters.
    (C) dogs always sense danger.
    (D) fires make dogs thirsty.

56. During an inspection of a building, it is *least* likely that a firefighter would need to check the
    (A) number and location of fire extinguishers.
    (B) storage areas for combustible material.
    (C) location of the main electric control switch.
    (D) location of public washrooms.

57. Which of these types of fires is most likely to have been started on purpose?
    (A) a fire in a frying pan
    (B) a fire in a mailbox
    (C) a fire in a child's bedroom
    (D) a fire in an attic

58. A firefighter chopping a hole in a roof to let smoke out of a building would probably be in *least* danger from the smoke when chopping with the wind blowing
    (A) from the firefighter's left.
    (B) toward the firefighter's face.
    (C) toward the firefighter's back.
    (D) from the firefighter's right.

59. More firefighters' hands are hurt in summer than in winter. This is most probably because
    (A) things they handle stay hot longer in summer.
    (B) their hands sweat and things slip more easily in summer.
    (C) fires burn hotter in summer.
    (D) they are more likely to remove their protective gloves in summer.

60. Normally, firefighters try to get as close as possible to the fire they are fighting in order to
    (A) find out what started the fire.
    (B) aim their hoses more accurately.
    (C) use as little hose as possible.
    (D) avoid working in thick smoke.

61. Which of these items would firefighters be *least* likely to need when putting out a fire at the scene of an automobile accident on a city street?
    (A) a portable ladder
    (B) a portable fire extinguisher
    (C) a stretcher
    (D) a hose

62. During a fire, a firefighter searching an apartment for victims finds an unconscious woman on her bed. It would probably be best for the firefighter to *first*
    (A) call for a doctor to help the woman.
    (B) give the woman first aid to make her conscious.
    (C) move the woman to a safe place.
    (D) stay with the woman in case she becomes conscious.

63. Firefighters often make their way through a burning room by crawling along the floor because
    (A) they are less likely to lose their balance.
    (B) it is easier for them to avoid flying sparks.
    (C) fire does not often spread along the floor.
    (D) there is less smoke at floor level.

64. In order to enforce the fire safety laws firefighters must inspect buildings and stores. It is *not* a good idea for firefighters to let owners of buildings and stores know when they are coming because
    (A) firefighters will waste valuable time if the owner breaks the appointment.
    (B) owners might try to hide fire hazards from the firefighters.
    (C) firefighters can make the inspection faster without an appointment.
    (D) owners would be angry if the firefighters were unable to keep the appointment.

65. Three fires occurred in an unused school building during one month. The firefighters believed that arsonists had started these fires. The fire department decided to take photographs of the crowds watching any more fires in this school building. This was a good idea mainly because
    (A) it would help the fire department's public relations program.
    (B) it would be useful in getting the city to tear down the building.
    (C) taking pictures forces the crowd to stand back out of danger.
    (D) people who set fires sometimes appear in the crowd watching the fire.

66. Firefighters probably would be in greatest personal danger from a fire in a
    (A) florist shop.
    (B) food store.
    (C) paint store.
    (D) bank.

67. Firefighters answering an alarm find a badly beaten and unconscious man lying next to the alarm box. Of the following, the *least* important thing the firefighters should do is
    (A) call the police.
    (B) administer first aid.
    (C) call for an ambulance.
    (D) question nearby residents.

68. At several fires in one neighborhood, people threw bricks and bottles at firefighters who were fighting fires. To try to solve this problem it would be best to
    (A) talk to neighborhood leaders to reduce these attacks.
    (B) refuse to answer alarms in the neighborhood.
    (C) turn the hoses on the people if it happens again.
    (D) allow firefighters to carry guns or other weapons.

69. Between fires, firefighters clean and check their equipment and vehicles. They probably do this because they want to
    (A) look busy when the public visits.
    (B) have more free time to watch TV or play cards.
    (C) make the neighborhood proud of the appearance of the equipment and vehicles.
    (D) keep the equipment and vehicles in good working order.

70. Storekeepers are most likely to be willing to follow fire safety rules if they
    (A) understand the reasons for them.
    (B) have plenty of time.
    (C) are ordered to follow them.
    (D) have just opened up a new store.

PART IV: UNDERSTANDING MECHANICAL DEVICES

Directions: Questions *71* through *74* test your ability to understand general mechanical devices. Pictures are shown and questions asked about the mechanical devices shown in the picture. Read each question and study the picture. Each question is followed by four choices. For each question, choose the one *best* answer (A, B, C or D). Then, on your SEPARATE ANSWER SHEET, in the row with the same number as the question, blacken the oval containing the same letter as your answer. Start marking your answers with question 71.

71. The reason for crossing the belt connecting these wheels is to
    (A) make the wheels turn in opposite directions.
    (B) make wheel 2 turn faster than wheel 1.
    (C) save wear on the belt.
    (D) take up slack in the belt.

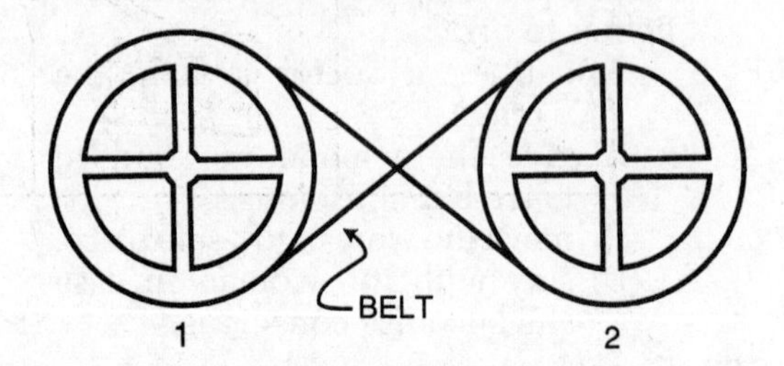

72. The purpose of the small gear between the two large gears is to
    (A) increase the speed of the larger gears.
    (B) allow the larger gears to turn in different directions.
    (C) decrease the speed of the larger gears.
    (D) make the larger gears turn in the same direction.

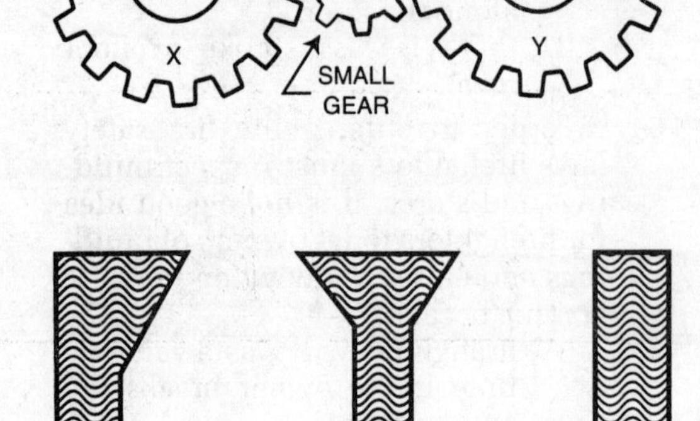

73. Each of these three-foot-high water cans has a bottom with an area of one square foot. The pressure on the bottom of the cans is
    (A) least in A.
    (B) least in B.
    (C) least in C.
    (D) the same in all.

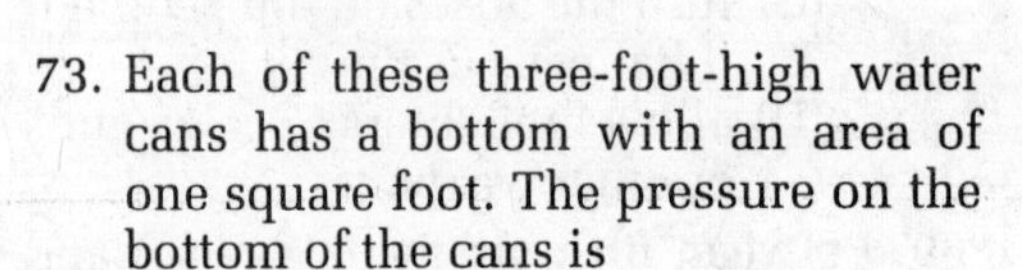

74. The reading on the scale should be
    (A) zero.
    (B) 10 pounds.
    (C) 13 pounds.
    (D) 26 pounds.

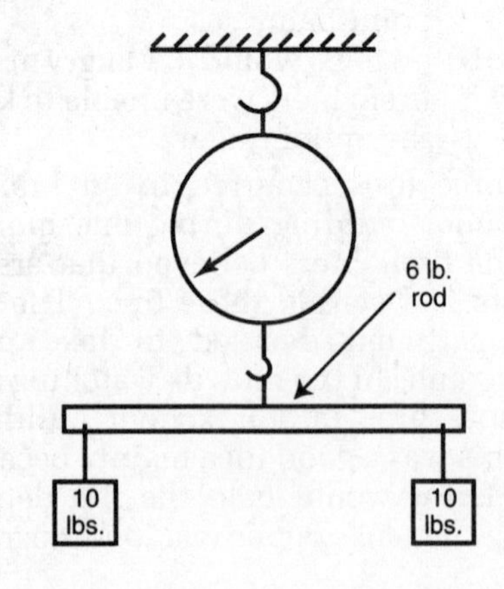

Directions: Questions *75* through *77* test knowledge of tools and how to use them. For each question, decide which one of the four things shown in the boxes labeled A, B, C or D normally is used with or goes best with the thing in the picture on the left. Then, on your SEPARATE ANSWER SHEET, in the row with the same number as the question, blacken the oval containing the same letter as your answer. Start marking your answers with question 75.

NOTE: All tools are *not* drawn to the same scale.

75.

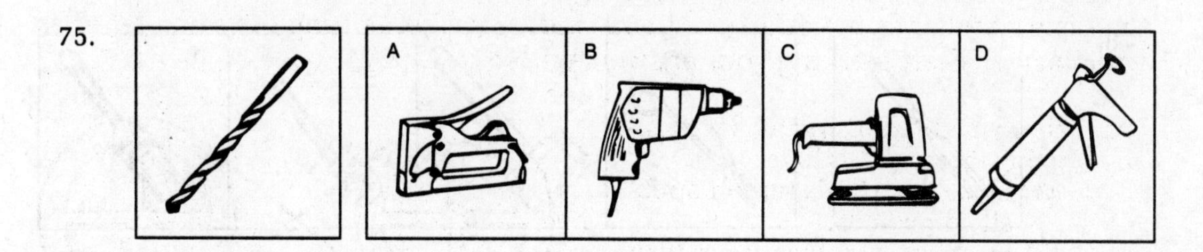

76.

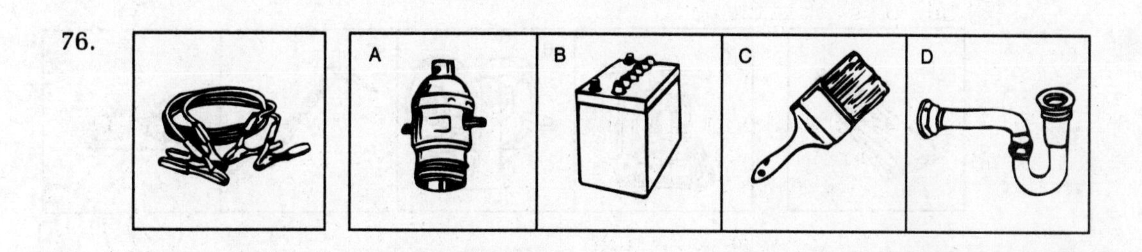

77.

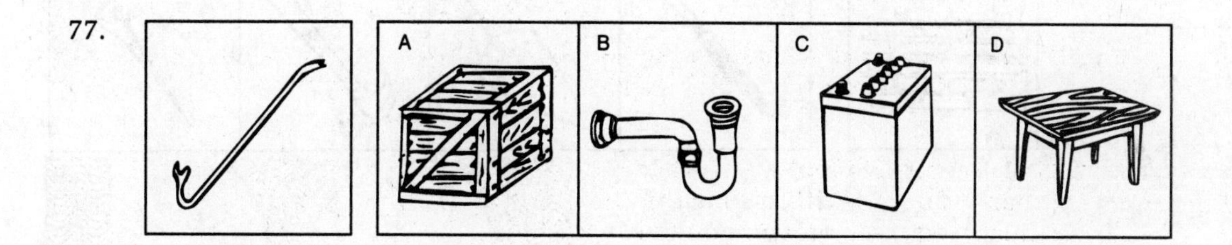

Directions: Questions *78* through *80* test knowledge of tools and how to use them. For each question, decide which one of the four things shown in the boxes labeled A, B, C or D normally is used with or goes best with the thing in the picture on the left. Then, on your SEPARATE ANSWER SHEET, in the row with the same number as the question, blacken the oval containing the same letter as your answer. Start marking your answers with question 78.

NOTE: All tools are *not* drawn to the same scale.

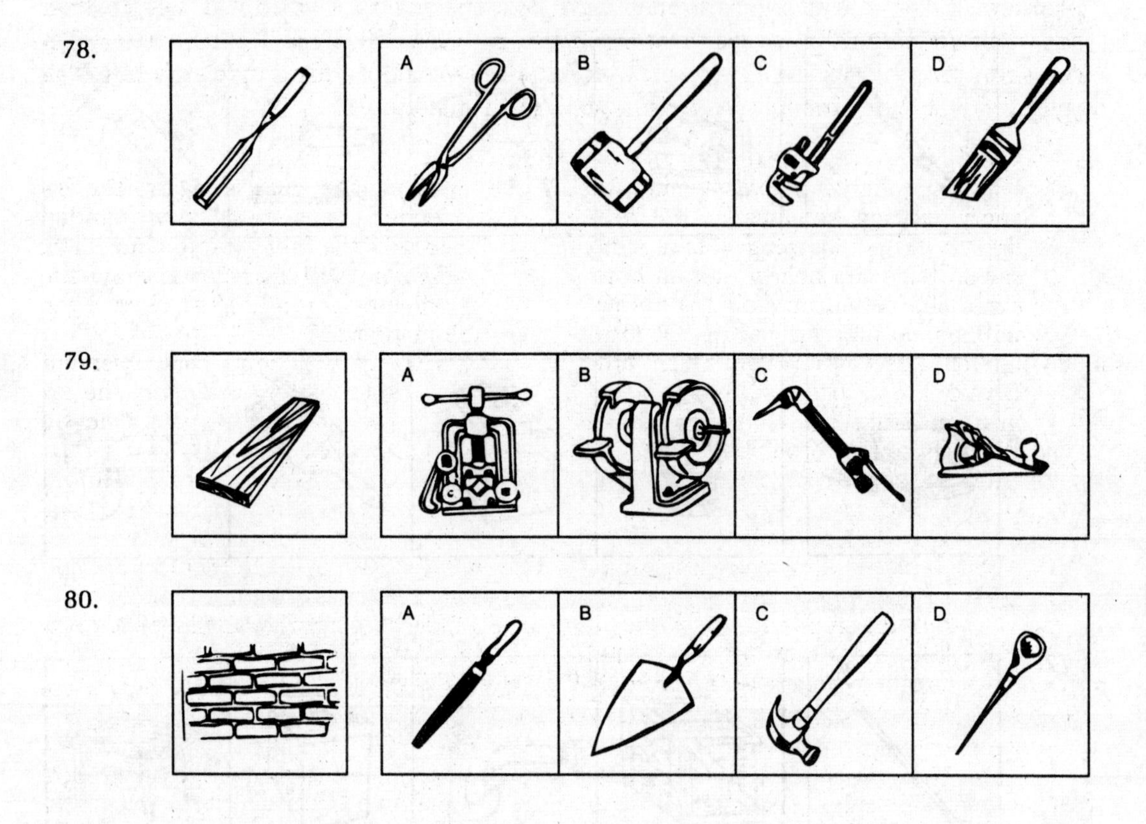

PART V: DEALING WITH OTHER PEOPLE

Directions: Firefighters must deal with the public during inspection of buildings, when groups visit the firehouse, and while they are fighting fires. As public servants they should treat people with courtesy and respect. However, at a fire, putting the fire out must be their first duty. Questions *81* through *100* present problem situations in which firefighters have to deal with other people. Read the problem, study the answer choices, and pick the one that you think would *best* solve the problem. Then, on your SEPARATE ANSWER SHEET, in the row with the same number as the question, blacken the oval containing the same letter as your answer. Start marking your answers with question 81.

81. When a firefighter arrives at a fire, an angry woman screams at the firefighter for not coming sooner. She says it has been nearly half an hour since she called and all her things will be burned up before the firefighter gets the fire out. The firefighter knows it has been only 6 minutes since the alarm came in. What should the firefighter do?
    (A) Ask her exactly what time she called and show her it was only 6 minutes ago.
    (B) Explain to her how firefighters respond to a fire and go on fighting the fire.
    (C) Tell her that times seems longer when people are worried and she is wrong.
    (D) Say nothing and get to work on the fire.

82. When inspecting a store, a firefighter sees some trash piled on a stair landing. This is a dangerous fire condition which violates the law. The owner says that the last inspector said it was okay because the people can use the elevator instead of the stairs. What should the firefighter do *first?*
    (A) Ignore the violation since firefighters should back each other up in dealing with the public.
    (B) Insist that the owner remove the trash since it is a violation of the law.
    (C) Ask the owner for the name of the previous inspector.
    (D) Try to find out why the owner wants to store trash on the stair landing.

83. While on fire inspection duty, a firefighter enters a restaurant during a very busy lunch time. The owner, who usually goes along on the inspection, says he is short-handed and asks the firefighter to come back when he will be able to show the firefighter around. What should the firefighter do?
    (A) Agree to come back later but inform the owner that the inspection will be more strict on the next visit.
    (B) Go ahead and inspect without the owner since it would be a good time to catch violations.
    (C) Tell the owner that he will be reported to the authorities.
    (D) Wait around and inspect when the crowd has gone.

84. Firefighters have almost put a fire out in an apartment building. A couple who live on the top floor ask a firefighter if they can go back to their apartment to bring out some clothing. What should the firefighter do?
    (A) Let the couple go back to their apartment since the fire is now almost put out.
    (B) Offer to go back and get the clothes for them.
    (C) Tell them that only one or the other, not both, can go back inside.
    (D) Tell them that they will be allowed back into the building when it is completely safe.

85. A man rushes into the firehouse telling the firefighters on duty that some children are throwing large rocks off a nearby overpass onto cars passing underneath. What should the firefighters do?
    (A) Go to the overpass with the man and stop the children.
    (B) Report it to their superior officer.

(C) Drive the man to the nearest police station.
(D) Tell him this is not a firefighter's problem.

86. During a dangerous fire, a citizen complains to a firefighter that another firefighter swore at her for being in the way. What should the firefighter do?
(A) Apologize to the citizen and explain in detail why the firefighter acted that way.
(B) Nothing, since the citizen is clearly wrong.
(C) Tell the woman to report the incident to the supervisor after the fire.
(D) Tell the firefighter to apologize or the citizen might make trouble.

87. Twenty firefighters in a firehouse want to get a color TV to replace their old set which is broken. The other two firefighters who want the set say they can't afford to pay their shares now. The best solution is for the twenty firefighters to
(A) suggest that the two take part-time jobs to pay their shares.
(B) forget about buying a set.
(C) suggest that the two borrow the money to pay their shares.
(D) buy the set without the contribution from the two and allow them more time to pay.

88. While inspecting a small store during business hours, a firefighter finds an exit door padlocked. According to fire safety laws, exit doors must not be locked while the store is open for business. The owner says burglars had once come in through that door and she has put the padlock key on a nail beside the door and shown it to all her employees. What should the firefighter do?
(A) Explain the fire safety law requirements and order the owner to remove the padlock.
(B) Get all the employees to show the firefighter they can open the locked door.
(C) Do nothing because the key is easy to see hanging from the nail.
(D) Tell the owner to give a key to each employee.

89. A fire truck is present at a fireworks display to put out any fires which may be started by the fireworks. Just when the fireworks begin, some children start climbing up on the truck. What should the firefighters do?
(A) Show the children around on the truck and explain how it works.
(B) Let only one or two of the children on the truck at a time so that the firefighters can watch them.
(C) Tell the children to get off the truck because they will be in the way if a fire starts.
(D) Move the truck away from the fireworks so the children will not be hurt.

90. During a fire a superior officer orders a firefighter to use a certain type of hose. The firefighter feels very strongly that a different type of hose should be used. In this situation the firefighter should
(A) use the different hose if the officer is not watching.
(B) ask the officer the reasons for the order.
(C) do what the officer says and ask for an explanation of the order after the fire is out.
(D) tell the officer that the different hose is better and explain why.

91. On a firefighter's first day on the job, a neighborhood resident visits the firehouse and complains to the firefighter that too much of the taxpayers' money is being spent on expensive firefighting equipment. The firefighter should say
(A) "I'm new on this job so let me get my superior officer to discuss this with you."
(B) "Since New York City is the biggest city, it needs the most expensive equipment."
(C) "Sir, you are entitled to your opinion even if you are wrong."
(D) "Why don't you write a letter of complaint and send it to the Mayor?"

92. A man who is drunk staggers into the firehouse and asks a firefighter for help in getting his car started. It

is parked less than a block away. What should the firefighter do?

(A) Go with him and see what is wrong with the car.
(B) Ask him how much he has had to drink.
(C) Tell him he shouldn't drive and suggest he take a cab.
(D) Offer to take him home.

93. A Scout troop visits a firehouse to see the equipment. The firefighters show them around the firehouse. Letting Scouts visit the firehouse is

(A) good, because firefighters should be kept busy at all times.
(B) bad, because firefighters should rest when not fighting fires.
(C) good, because it gives the firefighters a chance to teach the Scouts about fire safety.
(D) bad, because the Scouts can cause damage to the equipment in the firehouse.

94. During an inspection a building manager becomes angry and tells a firefighter that the fire department does a rotten job and inspections are a waste of time. The firefighter should

(A) tell the building manager to clean up his own place first before complaining about the fire department.
(B) suggest that they trade jobs for a few hours.
(C) try to find out why he feels this way.
(D) inform the building manager that the real problem is his bad attitude.

95. While putting fire fighting equipment back on the fire truck after an apartment fire, a firefighter is falsely accused by a woman of stealing ten dollars she had kept hidden in a sugar bowl in her apartment. The firefighter should

(A) ask her why firefighters would risk their lives for ten dollars.
(B) give her ten dollars to get her out of the way.
(C) tell her she had no business keeping money in a sugar bowl.
(D) deny it and tell her she may report it to a superior officer.

96. After firefighters had put out a fire in a store, the owner yelled to a firefighter that another firefighter had chopped a hole in the roof and had caused more damage than the fire. What should the firefighter do *first?*

(A) Tell the owner insurance pays for the damage.
(B) Tell the owner to talk to the firefighter who chopped the hole.
(C) Explain that it was necessary to make a hole or the firefighter wouldn't have done it.
(D) Suggest to the owner that he put his complaint in writing to the Fire Commissioner.

97. After giving a talk on fire prevention to a group of school children, a firefighter finds that the children have more questions than can be answered in the time allowed. The firefighter should

(A) answer as many questions as possible and try to arrange another visit.
(B) let only the teacher ask questions.
(C) stay with the children and answer all their questions.
(D) refuse to answer any questions.

98. Some neighborhood children have been coming to the firehouse several times a day. They mean to be friendly, but they are keeping the firefighters from doing their work. The firefighters should

(A) tell them not to come back to the firehouse anymore.
(B) call their parents and ask them to keep the children away.
(C) suggest that they start visiting the neighborhood police station.
(D) ask them to leave because the firefighters have a lot of work to do.

99. After a fire has been put out, a firefighter sees a newly appointed firefighter pulling down loose plaster from the ceiling, without wearing a helmet. The firefighter should
    (A) report the new firefighter to the officer in charge.
    (B) say nothing about what happened.
    (C) tell the new firefighter to put the helmet on because falling plaster can be dangerous.
    (D) use force, if necessary, to make the new firefighter wear the helmet.

100. While on duty, firefighters usually prepare their meals in the firehouse. Most of the firefighters in a certain firehouse like to take turns cooking meals. However, one of them offers to clean up the dishes rather than having to cook. The other firefighters should
    (A) tell the firefighter that cooking can be fun.
    (B) agree to the firefighter's offer.
    (C) let the firefighter make sandwiches instead of meals for dinner.
    (D) excuse the firefighter from cooking and doing the dishes.

END OF WRITTEN TEST

Instructions for the End of the Test

RECORD YOUR ANSWERS IN THIS TEST BOOKLET AS WELL AS ON THE ANSWER SHEET BEFORE THE LAST SIGNAL—This is necessary in the event that you wish to protest the proposed key answers.

ANSWER SHEET COLLECTION—When finished, remain seated and summon the monitor to collect your answer sheet. You are responsible for seeing that the monitor collects your answer sheet. You should make a note of your T.P. No. (Test Paper Number) and include this number and your Social Security number in any correspondence to the Department of Personnel with respect to this examination. Leave the building quickly and quietly.

The Department of Personnel, in establishing key answers to this test, reserves the right to determine which of the answers listed for each question is to be deemed and accredited as acceptable, and whether more than one of the answers listed for each question is to be deemed acceptable and accredited as such. The Department further reserves the right to cancel and annul any question whenever, upon inquiry, it deems that none of the listed answers to the question can properly be considered acceptable.

Please do not telephone this Department to request information on the progress of the rating of this test. Such information will not be given. All candidates will be notified individually by mail of their rating after all papers have been rated. Notify this Department promptly in writing of any change in your address to ensure prompt delivery of such notice.

ANSWER KEY

1. **C**	21. **B**	41. **D**	61. **A**	81. **D**
2. **D**	22. **D**	42. **A**	62. **C**	82. **B**
3. **A**	23. **B**	43. **C**	63. **D**	83. **B**
4. **D**	24. **B**	44. **B**	64. **B**	84. **D**
5. **B**	25. **A**	45. **C**	65. **D**	85. **B**
6. **A**	26. **C**	46. **A**	66. **C**	86. **C**
7. **D**	27. **C**	47. **D**	67. **D**	87. **D**
8. **B**	28. **A**	48. **C**	68. **A**	88. **A**
9. **C**	29. **D**	49. **B**	69. **D**	89. **C**
10. **A**	30. **D**	50. **B**	70. **A**	90. **C**
11. **C**	31. **B**	51. **A**	71. **A**	91. **A**
12. **A**	32. **A**	52. **D**	72. **D**	92. **C**
13. **B**	33. **B**	53. **A**	73. **D**	93. **C**
14. **B**	34. **C**	54. **C**	74. **D**	94. **C**
15. **D**	35. **A**	55. **B**	75. **B**	95. **D**
16. **C**	36. **A**	56. **D**	76. **B**	96. **C**
17. **D**	37. **D**	57. **B**	77. **A**	97. **A**
18. **A**	38. **C**	58. **C**	78. **B**	98. **D**
19. **C**	39. **D**	59. **D**	79. **D**	99. **C**
20. **A**	40. **B**	60. **B**	80. **B**	100. **B**

DIAGNOSTIC PROCEDURE

Use the following diagnostic chart to determine how well you have done and to identify your areas of weakness.

Enter your number of correct answers for each section in the appropriate box in the column headed "Your Number Correct." The column immediately to the right will indicate how well you did on each section of the test.

Below the chart you will find directions for the chapter(s) in the guide you should review to strengthen your weak area(s).

Section Number	Question Number	Area	Your Number Correct	Scale	
One	1–10	Recall		10 right	Excellent
				9 right	Good
				8 right	Fair
				Under 8 right	Poor
Two	11–45	Understand Job Information		33–35 right	Excellent
				31–32 right	Good
				30 right	Fair
				Under 30 right	Poor
Three	46–70	Judgment and Reasoning		24–25 right	Excellent
				23 right	Good
				22 right	Fair
				Under 22 right	Poor
Four	71–80	Mechanical Devices and Tools		10 right	Excellent
				9 right	Good
				8 right	Fair
				Under 8 right	Poor
Five	81–100	Dealing with People		19–20 right	Excellent
				18 right	Good
				17 right	Fair
				Under 17 right	Poor

1. If you are weak in Section One, concentrate on Chapter 6.
2. If you are weak in Section Two, concentrate on Chapter 7.
3. If you are weak in Section Three, concentrate on Chapter 8.
4. If you are weak in Section Four, concentrate on Chapters 9 and 10.
5. If you are weak in Section Five, concentrate on Chapter 11.

Note: Consider yourself weak in a section if you receive other than an "Excellent" rating.

ANSWER EXPLANATIONS

1. **C** The dining room has three doorways but no doors. The kitchen has a doorway but no door; however, it is not included as an answer choice.
2. **D** This question is based on travel distance, and the room that requires a person leaving the bathroom to travel the greatest distance is the kitchen.
3. **A** Of the choices offered, the kitchen is the only room with a window from the fire escape. Bedroom 3 also has a window from the fire escape, but it is not given as a choice.
4. **D** Because the chimney is shared by both apartments, a possibility of fire extension exists.
5. **B** This is the safest route, leading away from the fire and toward the main entrance of the apartment. Choice C would put the persons in close proximity to the fire and increase the risk of injury. Choices A and D are secondary means of escape. The best choice is the door that leads to a known safe area.
6. **A** The only way out of the bathroom is through the door and into the hall; from the hall there are two ways of escape.
7. **D** The fire and smoke in the hall would not permit escape via that route. In this case the only way out of bedroom 2 would be through the window and down a ladder. Persons in the dining room (A), kitchen (B), and living room (C) can get access to the living room door or the fire escape without passing through the hall.
8. **B** The four ways are (1) through the kitchen, (2) through the hall, (3) through the living room, and (4) through the window.
9. **C** The kitchen is an isolated room leading only to the dining room and the fire escape.
10. **A** This kind of question can be confusing; it asks you to enter the apartment from the fire escape and then go to the kitchen. The obvious way is the direct route—fire escape to kitchen, but this is not offered as a choice. At this point you must fight the temptation to say that the question is wrong. You have to look at it closely. It is possible to enter the apartment from the fire escape through the window of bedroom 3 and go through the hall into the dining room and then the kitchen. This is the route spelled out in choice A.
11. **C** The second sentence makes this statement. Choices A and B may be true but are not stated directly in the passage. You must always choose the best answer on the basis of what you have just read.
12. **A** An arsonist is defined in the opening sentence as one who deliberately sets fires. Choices B, C and D may be correct for some arsonists but not for all. The best choice is the one that fits all occurrences.
13. **B** This is clearly stated in the last sentence. Choices A, C, and D are not mentioned.
14. **B** This information is found in the fourth sentence.
15. **D** This choice is derived from looking at the several sentences that show who the arsonist may be.
16. **C** This information is found in the third sentence.
17. **D** The fourth sentence states that hot gases can raise room temperature sufficiently for flames to appear.
18. **A** The fifth sentence tells us that this can occur. Choice B—this is incorrect; soaking with water will prevent the generation of gases that will burn. Choice C—the last two sentences state that ventilation removes the hot gases. Choice D—there is no mention of the effect of cold on a fire; damp objects will not give off gases readily.
19. **C** This information is found in the last two sentences. Choices A, B, and D may have some degree of truth, but you must base your answer only on the material presented to you.
20. **A** This is the thrust of the last two sentences. Choice B—ventilation does not extinguish the fire; water does. Choices C and D—these may be true but are not mentioned in the passage.

21. B This information is found in the fourth sentence; it tells us that in a closed room the temperature can rise enough for the room to burst into flame.

22. D This is the theme of this passage. By providing rapid escape routes, firefighters relieve overcrowding of fire escapes.

23. B The correct answer is found in the second sentence. The two ladders are the portable ladder and the drop ladder. Four ladders are mentioned, but only two are used from the first level of the fire escape.

24. B This requires some deductive reasoning to find the one best answer. The passage tells us that the "gooseneck" ladder hooks over the roof. We know that the roof may or may not be above the first or above the second floor, but it is always above the top floor (level). Choice A is incorrect; a portable ladder is placed opposite the drop ladder (see the second sentence).

25. A This is a repetition of question 22 in a different form. The thrust of the passage, overcrowded fire escapes, is given in the first sentence, as is often the case.

26. C In this passage the author leads us to the correct answer by the use of such terms as "disorganized," "the first step," "start," and "first call out." All of these indicate a strong need for organization. The other selections are mentioned in the passage only once. Keep in mind that the main point or theme is the recurring thought.

27. C This information is found in the opening sentence; these victims may or may not be unconscious (choice A), trapped under rubble (choice B), or injured (choice D). However, since they are easily seen and heard, they can be immediately rescued.

28. A The need for periods of silence is stated in the last sentence.

29. D The fourth sentence tells us to start from the outer edge and work toward the center. Don't be misled by the dropping of the word "outer." In this case "edge" and "outer edge" mean the same thing.

30. D This is stated in the third sentence.

31. B The answer to this question is not clearly stated in the passage; rather, it must be found by elimination of the incorrect choices. The fifth sentence tells us that victims are sometimes located by calling out or tapping on pipes, that is, through the use of sound. Now let's look at each selection and see how it agrees with this reasoning. Choice D is incorrect; pipes normally are covered at least partially by rubble. Choice C may be true, but the purpose of the calling out or tapping is to locate the victim, not to signal other firefighters. Choice A is similar to choice C in that this is a method for sending messages to other firefighters and is not appropriate for locating a victim. Choice B states that sound travels, and hence we can assume that it is possible for a victim to hear it. Since we have eliminated A, C, and D, choice B is the best selection.

32. A This is clearly stated in the third sentence.

33. B Since there is no clear-cut answer to this question, you must eliminate the poor choices first. Choice A—cobwebs are organic material and are destroyed easily. Choice C—spiders have no idea when or where a fire will occur. Choice D—cobwebs are lacy and open and will not fill a void. By elimination choice B is the best choice.

34. C The seventh sentence tells us that if there is blistering paint, one sign of spreading fire, the partition must be opened. This would be done by chopping a hole; a small hole is often sufficient.

35. A In the sixth sentence, item 3, we are told that a hot partition is identified by touch.

36. A How a fire spreads is the recurring idea of the passage.

37. D This firefighter is inside the fire station, where a guide is *not* needed, and therefore is in violation of the directive found in the first sentence.

38. C Firefighters A and B are clearly in the street and hence are responsible for the stopping of cars. Firefighters C and E are responsible for stopping pedestrian traffic.

39. **D** This information is found at the end of sentence 4.
40. **B** Sentence 5 gives this information.
41. **D** This passage tells us that many people do not know what fire fighters do between fires and that fire fighters could do something about this lack of knowledge; however, it does not tell us what they do. If you chose A, B, or C you probably were using information or impressions you already had, not information contained in the passage.
42. **A** The third sentence states very clearly: "Fire fighters can help change this image . . . ," and the fifth sentence reinforces this by stating that "fire departments seldom let the press know . . . ."
43. **C** The fourth sentence tells us that 90 percent of the rescues occur at small fires. There is no justification in the passage for any other choice.
44. **B** The fifth sentence states that fire departments seldom let the press know about fire fighters who have performed acts of bravery at routine fires.
45. **C** The title should clearly state the theme of the passage—in this case, the need to correct the public perception of what fire fighters do. The title that best fits this concept is C—Improving the Fire Department's Public Image.
46. **A** A door that opens out allows many people to pass through it quickly. Many disastrous fires have shown that the doors in choices B, C, and D become jammed when large crowds of people attempt to use them quickly.
47. **D** To prevent the spread of fire we must provide a barrier. The closed door will help to cut the fire off. Choice A would promote, not delay, the spread of fire. The windows in choice B would succumb to the fire much sooner than would the door. Choice C would reduce the overall size of the fire but would not prevent the fire from spreading from one room to the other.
48. **C** This question calls for you to consider the nature of the occupancy and the condition of the occupants. Choices A, B, and D relate to places with occupants who generally are in good physical health and are able to move about without help. On the other hand, choice C, a nursing home, houses the elderly and infirm, many of whom cannot move about freely without help.
49. **B** Industrial plants are often shut down at night and are in neighborhoods where few people live or visit except when the building is open. Choice A is incorrect; the opposite is generally true because of the reduced traffic load at night. Choice C is incorrect; arson is committed at all hours of the day or night. Choice D may be true but is not related to the question and hence is not the best choice.
50. **B** By referring the reporter to the supervisor, the fire fighter properly directs the reporter to the person who is best able to give accurate and appropriate answers to questions.
51. **A** Calling for the ladder alerts other fire fighters to the problem and allows for an effective rescue effort. Choice B is incorrect; help may not arrive because the other fire fighters may not know about the problem. Choice C would be foolish and dangerous. Choice D is incorrect; heat rises and will quickly spread to the floor above. Also, the stairs are already on fire and there would be little chance of moving up.
52. **D** The question of suicide would be a factor in determining the cause of the accident, but not in deciding what first aid to apply. Choices A, B, and C all relate to determining possible first-aid needs.
53. **A** The image in the mirror is deceptive and can cause the fire fighter to lose valuable time and water fighting fire where it doesn't exist. This of course would allow the real fire to grow and weaken the building. Choices B and C are possible but do not apply to all mirrors. Note in choice A the use of the word "may," while in choices B and C the verb is "are." Choice D is a false statement.
54. **C** As was mentioned early in this book, fire fighting is a group effort. If a fire fighter cannot physically

perform the tasks required of him, the probability of another firefighter getting hurt is greatly increased and the probability of a successful fire fighting operation is reduced.

55. **B** This question asks you to select the "most probable" answer; knowing the exact reason is not necessary. For each selection, ask, "Is this probable?" If the answer is yes, then ask, "Is it more or less probable than the preceding choice?" Then select the best choice—B, in this case.

56. **D** The purposes of fire-prevention inspection are to reduce the chance that fire will occur and to ensure, if it does, that the occupants can escape safely. Choices A, B, and C clearly relate to these goals. Choice D does not.

57. **B** The fires described in A, C, and D all have many possible accidental causes. There are, however, an extremely limited number of accidental ways of starting a fire in a mailbox. Since the question involves a fire that was started on purpose, the mailbox fire (B) is the best choice.

58. **C** In this position the wind will blow the smoke away from the firefighter and provide fresh air and increased visibility.

59. **D** Because the weather is hotter in the summer, and because a firefighter's hands sweat more, firefighters tend to remove their gloves more often at that time of year. As a result the hands have no protective covering and the opportunity for hand injuries is increased.

60. **B** By getting close, the firefighters can put the water directly on the burning materials and rapidly extinguish the fire, leading to quick control and less damage. Choice A—this is done after the fire is extinguished. Choice C—it takes more, not less, hose to get to the site of the fire. Choice D—firefighters must pass through the smoke to get close to the fire.

61. **A** An automobile is a low object, and a ladder would have very limited or no use. Choices B and D may be used to extinguish a fire or wash away spilled flammable liquids. Choice C, "stretcher," may be used to transport an injured person.

62. **C** The first action would be to ensure that the person was safe from the fire, dangerous smoke, and gases. This is accomplished by moving the woman. Choice D is in contradiction with the above and could increase the danger for the woman and the firefighter. Choices A and B should be done immediately after the removal of the victim.

63. **D** Smoke and heat rise. Therefore it is safer and cooler, and there is greater visibility, at the floor level.

64. **B** The objective of fire prevention inspections is the correction of fire hazards and unsafe practices. A person who is knowingly performing an illegal or improper activity often seeks to conceal this from firefighters performing inspections and then resumes the activity after the firefighters leave.

65. **D** Arsonists have demonstrated a consistent pattern of returning to the scene to view the fire. They are often most willing to help firefighters and to talk about the fire.

66. **C** Paint shops contain many flammable and combustible substances that can explode and spread the fire rapidly. The occupancies in choices A, B, and D represent relatively low hazards.

67. **D** Attending to the person's well-being and determining the cause of his injuries are most important. Choices B and C relate to the first of these concerns. Since this is not a fire incident, determination of the cause is a function of the police department. Therefore the firefighters should call the police (choice A), and they will question the residents.

68. **A** By talking with neighborhood leaders, mutual misunderstanding between the firefighters and the community can be resolved. The leaders can then convince their followers to cease the hostile actions. Choice B would lead to

larger fires, increased numbers of injuries and deaths, and greater civil unrest. Choice C would lead to greater hostility between the citizens and the fire fighters and could result in serious injury. Choice D would also lead to serious injuries and greater civil unrest.

69. **D** Fires require quick response. Clean equipment in good working order is essential for immediate, effective response to fires.

70. **A** When storekeepers understand the reasons behind the rules, they can better appreciate how the rules help to protect their business.

71. **A** Crossing the belts allows the wheels to transmit power in opposite directions. When the belt is straight, power goes in the same direction.

72. **D** When two sprocket gears mesh, a change in the direction of one produces an opposite change in the other. When you draw an arrow to indicate the direction, make sure the arrow goes at least halfway around the gear. This will give you a much clearer picture of what is happening.

73. **D** Pressure from fluid is independent of the shape of the container. It depends on the height of the fluid and the density of the fluid. In this question all the cans are the same height (3 ft.), and water would have the same density in each can.

74. **D** To determine the weight you must add all the weights attached to the scale. The rod weighs 6 lb., and each of the two weights attached to the rod weighs 10 lb.: 10 lb. + 10 lb. + 6 lb. = 26 lb.

75. **B** The picture at the left shows a high-speed drill bit, and the question asks for the tool with which this item would be used. The electric drill (B) uses a high-speed drill bit. Picture A is a heavy-duty staple gun. Picture C is a hand-held vibrating electric sander. Picture D is a caulking gun.

76. **B** The picture at the left shows a set of booster cables, also known as jumper cables. These are used to transfer electric energy from a wet cell battery (B). Picture A is an electric light bulb outlet. Picture C is a brush used for painting. Picture D is a U-trap used with plumbing fixtures.

77. **A** The picture at the left shows a crowbar, used for prying. It is most appropriately used for picture A, a wooden crate. B is a plumbing trap, C a wet cell battery, and D a table.

78. **B** The picture at the left shows a wood chisel, which would be used in conjunction with the mallet (B). Choice A is a pair of tin snips. Choice C is a pipe wrench. Choice D is a paint brush.

79. **D** The picture at the left shows a flat board and the tool most appropriate is the hand plane (D). Picture A is a pipe vise. Picture B is a power grinding machine. Picture C is an acetylene torch.

80. **B** The picture at the left shows a brick wall, and the most closely related tool is the hand trowel (B). Choice A is a hand file. Choice C is a claw hammer. Choice D is a scratch awl.

81. **D** Fire fighters must set priorities. Explaining to the woman the time sequence of the response to the fire is important but must be postponed until the fire has been extinguished. After the fire is under control, a fire officer should tell the woman the actual response time and explain why she perceived the passing of time so differently.

82. **B** Ensuring that fire hazards are removed and fire safety laws are obeyed is the primary task of every fire fighter while on fire-prevention inspection.

83. **B** By inspecting the restaurant at a time when it is very busy you will be viewing the occupancy under actual operating conditions. The possibility of hazardous conditions occurring is greater as the restaurant becomes very busy. By inspecting at this time you may discover such hazards and correct them before a serious injury occurs.

84. D Firefighters are responsible for the safety of the occupants and their possessions. To allow an occupant into the building before it is safe would be an improper action. By telling the couple that they will be allowed back into the building when it is safe, the firefighter fulfills his responsibilities and lets them know what to expect.

85. B Once notified, the superior officer can take appropriate action. The officer's actions could include notifying the police department or responding with the unit.

86. C It is important to respond to citizens' complaints; however, control and extinguishment of the fire must be accomplished first. Directing the woman to inform the superior after the fire is under control puts the problem into the proper perspective.

87. D This is a human relations problem, in which the rights, positions, and financial conditions of all persons must be respected. However, if the group of 20 can afford to purchase the TV it has the right to do so. The other two firefighters have expressed an intent to pay and should be allowed more time to do so.

88. A Compliance with the law is mandatory. Firefighters do not have the authority to grant variances. By explaining the concepts behind the law and the dangers presented by the lock and key arrangement, the firefighter should be able to gain the owner's understanding and cooperation.

89. C This is necessary because of the danger involved in a fireworks display. Should something go wrong, the firefighters and the apparatus must be ready to go into operation immediately. However, when telling the children to get off the truck, the firefighters should invite them to come to the fire station to visit and to look at the fire equipment.

90. C During a fire, compliance with the orders and directions of a fire officer is mandatory. While the fire is in progress is not the time to debate operational procedures. However, after the fire is under control, the firefighter would be right to discuss the order with the officer and learn the reasons for his choice of hose.

91. A This is a common occurrence, and the firefighter in choice A did the right thing. The officer is better prepared to explain the need for effective fire protection in the community. Choices B, C, and D are antagonistic and disrespectful to citizens who feel that they have a legitimate complaint.

92. C Permitting a person who is obviously intoxicated to drive a car is a form of moral negligence. By instructing the person to take a cab, you may very well have saved his life.

93. C Teaching fire safety is one of the many roles of a firefighter.

94. C Often people misinterpret or do not completely understand the role of the fire service. By finding out why the manager is so hostile, the firefighter will be better equipped to explain why the laws and the fire service are important.

95. D Accusations of theft and other wrongdoing should be directed to superiors for investigation. In this case we know from the question that the accusation is false, and the superior officer would have the duty and responsibility of proving the firefighter's innocence.

96. C This is a common problem. After the fire has been extinguished and the smoke has lifted, there may appear to have been no need for the hole. However, during the fire, conditions warranted chopping a hole in the roof to vent the smoke and heat. Explaining to the store owner that this action reduces the *overall* damage will often decrease his anger.

97. A Talking to school children is an important and pleasant part of a firefighter's duties. By offering to set up another visit, the firefighter shows a willingness to answer additional questions at another time.

98. D When the firefighters explain the need to do their work, the children understand the problem and do not feel offended by the request to leave.

99. **C** There is a time to tell a supervisor and a time to approach a fellow firefighter directly. The general rule is as follows: If the improper act is immediately dangerous, you should let the person know about it without going through the chain of command. If the action is an infraction that is not immediately dangerous, you generally should notify the firefighter's supervisor.

100. **B** Preparing meals and cleaning up after them are voluntary activities in fire stations. Within a group individuals must be allowed to contribute according to their own strengths and capabilities. A firefighter who cannot cook but is willing to clean up should do that.

CHAPTER 13

Practice Examination Two

Official Examination
City of New York

In this chapter you will find the second official examination that you should take for practice.

Be sure that when you take this examination you allow yourself a 3½ hour uninterrupted time period. Do not try to take the examination in parts; if you do, you will defeat its purpose. By taking the examination at one sitting you will become familiar with the test conditions and with your personal idiosyncrasies about sitting and working for this period of time.

Before Taking The Examination

First, go back and review quickly the test-taking strategies outlined in Chapter 3. Then, when you begin the examination, be sure to read and follow all instructions. Time for reading the instructions has been figured into the examination. Read each question carefully, and answer only what is asked of you. Select the answer that is the one *best* choice of those provided, and then record your selection on the answer sheet.

The answer sheet precedes the examination. The Diagnotic Procedure and Answer Explanations are given at the end of the chapter.

ANSWER SHEET
PRACTICE EXAMINATION TWO

Follow the instructions given in the test. Mark only your answers in the ovals below.

WARNING: Be sure that the oval you fill is in the same row as the question you are answering. Use a No. 2 pencil (soft pencil).

BE SURE YOUR PENCIL MARKS ARE HEAVY AND BLACK. ERASE COMPLETELY ANY ANSWER YOU WISH TO CHANGE.

START HERE DO NOT make stray pencil dots, dashes or marks.

1 Ⓐ Ⓑ Ⓒ Ⓓ	2 Ⓐ Ⓑ Ⓒ Ⓓ	3 Ⓐ Ⓑ Ⓒ Ⓓ
4 Ⓐ Ⓑ Ⓒ Ⓓ	5 Ⓐ Ⓑ Ⓒ Ⓓ	6 Ⓐ Ⓑ Ⓒ Ⓓ
7 Ⓐ Ⓑ Ⓒ Ⓓ	8 Ⓐ Ⓑ Ⓒ Ⓓ	9 Ⓐ Ⓑ Ⓒ Ⓓ
10 Ⓐ Ⓑ Ⓒ Ⓓ	11 Ⓐ Ⓑ Ⓒ Ⓓ	12 Ⓐ Ⓑ Ⓒ Ⓓ
13 Ⓐ Ⓑ Ⓒ Ⓓ	14 Ⓐ Ⓑ Ⓒ Ⓓ	15 Ⓐ Ⓑ Ⓒ Ⓓ
16 Ⓐ Ⓑ Ⓒ Ⓓ	17 Ⓐ Ⓑ Ⓒ Ⓓ	18 Ⓐ Ⓑ Ⓒ Ⓓ
19 Ⓐ Ⓑ Ⓒ Ⓓ	20 Ⓐ Ⓑ Ⓒ Ⓓ	21 Ⓐ Ⓑ Ⓒ Ⓓ
22 Ⓐ Ⓑ Ⓒ Ⓓ	23 Ⓐ Ⓑ Ⓒ Ⓓ	24 Ⓐ Ⓑ Ⓒ Ⓓ
25 Ⓐ Ⓑ Ⓒ Ⓓ	26 Ⓐ Ⓑ Ⓒ Ⓓ	27 Ⓐ Ⓑ Ⓒ Ⓓ
28 Ⓐ Ⓑ Ⓒ Ⓓ	29 Ⓐ Ⓑ Ⓒ Ⓓ	30 Ⓐ Ⓑ Ⓒ Ⓓ
31 Ⓐ Ⓑ Ⓒ Ⓓ	32 Ⓐ Ⓑ Ⓒ Ⓓ	33 Ⓐ Ⓑ Ⓒ Ⓓ
34 Ⓐ Ⓑ Ⓒ Ⓓ	35 Ⓐ Ⓑ Ⓒ Ⓓ	36 Ⓐ Ⓑ Ⓒ Ⓓ
37 Ⓐ Ⓑ Ⓒ Ⓓ	38 Ⓐ Ⓑ Ⓒ Ⓓ	39 Ⓐ Ⓑ Ⓒ Ⓓ
40 Ⓐ Ⓑ Ⓒ Ⓓ	41 Ⓐ Ⓑ Ⓒ Ⓓ	42 Ⓐ Ⓑ Ⓒ Ⓓ
43 Ⓐ Ⓑ Ⓒ Ⓓ	44 Ⓐ Ⓑ Ⓒ Ⓓ	45 Ⓐ Ⓑ Ⓒ Ⓓ
46 Ⓐ Ⓑ Ⓒ Ⓓ	47 Ⓐ Ⓑ Ⓒ Ⓓ	48 Ⓐ Ⓑ Ⓒ Ⓓ
49 Ⓐ Ⓑ Ⓒ Ⓓ	50 Ⓐ Ⓑ Ⓒ Ⓓ	51 Ⓐ Ⓑ Ⓒ Ⓓ
52 Ⓐ Ⓑ Ⓒ Ⓓ	53 Ⓐ Ⓑ Ⓒ Ⓓ	54 Ⓐ Ⓑ Ⓒ Ⓓ
55 Ⓐ Ⓑ Ⓒ Ⓓ	56 Ⓐ Ⓑ Ⓒ Ⓓ	57 Ⓐ Ⓑ Ⓒ Ⓓ
58 Ⓐ Ⓑ Ⓒ Ⓓ	59 Ⓐ Ⓑ Ⓒ Ⓓ	60 Ⓐ Ⓑ Ⓒ Ⓓ
61 Ⓐ Ⓑ Ⓒ Ⓓ	62 Ⓐ Ⓑ Ⓒ Ⓓ	63 Ⓐ Ⓑ Ⓒ Ⓓ
64 Ⓐ Ⓑ Ⓒ Ⓓ	65 Ⓐ Ⓑ Ⓒ Ⓓ	66 Ⓐ Ⓑ Ⓒ Ⓓ
67 Ⓐ Ⓑ Ⓒ Ⓓ	68 Ⓐ Ⓑ Ⓒ Ⓓ	69 Ⓐ Ⓑ Ⓒ Ⓓ
70 Ⓐ Ⓑ Ⓒ Ⓓ	71 Ⓐ Ⓑ Ⓒ Ⓓ	72 Ⓐ Ⓑ Ⓒ Ⓓ
73 Ⓐ Ⓑ Ⓒ Ⓓ	74 Ⓐ Ⓑ Ⓒ Ⓓ	75 Ⓐ Ⓑ Ⓒ Ⓓ
76 Ⓐ Ⓑ Ⓒ Ⓓ	77 Ⓐ Ⓑ Ⓒ Ⓓ	78 Ⓐ Ⓑ Ⓒ Ⓓ
79 Ⓐ Ⓑ Ⓒ Ⓓ	80 Ⓐ Ⓑ Ⓒ Ⓓ	81 Ⓐ Ⓑ Ⓒ Ⓓ
82 Ⓐ Ⓑ Ⓒ Ⓓ	83 Ⓐ Ⓑ Ⓒ Ⓓ	84 Ⓐ Ⓑ Ⓒ Ⓓ
85 Ⓐ Ⓑ Ⓒ Ⓓ	86 Ⓐ Ⓑ Ⓒ Ⓓ	87 Ⓐ Ⓑ Ⓒ Ⓓ
88 Ⓐ Ⓑ Ⓒ Ⓓ	89 Ⓐ Ⓑ Ⓒ Ⓓ	90 Ⓐ Ⓑ Ⓒ Ⓓ
91 Ⓐ Ⓑ Ⓒ Ⓓ	92 Ⓐ Ⓑ Ⓒ Ⓓ	93 Ⓐ Ⓑ Ⓒ Ⓓ
94 Ⓐ Ⓑ Ⓒ Ⓓ	95 Ⓐ Ⓑ Ⓒ Ⓓ	96 Ⓐ Ⓑ Ⓒ Ⓓ
97 Ⓐ Ⓑ Ⓒ Ⓓ	98 Ⓐ Ⓑ Ⓒ Ⓓ	99 Ⓐ Ⓑ Ⓒ Ⓓ
100 Ⓐ Ⓑ Ⓒ Ⓓ		

City of New York Department of Personnel Examination No #1162 Firefighter

Firefighters must be able to find their way in and out of buildings that are filled with smoke. They must learn the floor plan quickly for their own safety and to help fight the fire and remove victims.

Look at this floor plan of an apartment. There is an apartment on each side of this one. It is on the *fifth floor* of the building.

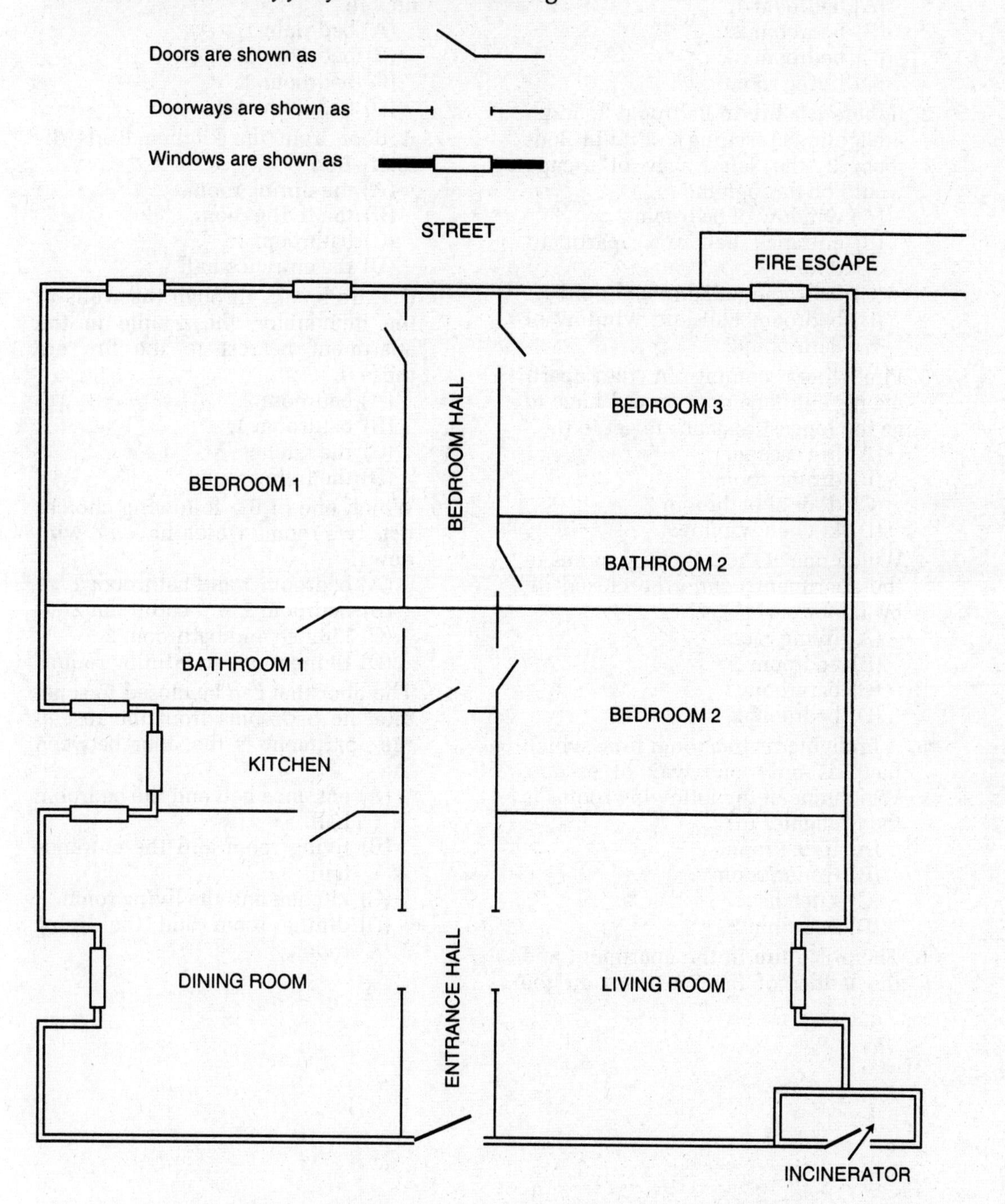

CLOSE THIS BOOKLET WHEN THE THIRD BELL RINGS

DO NOT WRITE UNTIL THE FOURTH BELL RINGS

*The New York City Department of Personnel makes no commitment, and no implication is to be drawn, as to the content, style, or format of any future examination for the position of Firefighter.

Answer questions *1* through *10* solely on the basis of the Floor Plan.

1. A person escaping a fire in the apartment can get on the fire escape by going through the window of
   (A) bedroom 1.
   (B) bedroom 2.
   (C) bedroom 3.
   (D) living room.
2. If there is a fire in Bedroom 3, and a firefighter is rescuing a child in Bedroom 2, the safest way of escape would be through the
   (A) window of bedroom 2.
   (B) entrance hall and apartment door.
   (C) bedroom hall and bedroom 1.
   (D) bedroom hall and window of bathroom 2.
3. Firefighters coming in the apartment's entrance door would have to go the longest distance to get to the
   (A) fire escape.
   (B) dining room.
   (C) door of bathroom 2.
   (D) kitchen window.
4. Which one of the following rooms in the apartment *cannot* be closed off by a door?
   (A) living room
   (B) bedroom 1
   (C) bathroom 1
   (D) bedroom 2
5. A firefighter is in a room from which there is only one way of escape. Which one of the following rooms is the firefighter in?
   (A) living room
   (B) dining room
   (C) kitchen
   (D) bedroom 2
6. There is a fire in the apartment and the ladder of the fire truck in the street cannot be placed against the fire escape. The ladder should be raised from the street to reach a window in
   (A) bedroom 1.
   (B) bedroom 2.
   (C) bedroom 3.
   (D) bedroom hall.
7. A door from the kitchen leads directly into
   (A) the dining room.
   (B) the living room.
   (C) bathroom 1.
   (D) the entrance hall.
8. If a fire breaks through the walls of the incinerator, the people in the apartment nearest to the fire are those in
   (A) bedroom 2.
   (B) bathroom 1.
   (C) the kitchen.
   (D) the living room.
9. Which one of the following choices lists two rooms which have *no* windows?
   (A) bedroom 1 and bathroom 1
   (B) bedroom 2 and bathroom 2
   (C) kitchen and bathroom 2
   (D) living room and dining room
10. The door that can be closed to separate the bedrooms from the rest of the apartment is the door between the
    (A) entrance hall and the bedroom hall.
    (B) living room and the entrance hall.
    (C) kitchen and the living room.
    (D) dining room and the living room.

Questions *11* through *13* should be answered solely on the basis of the passage below.

Automatic sprinkler systems are installed in many buildings. They extinguish or keep from spreading 96 percent of all fires in areas they protect. Sprinkler systems are made up of pipes which hang below the ceiling of each protected area and sprinkler heads which are placed along the pipes. The pipes are usually filled with water and each sprinkler head has a heat sensitive part. When the heat from the fire reaches the sensitive part of the sprinkler head, the head opens and showers water upon the fire in the form of spray. The heads are spaced so that the fire is covered by overlapping showers of water from the open heads.

11. Automatic sprinkler systems are installed in buildings to
    (A) prevent the build up of dangerous gases.
    (B) eliminate the need for fire insurance.
    (C) extinguish fires or keep them from spreading.
    (D) protect 96 percent of the floor space.

12. If more than one sprinkler head opens, the area sprayed will be
    (A) flooded with hot water.
    (B) overlapped by showers of water.
    (C) subject to less water damage.
    (D) about 1 foot per sprinkler head.

13. A sprinkler head will open and shower water when
    (A) it is reached by heat from a fire.
    (B) water pressure in the pipes gets too high.
    (C) it is reached by sounds from a fire alarm.
    (D) water temperature in the pipes gets too low.

<table>
<tr><td colspan="4">BUILDING INSPECTION FORM</td></tr>
<tr><td colspan="4">DIVISION (1) | BATTALION (2) | COMPANY (3)</td></tr>
<tr><td rowspan="2">BUILDING INFORMATION</td><td colspan="2">Name of Business
(4)</td><td>Address
(5)</td></tr>
<tr><td colspan="2">Type of Business
(6)</td><td>Occupancy Code Number
(7)</td></tr>
<tr><td rowspan="2">CONDITION OF EXITS</td><td>Number of Exits (8)</td><td>Exits Obstructed
(9)</td><td>Exits Unlocked
(10)</td></tr>
<tr><td>Exit Signs
(11)</td><td>Exit Sign Lights
(12)</td><td>Fire Doors
(13)</td></tr>
<tr><td rowspan="3">HOUSEKEEPING CONDITIONS</td><td colspan="2">Rubbish Receptacles
(14)</td><td>No Smoking Signs
(15)</td></tr>
<tr><td colspan="3">Clearance of Stock in Feet from Sprinkler Heads
(16)</td></tr>
<tr><td>Electrical Wiring
(17)</td><td>Switches
(18)</td><td>Junction Box
(19)</td></tr>
<tr><td>CONDITION OF FIRE EXTINGUISHERS</td><td>Charged
(20)</td><td>Placement
(21)</td><td>Date of Last Inspection
(22)</td></tr>
<tr><td rowspan="2">CONDITION OF AUTOMATIC SPRINKLER SYSTEM</td><td>Color of Siamese
(23)</td><td>Main Control Valve
(24)</td><td>Shut-off Sign
(25)</td></tr>
<tr><td colspan="2">Certificate of Fitness
(26)</td><td>Date of Last Inspection
(27)</td></tr>
<tr><td rowspan="2">SPECIAL CONDITIONS</td><td colspan="2" rowspan="2">Rubbish/Obstructions
(28)</td><td>Certificate of Occupancy (29)</td></tr>
<tr><td>Heavy Load Signs
(30)</td></tr>
<tr><td rowspan="2">FIRE DEPARTMENT INFORMATION</td><td colspan="2">Inspector
Name ________ (31)
Signature ________</td><td>Rank (32) | Date (33)</td></tr>
<tr><td colspan="2">Officer
Name ________ (34)
Signature ________</td><td>Rank (35) | Date (36)</td></tr>
</table>

Answer questions *14* through *28* solely on the basis of the following facts and the Building Inspection Form. Each box on the form is numbered. Read the facts and review the form before answering the questions.

Firefighters are required to inspect all buildings within their assigned area of the city. They check conditions within the building for violations of fire safety laws. While inspecting a building, they must fill out a Building Inspection Form as a record of the conditions they observed.

On June 12, 1982, Firefighter Edward Gold, assigned to Engine Company 82, is ordered by Captain John Bailey to inspect the building at 1400 Compton Place as part of the engine company's monthly building inspection duty. The building is a one-story brick warehouse where books of the S & G Publishing Company are stored before shipment to stores.

Firefighter Gold enters the warehouse through the main entrance door in the front of the building. Though an exit sign is present above the door, the sign is unlit because of a burned out bulb. There is a small office to one side of the main entrance area where Firefighter Gold goes to meet the warehouse manager, Mr. Stevens. The firefighter explains the purpose of the inspection.

Firefighter Gold tells the manager that he will check the automatic sprinkler system first because, if a fire got started in a warehouse full of stored books, the fire could spread rapidly. He asks Mr. Stevens for the Certificate of Fitness issued to the company employee certified to maintain the sprinkler system in working order. The Certificate is dated June 1, 1979, and Gold observes that it has expired. The manager promises to have the Certificate renewed as soon as possible.

The firefighter wants to locate the main control valve of the sprinkler system. He asks Mr. Stevens to go with him and show him its location. Gold and the manager leave through an office door which leads into the main working area of the warehouse. They locate the main sprinkler control valve on the wall in a corner of the work area behind high shelves stocked with books. The firefighter observes that the main control valve is sealed in the open position. Gold next climbs a ladder lying against the storage shelves and measures the distance between the top of the stack of books on the highest shelf and the sprinkler heads suspended on pipes below the ceiling. The distances is three feet.

Firefighter Gold next inspects the remaining exits from the building. A large fire door leads out to the loading dock in the rear of the warehouse. A small door on the side of the warehouse that is used by employees when they leave for the day is partially obstructed by cartons. Lighted exit signs can be clearly seen above both doors. During working hours, only the main entrance door and the fire door to the loading dock are unlocked. Mr. Stevens says he keeps the side door locked to keep employees from leaving early and only unlocks it at closing time.

Firefighter Gold and the manager then walk through the main work area. Gold observes that fireproof rubbish receptacles are placed at frequent intervals. However, they are not covered and the contents are overflowing, resulting in several piles of litter on the floor. "No Smoking" signs are on the walls of the work area, but are difficult to see behind the rows of high storage shelves.

The two fire extinguishers in the work area are found lying on the floor rather than hung on wall racks. The two other fire extinguishers in the warehouse, one in the office and one in the employee lounge, are both correctly hung on wall racks.

All four fire extinguishers are fully charged. According to their tags, they were last inspected on March 11, 1982.

Firefighter Gold continues the inspection by checking on the electrical wiring which appears to be generally in good condition. However, four switch boxes lack covers. The main junction box has a cover, but it cannot be closed because the cover is corroded.

The inspection is now complete, so Firefighter Gold thanks Mr. Stevens for his cooperation and leaves the building. Gold checks that all required information is entered on the Building Inspection Form, including information concerning building violations. Firefighter Gold signs and dates the Building Inspection Form and then submits it to Captain Bailey for his review. After reviewing Firefighter Gold's report, Captain Bailey signs the Building Inspection Form.

14. Which one of the following should be entered in Box 3?
    (A) Ladder Company 79
    (B) Engine Company 12
    (C) Ladder Company 140
    (C) Engine Company 82
15. Which one of the following should be entered in Box 4?
    (A) G & R Printing Company
    (B) S & G Printing Company
    (C) R & G Publishing Company
    (D) S & G Publishing Company
16. Which one of the following should be entered in Box 8?
    (A) 2
    (B) 3
    (C) 4
    (D) 5
17. Which one of the following should be entered in Box 9?
    (A) office door
    (B) side door
    (C) main door
    (D) fire door
18. Which one of the following should be entered in Box 10?
    (A) fire door and main door
    (B) side door and office door
    (C) fire door and side door
    (D) main door and cellar door
19. The entry in Box 12 should show that replacement bulbs are needed for
    (A) one light.
    (B) two lights.
    (C) three lights.
    (D) all lights.
20. The entry in Box 14 should show that covers are missing from
    (A) two of the rubbish receptacles.
    (B) three of the rubbish receptacles.
    (C) four of the rubbish receptacles.
    (D) all of the rubbish receptacles.
21. Which one of the following should be entered in Box 16?
    (A) one and one-half feet
    (B) two feet
    (C) two and one-half feet
    (D) three feet
22. Which one of the following should be entered in Box 19?
    (A) faulty circuits
    (B) exposed wiring
    (C) corroded cover
    (D) good condition
23. Which one of the following entries about the placement of fire extinguishers should appear in Box 21?
    (A) one on the floor, three hung on wall racks
    (B) two on the floor, two hung on wall racks
    (C) three on the floor, one hung on wall rack
    (D) four hung on wall racks
24. Which one of the following should be entered in Box 22?
    (A) June 1, 1979
    (B) May 21, 1981
    (C) March 11, 1982
    (D) May 1, 1982
25. The entry in Box 24 should show that the position of the main control valve is
    (A) open.
    (B) half-open.
    (C) one-third closed.
    (D) closed.
26. Which one of the following should be entered in Box 26?
    (A) expired
    (B) missing from file
    (C) never issued
    (D) current
27. Which one of the following should be entered in Box 28?
    (A) ceiling plaster cracked
    (B) rubbish piles litter work floor
    (C) second floor stairway blocked
    (D) open paint cans on loading dock
28. Which one of the following should be entered in Box 34?
    (A) John Bailey
    (B) Edward Gold
    (C) John Gold
    (D) Edward Bailey

Questions *29* through *32* should be answered solely on the basis of the passage below.

About 48 percent of all reported fires are false alarms. False alarms add more risk of danger to firefighters, citizens, and property, as well as waste the money and time of the fire department. When the first firefighters are called to a reported fire, they do not know if the alarm is for a real fire or is a false alarm. Until they have made sure that the alarm is false, they must not respond to a new alarm even if a real fire is burning and people's lives and property are in danger. If they do not find a fire or an emergency at the original location, then the firefighters radio the fire department that they have been called to a false alarm. The fire department radios back and tells the firefighters that they are in active service again and tells them where to respond for the next alarm. If that location is far from that of the false alarm, then the distance and the time it takes to get to the new location are increased. This means that firefighters will arrive later to help in fighting the real fire and the fire will have more time to burn. The fire will be bigger and more dangerous just because someone called the firefighters to a false alarm. In addition, each time the firefighters ride to the location of a false alarm, there is additional risk of unnecessary accidents and injuries to them and to citizens.

29. The main point of the passage is that false alarms
    (A) seldom interrupt other activities in the firehouse.
    (B) occur more often during the winter.
    (C) are rarely turned in by children.
    (D) add more risk of danger to life and property.

30. When firefighters are called to a false alarm, they must not respond to other alarms until they
    (A) turn in a written report to the fire department.
    (B) take a vote and all agree to go.
    (C) are put back into active service by the fire department.
    (D) decide on the quickest route.

31. Before firefighters get to the location of a reported fire, they
    (A) finish eating their lunch at the firehouse.
    (B) do not know if the alarm is real or false.
    (C) search the neighborhood for the person who made the report.
    (D) do not know if the alarm is from an alarm box or telephone.

32. The passage states that false alarms
    (A) shorten travel time to real fires.
    (B) give firefighters needed driving practice.
    (C) save money on fuel for the fire department.
    (D) account for about 48 percent of reported fires.

33. An elderly man staggers into the firehouse and tells the firefighters on duty that he is having trouble breathing. Of the following, it would be best for the firefighters to
    (A) send the elderly man away as his staggering shows that he has been drinking too much.
    (B) place the elderly man in a chair and quickly call for assistance.
    (C) tell the elderly man to go to the hospital and see a doctor.
    (D) help the elderly man leave the firehouse as this is not a problem that firefighters should handle.

34. As firefighters travel to and from their firehouse they usually look around the neighborhood in order to spot dangerous conditions. If they spot a dangerous condition, firefighters will take action to correct it. They do this because they want to prevent fires. While on the way to work overtime at a nearby firehouse, a firefighter passes a local gas station and spots a leaking gasoline pump. Which one of the following is the most appropriate course of action for the firefighter to take?
    (A) Stop at the gas station and make sure that the leak is actually gasoline by lighting a match to it.
    (B) Continue on to work because the gas station attendant will take care of the leak.

(C) Stop at the gas station and tell the gas station attendant to make sure the leak is repaired.
(D) Call the Mayor's Office to complain that the leaking gasoline is polluting the area.

35. A newly appointed firefighter is assigned to go with an experienced firefighter to inspect a paint store. The paint store owner refuses to allow the inspection, saying that she is closing the store early that day and going on vacation. The new firefighter demands rudely that the inspection be allowed, even though it would be permissible to delay it. Of the following, it would be best for the experienced firefighter to
(A) repeat the demand that the inspection be allowed and quote the law to the store owner.
(B) tell the new firefighter that it would be best to schedule the inspection after the store owner's vacation.
(C) tell the store owner to step aside, and instruct the new firefighter to enter the store and begin the inspection.
(D) tell the new firefighter to forget about the inspection because the store owner is uncooperative.

Questions *36* through *43* are to be answered by referring to the items pictured below. The sizes of the items shown are *not* their actual sizes. Each item is identified by a number. For each question, select the answer which gives the identifying number of the item that best answers the question.

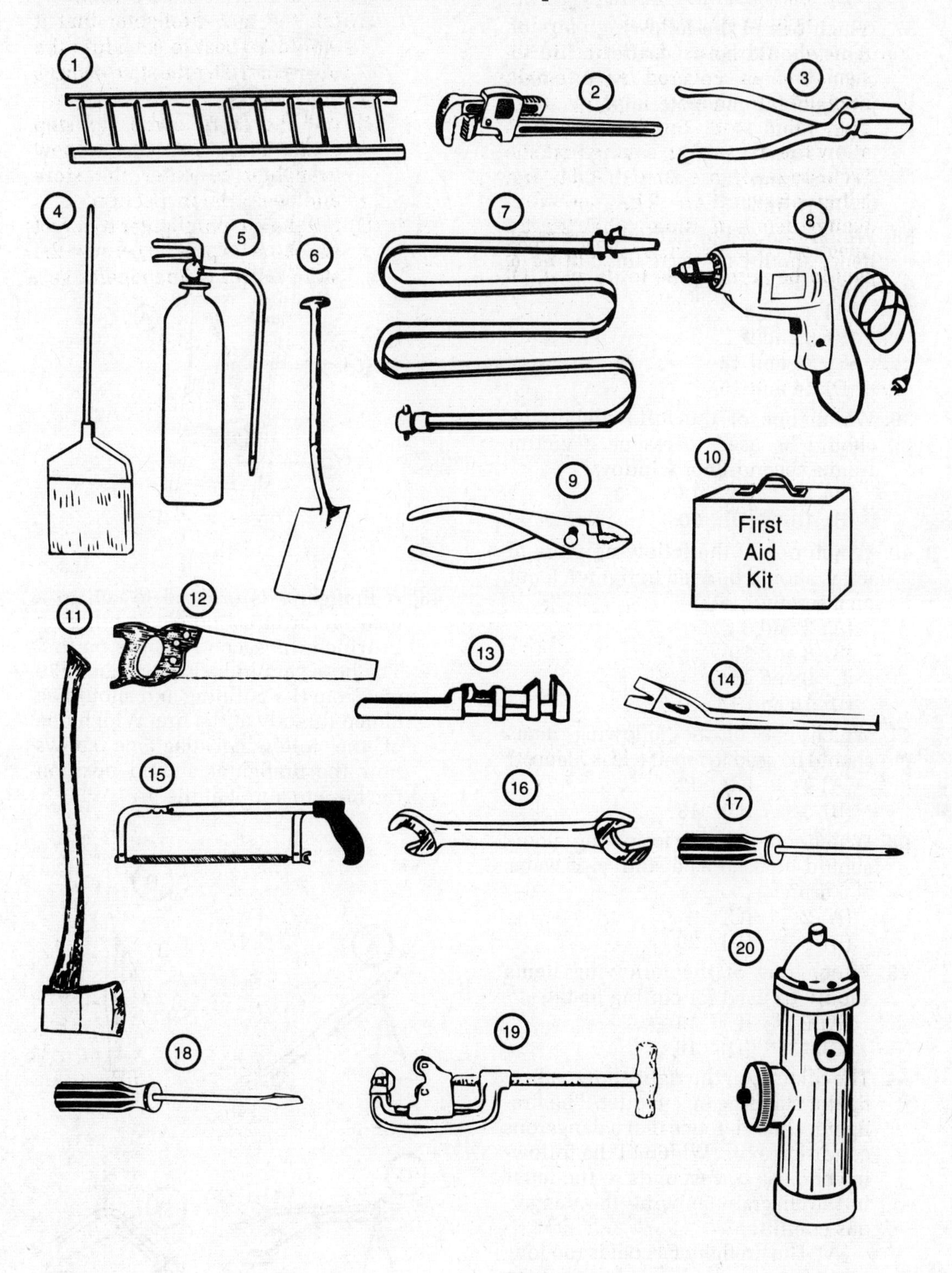

36. Which one of the following items should be connected to a hydrant and used to put out a fire?
    (A) 5 (C) 8
    (B) 7 (D) 17
37. Which one of the following pairs of items should be used after a fire to clean a floor covered with small pieces of burned materials?
    (A) 1 and 14
    (B) 4 and 6
    (C) 10 and 12
    (D) 11 and 13
38. Which one of the following pairs of items should be used for cutting a branch from a tree?
    (A) 2 and 3
    (B) 8 and 9
    (C) 11 and 12
    (D) 14 and 15
39. Which one of the following items should be used to rescue a victim from a second-floor window?
    (A) 1 (C) 15
    (B) 10 (D) 20
40. Which one of the following pairs of items should be used to tighten a nut on a screw?
    (A) 2 and 3
    (B) 8 and 19
    (C) 9 and 14
    (D) 16 and 18
41. Which one of the following items should be used to repair a leaky faucet?
    (A) 4 (C) 12
    (B) 5 (D) 13
42. Which one of the following items should be used as a source of water at a fire?
    (A) 2 (C) 9
    (B) 6 (D) 20
43. Which one of the following items should be used for cutting metal?
    (A) 6 (C) 15
    (B) 13 (D) 18
44. The picture on the right shows a firefighter standing on a ladder. The firefighter should notice that a dangerous condition exits. Which of the following choices corresponds to the letter in the diagram showing the dangerous condition?
    (A) The firefighter's coat is too long for safe climbing of the ladder.
    (B) A helmet keeps the firefighter from seeing what is going on.
    (C) The firefighter's feet are on the ladder rung.
    (D) The ladder rung is missing.

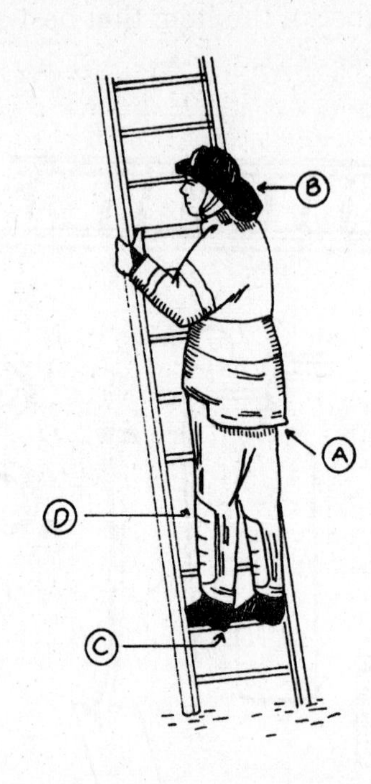

45. A firefighter is ordered to set up a hose on the street outside a building in which the second floor is on fire. The hose should be located about 30 feet from the building and should be aimed directly at the fire. Which one of the following diagrams shows how the firefighter should position the hose to aim it at the fire?

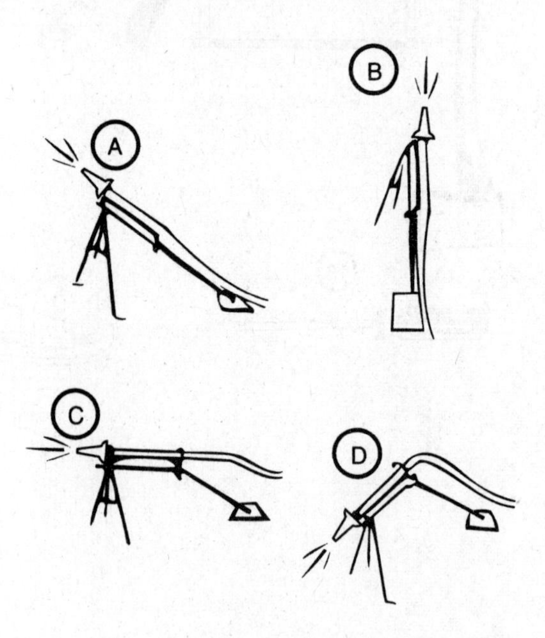

Questions *46* through *52* are to be answered solely on the basis of the following information.

The portable power saw lets the firefighter cut through various materials so that a fire can be reached. It can be dangerous, however, if it is not properly used or if it has not been inspected and tested to insure that it is in serviceable condition. The parts of the saw should be clean and free of foreign material, especially the exhaust port and spark arrestor, the carburetor enclosure, the cooling fins, the spark plugs, and the V-belt pulley if the saw has one.

The saw should be checked to make sure it has both air and fuel filters. It should never be run without an air filter. The V-belt pulley, if present, must be checked to make sure it is not too tight or too loose. If too loose, it could cause slipping. If too tight, the blade might turn when the engine idles, there might be damage to the clutch bearing, or the motor might stall when the blade is stopped. All nuts, bolts, and screws should be checked for tightness.

The saw may use carbide-tipped blades, aluminum oxide blades, or silicon carbide blades. Carbide-tipped blades should be returned for replacement when two or more tips are broken or missing or when the tips are worn down to the circumference of the blade. Aluminum oxide and silicon carbide blades should be replaced when they are cracked, badly nicked, or when worn down to an eight-inch diameter or less.

46. The principal reason for inspecting power saws is to make sure that
(A) they are clean.
(B) they are in serviceable condition.
(C) the pulley is not too tight or too loose.
(D) the blades are replaced.

47. What does the passage mean when it says the saw should be kept free of foreign material?
(A) Only American-made parts should be used.
(B) The saw should not be used on material that might damage it.
(C) Both air and fuel filters should be used.
(D) Anything that does not belong on the saw or in it should be removed.

48. Some saws are made to work *without* which one of the following items?
(A) an air filter
(B) a fuel filter
(C) a V-belt pulley
(D) blades

49. If the V-belt pulley on a power saw is too loose, it is most likely to cause
(A) the blade to turn when the engine idles.
(B) damage to the clutch bearing.
(C) the motor to stall when the blade is stopped.
(D) slipping.

50. The passage says that a power saw should *never* be run without
(A) an air filter.
(B) a fuel filter.
(C) a V-belt pulley.
(D) a blade.

51. Which of the following blades should be replaced when two or more tips are missing?
(A) Both aluminum oxide and carbide-tipped blades.
(B) Carbide-tipped blades only.
(C) Both silicon carbide and aluminum oxide blades.
(D) Silicon carbide blades only.

52. Which of the following blades should be replaced when worn down to an eight-inch diameter or less?
(A) Both aluminum oxide and carbide-tipped blades.
(B) Carbide-tipped blades only.
(C) Both silicon carbide and aluminum oxide blades.
(D) Silicon carbide blades only.

53. Fire engines use diesel motors to make them run. Diesel motors have devices called air cleaners which keep dirt from the inside of the motor. To make sure that the air cleaners are cleaned or replaced when necessary, an indicator on the fire engine will display a red color if the air cleaner has become too dirty. Each time the lubricating oil in the motor is changed,

or whenever the indicator shows red, the air cleaners must be inspected and cleaned or replaced.

Of the following, the most accurate statement concerning air cleaners on fire engine diesel motors is that they should be

(A) cleaned every day.
(B) replaced only when the oil is changed.
(C) inspected and cleaned only when the oil is changed.
(D) inspected and cleaned or replaced when the indicator shows red.

54. Firefighter Green must check the supply of air tank cylinders at the beginning of each tour of duty. There must be ten air tank cylinders always full of air and ready to be exchanged for used, empty air tank cylinders. At the start of a new tour of duty, Firefighter Green finds that out of twenty cylinders present, only five cylinders are full and ready to be exchanged. What is the minimum number of used empty cylinders that Firefighter Green must replace with full cylinders?

(A) 5
(B) 10
(C) 15
(D) 20

Questions *55* through *58* should be answered solely on the basis of the passage below.

Fires in vacant buildings are a major problem for firefighters. People enter vacant buildings to remove building material or they damage stairs, floors, doors and other parts of the building. The buildings are turned into dangerous structures with stairs missing, holes in the floors, weakened walls and loose bricks. Children and arsonists find large amounts of wood, paper and other combustible materials in the buildings and start fires which damage and weaken the buildings even more. Firefighters have been injured putting out fires in these buildings due to these dangerous conditions. Most injuries caused while putting out fires in vacant buildings could be eliminated if all of these buildings were repaired. All such injuries could be eliminated if the buildings were demolished. Until then, firefighters should take extra care while putting out fires in vacant buildings.

55. The problem of fires in vacant buildings could be solved by

(A) repairing buildings.
(B) closing up the cellar door and windows with bricks and cement.
(C) arresting suspicious persons before they start the fires.
(D) demolishing the buildings.

56. Firefighters are injured putting out fires in vacant buildings because

(A) there are no tenants to help fight the fires.
(B) conditions are dangerous in these buildings.
(C) they are not as careful when nobody lives in the buildings.
(D) the water in the buildings has been turned off.

57. Vacant buildings often have

(A) occupied buildings on either side of them.
(B) safe empty spaces where neighborhood children can play.
(C) combustible materials inside them.
(D) strong walls and floors that cannot burn.

58. While firefighters are putting out fires in vacant buildings, they should

(A) be extra careful of missing stairs.
(B) arrest the arsonists who start the fires.
(C) learn the reasons why the fires are set.
(D) help to repair the buildings.

Answer questions *59* through *63* solely on the basis of the following information and map.

A firefighter may be required to assist civilians who seek travel directions or referral to city agencies and facilities.

The following is a map of part of a city, where several public offices and other institutions are located. Each of the squares represents one city block. Street names are as shown. If there is an arrow next to the street name, it means the street is one way only in the direction of the arrow. If there is no arrow next to the street name, two-way traffic is allowed.

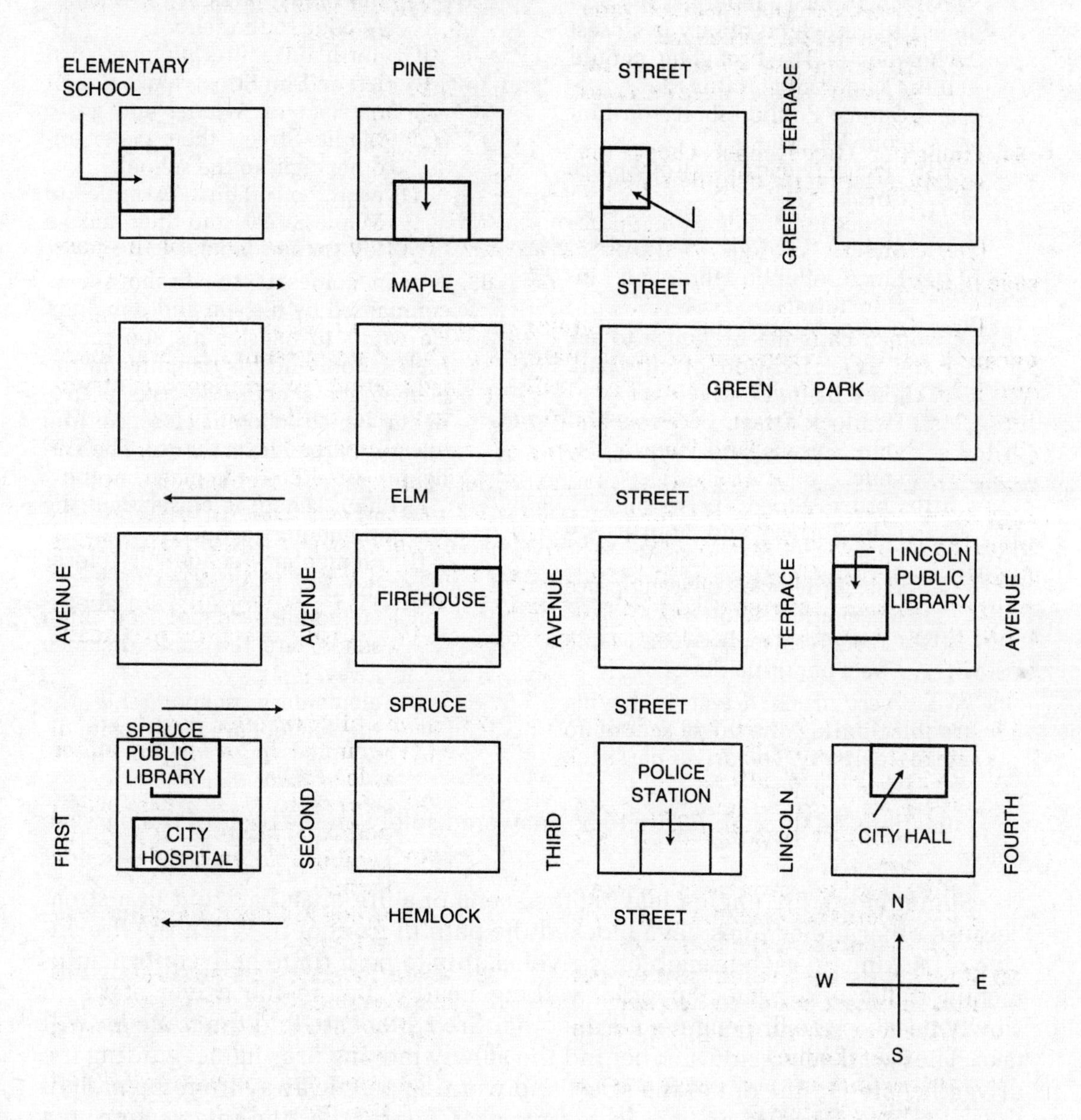

59. A woman whose handbag was stolen from her in Green Park asks a firefighter at the firehouse where to go to report the crime. The firefighter should tell the woman to go to the
    (A) police station on Spruce St.
    (B) police station on Hemlock St.
    (C) city hall on Spruce St.
    (D) city hall on Hemlock St.

60. A disabled senior citizen who lives on Green Terrace telephones the firehouse to ask which library is closest to her home. The firefighter should tell the senior citizen it is the
    (A) Spruce Public Library on Lincoln Terrace.
    (B) Lincoln Public Library on Spruce Street.
    (C) Spruce Public Library on Spruce Street.
    (D) Lincoln Public Library on Lincoln Terrace.

61. A woman calls the firehouse to ask for the exact location of city hall. She should be told that it is on
    (A) Hemlock Street, between Lincoln Terrace and Fourth Avenue.
    (B) Spruce Street, between Lincoln Terrace and Fourth Avenue.
    (C) Lincoln Terrace, between Spruce Street and Elm Street.
    (D) Green Terrace, between Maple Street and Pine Street.

62. A delivery truck driver is having trouble finding the high school to make a delivery. The driver parks the truck across from the firehouse on Third Avenue facing north and goes into the firehouse to ask directions. In giving directions the firefighter should tell the driver to go
    (A) north on Third Avenue to Pine Street and then make a right to the school.
    (B) south on Third Avenue, make a left on Hemlock Street, and then make a right on Second Avenue to the school.
    (C) north on Third Avenue, turn left on Elm Street, make a right on Second Avenue and go to Maple Street, then make another right to the school.
    (D) north on Third Avenue to Maple Street, and then make a left to the school.

63. A man comes to the firehouse accompanied by his son and daughter. He wants to register his son in the high school and his daughter in the elementary school. He asks a firefighter which school is closer for him to walk to from the firehouse. The firefighter should tell the man that the
    (A) high school is closer than the elementary school.
    (B) elementary school is closer than the high school.
    (C) elementary school and high school are the same distance away.
    (D) elementary school and the high school are in opposite directions.

Questions *64* through *68* are to be answered solely on the basis of the passage below.

Sometimes a fire engine leaving the scene of a fire must back out of a street because other fire engines have blocked the path in front of it. When the fire engine is backing up each firefighter is given a duty to perform to help control automobile traffic and protect people walking nearby. Before the driver starts to slowly back up the fire engine, all the other firefighters are told the route he will take. They walk alongside and behind the slowly moving fire engine, guiding the driver, keeping traffic out of the street and warning people away from the path of the vehicle. As the fire engine, in reverse gear, approaches the intersection, the driver brings it to a full stop and waits for his supervisor to give the order to start moving again. If traffic is blocking the intersection, two firefighters enter the intersection to direct traffic. They clear the cars and people out of the intersection, making way for the fire engine to back into it. The driver then goes forward, turning into the intersection. Two other firefighters keep cars and people away from the front of the fire engine as it moves. Because of the extra care needed to control cars and protect people in the streets when a fire engine is backing up, it is better to drive a fire engine forward whenever possible.

64. A fire engine is leaving the scene of a fire. The street in front of it is blocked by people and other fire engines. Of the following, it would be best for the driver to
    (A) put on the siren to clear a path.
    (B) back out of the street slowly.
    (C) drive on the sidewalk around the other fire engines.
    (D) move the other fire engines out of the way.

65. Firefighters walk alongside and behind the fire engine when it is backing up in order to
    (A) strengthen their legs and stay physically fit.
    (B) look around the neighborhood for fires.
    (C) ensure that the engine moves slowly.
    (D) control traffic, protect people and assist the driver.

66. A fire engine going in reverse approaches an intersection blocked with cars and trucks. The driver should
    (A) go forward, and then try to back into the intersection at a different angle.
    (B) slowly enter the intersection as the firefighters guiding the driver give the signal to move.
    (C) back up through the intersection without stopping.
    (D) stop, then enter the intersection only when the supervisor gives the signal to move.

67. The passage states that the two firefighters who first enter the intersection
    (A) clear the intersection of cars and people.
    (B) direct the cars past the fire engine when the engine is in forward gear.
    (C) see if the traffic signal is working properly.
    (D) set up barriers to block any traffic.

68. The diagram below shows a fire engine backing slowly out of Jones Street. The letters indicate where firefighters are standing. Which firefighter is *not* in the correct position?
    (A) Firefighter D
    (B) Firefighter E
    (C) Firefighter A
    (D) Firefighter C

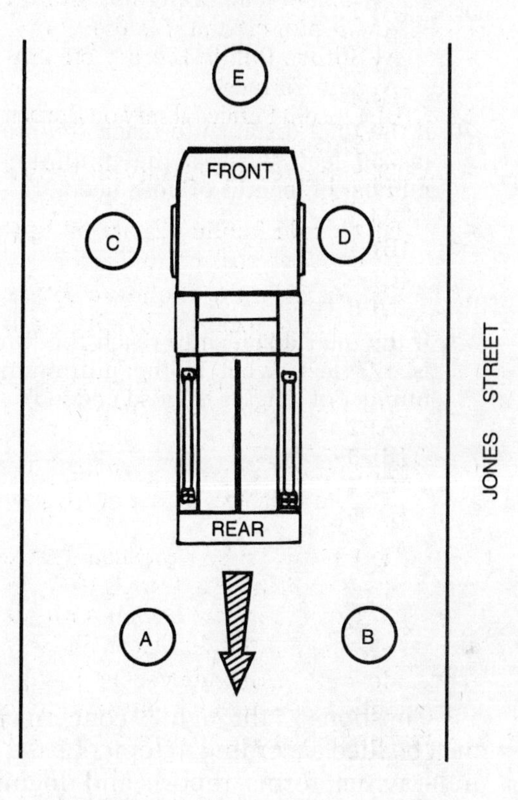

Questions *69* through *74* are to be answered solely on the basis of the following information.

In order to extinguish fires, firefighters must pull enough hose from the fire engine to reach the fire. Each length of hose is 50 feet long. The lengths of hose are attached together so that the water can go from the pump on the fire engine to a position where it will extinguish the fire.

69. If the total distance to reach the fire is 50 feet, what is the minimum number of lengths of hose needed?
    (A) 1
    (B) 2
    (C) 3
    (D) 4

70. If the total distance to reach the fire is 250 feet, what is the minimum number of lengths of hose needed?
    (A) 3
    (B) 4
    (C) 5
    (D) 6

71. If the total distance to reach the fire is 175 feet, what is the minimum number of lengths of hose needed?
    (A) 2
    (B) 3
    (C) 4
    (D) 5

72. If the total distance to reach the fire is 125 feet, what is the minimum number of lengths of hose needed?
    (A) 2
    (B) 3
    (C) 4
    (D) 5

73. If the total distance to reach the fire is 315 feet, what is the minimum number of lengths of hose needed?
    (A) 3
    (B) 5
    (C) 6
    (D) 7

74. If the total distance to reach the fire is 230 feet, what is the minimum number of lengths of hose needed?
    (A) 4
    (B) 5
    (C) 6
    (D) 7

Questions *75* through *79* concern various forms, reports, or other documents that must be filed according to topic. Listed below are four topics numbered 1 through 4, under which forms, reports and documents may be filed. In each question choose the topic under which the form, report or document concerned should be filed.

1. Equipment and supplies
2. Fire Prevention
3. Personnel
4. Training

75. Under which topic would it be most appropriate to file a letter on a heroic act performed by a member of the fire company?
    (A) 1 (C) 3
    (B) 2 (D) 4

76. Under which topic should a firefighter look for information about the fire company's new portable ladder?
    (A) 1 (C) 3
    (B) 2 (D) 4

77. Under which topic should a firefighter locate a copy of the fire company's fire prevention building inspection schedule for the current year?
    (A) 1 (C) 3
    (B) 2 (D) 4

78. Under which topic should a firefighter file a copy of a report on company property which has been damaged?
    (A) 1
    (B) 2
    (C) 3
    (D) 4

79. Under which topic should a firefighter be able to locate a roster of firefighters assigned to the company?
    (A) 1
    (B) 2
    (C) 3
    (D) 4

Questions *80* and *81* should be answered solely on the basis of the passage below.

Firefighters inspect many different kinds of places to find fire hazards and have them removed. During these inspections the firefighters try to learn as much as possible about the place. This knowledge is useful should the firefighters have to fight a fire at some later date at that location. When inspecting subways, firefighters are much concerned with the effects a fire might have on the passengers because, unless they have been trapped in a subway car during a fire, most subway riders do not think about the dangers involved in a fire in the subway. During a fire, the air in cars crowded with passengers may become intensely hot. The cars may fill with dense smoke. Lights may dim or go out altogether, leaving the passengers in darkness. Ventilation from fans and air conditioning may stop. The train may be stuck and unable to be moved through the tunnel to a station. Fear may send the trapped passengers into a panic. Firefighters must protect the passengers from the fire, heat and smoke, calm them down, get them out quickly to a safe area, and put out the fire. To do this firefighters may have to climb from street level down into the subway tunnel to reach a train stopped inside the tunnel. Before actually going on the tracks, they must be sure that the 600 volts of live electricity carried by the third rail is shut off. They may have to stretch fire hose a long distance down subway stairs, on platforms, and along the subway tracks to get the water to the fire and put it out. Subway fires are difficult to fight because of these special problems, but preparing for them in advance can help save the lives of both firefighters and passengers.

80. During a subway fire a train is stuck in a tunnel. Firefighters have been ordered into the tunnel. Before firefighters actually step down on the tracks they must be sure that
    (A) all the passengers have been removed from the burning subway cars to a safe place.
    (B) they have stretched their fire hose a long distance to put water on the fire.
    (C) live electricity carried by the third rail is shut off.
    (D) the train is moved from the tunnel to the nearest station.

81. According to the above passage, fire in the subway may leave passengers in subway cars in darkness. This occurs mainly because
    (A) the lights may go out.
    (B) air in the cars may become very hot.
    (C) ventilation may stop.
    (D) people may panic.

82. Firefighters must check gauges on fire engines so that defects are discovered and corrected. Some fire engines are equipped with gauges called "chargicators" which indicate whether or not the electrical system is operating properly. When the fire engine's motor is running, the chargicator of a properly operat-

ing electrical system will show a reading of 13.5 to 14.2 volts on the scale.

Which one of the following gauges shows a properly operating electrical system?

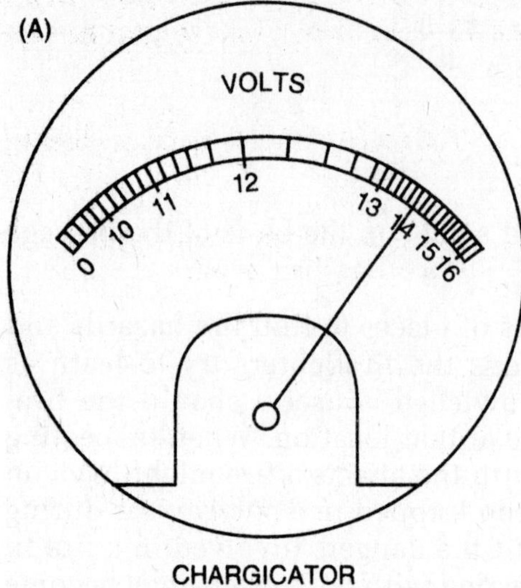

(B)

VOLTS

0 10 11 12 13 14 15 16

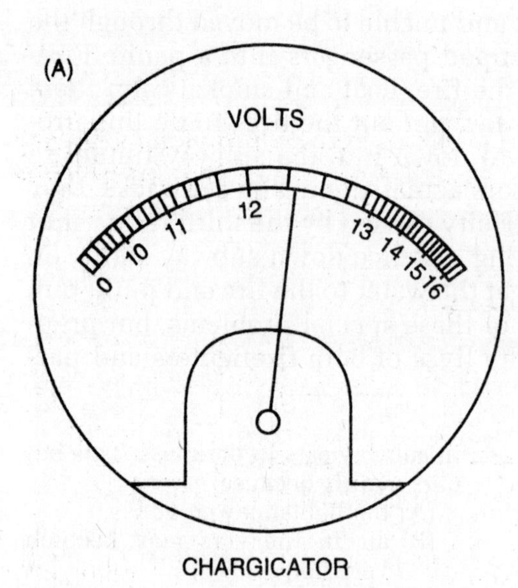

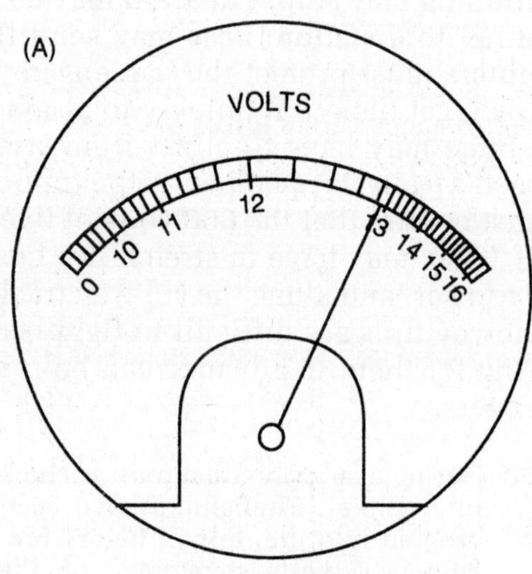

Answer questions *83* and *84* solely on the basis of the passage and diagram below.

Firefighters breathe through an air mask to protect their lungs from dangerous smoke when fighting fires. The air for the mask comes from a cylinder which the firefighter wears. A full cylinder contains 45 cubic feet of air when pressurized to 4500 pounds per square inch.

83. A gauge that firefighters read to tell how much air is left in the cylinder is pictured in the diagram below.

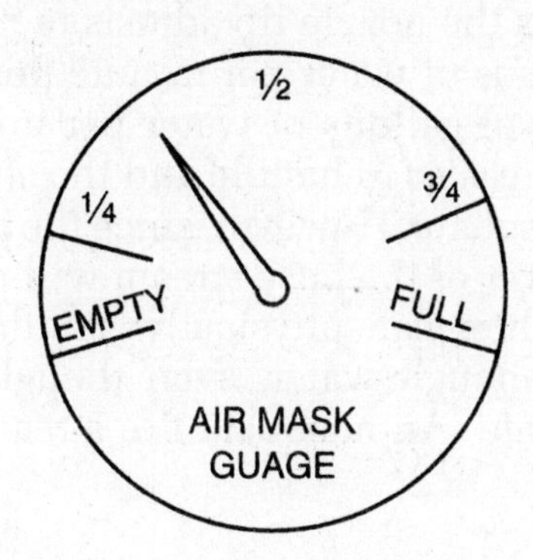

The gauge indicates that the cylinder is
(A) full.
(B) empty.
(C) more than ¾ full.
(D) less than ½ full.

84. A gauge which is part of the cylinder shows the pressure of the air in the cylinder in hundreds of pounds per square inch. Which of the following diagrams shows a cylinder which is more than half full?

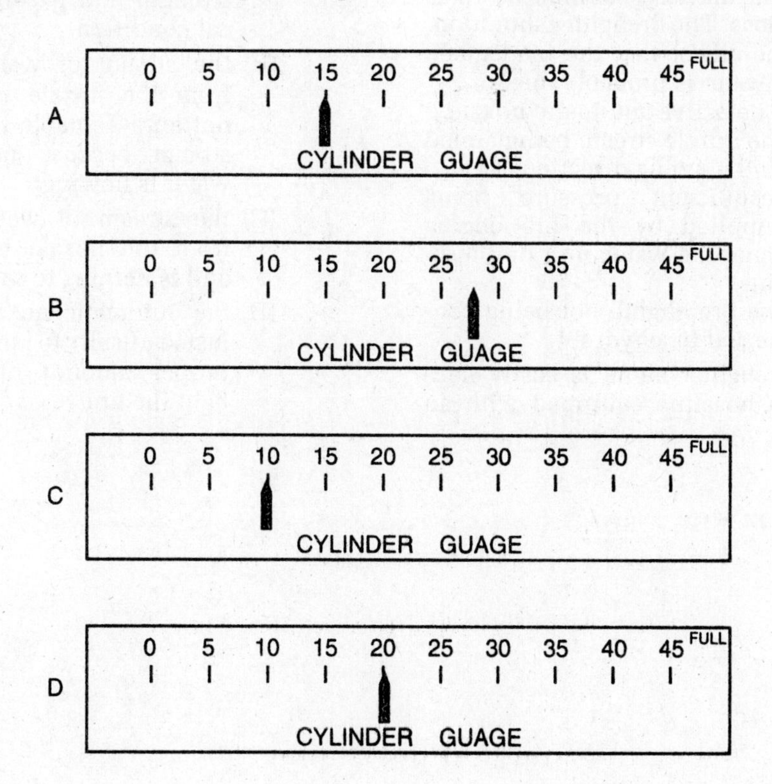

Questions *85* and *86* are to be answered solely on the basis of the following passage.

The Fire Department uses a firehose nozzle with an automatically adjusting tip. The automatically adjusting nozzle tip keeps the water pressure at the tip constant even though the amount of water being pumped through the hose from the fire engine may vary. A partial loss of water in the hoseline does not result in the stream of water from the nozzle falling short of the target. A partial loss of water is caused by a kink in the hose somewhere between the fire engine pumping the water and the nozzle or by insufficient pressure being supplied by the fire engine pumping water into the hoseline.

The danger of this automatic nozzle is that as the nozzle tip adjusts to maintain constant water pressure, the number of gallons of water per minute flowing out of the nozzle is reduced. When the number of gallons of water per minute flowing from the nozzle is reduced, the nozzle is easier to handle and the stream of water coming from the nozzle appears to be adequate. However, since the number of gallons of flow is reduced, the cooling power of the hose stream will probably not be enough to fight the fire. If a firefighter can physically handle the hoseline alone, the nozzle is not discharging enough water, even though the stream coming out of the nozzle appears adequate. An adequate fire stream requires two firefighters to handle the hoseline.

85. An officer tells a firefighter to find out why enough water is not coming out of a hoseline equipped with an automatic nozzle. The firefighter follows the hoseline from the nozzle back to the fire engine pumping the water into the hose, but finds no kinks in the hose. The firefighter should inform the officer that the inadequate flow of water is probably due to
    (A) a defective automatic nozzle.
    (B) the nozzle stream being aimed in the wrong direction.
    (C) insufficient pressure being supplied by the fire engine pumping water into the hoseline.
    (D) the fire engine not being connected to a hydrant.

86. One firefighter alone is easily handling a hoseline equipped with an automatic nozzle. The hoseline's stream is reaching the fire. According to the passage, the firefighter should probably conclude that:
    (A) being able to handle the hoseline alone indicates extreme strength and excellent physical condition.
    (B) the stream of water coming from the nozzle is probably not an acceptable fire fighting stream because not enough water is flowing.
    (C) the stream of water coming from the nozzle is adequate and is helping to save water.
    (D) the automatic nozzle has adjusted itself to provide the proper amount of water to fight the fire.

Answer questions *87* through *90* solely on the basis of the following facts and diagrams.

The gauges shown below in Diagrams I and II represent gauges on a fire engine's pump control panel at the scene of a fire. Diagram I gives the readings at 10 A.M. and Diagram II gives the readings at 10:15 A.M. Each diagram has one gauge labeled "Incoming" and one gauge marked "Outgoing." The "Incoming" gauges show the pressure in pounds per square inch (psi) of the water coming into the pumps on the fire engine from a hydrant. The "Outgoing" gauges show the pressure in pounds per square inch (psi) of the water leaving the pumps on the fire engine. The pumps on the fire engine raise the pressure of the water coming from the hydrant to the higher pressures needed in the fire hoses.

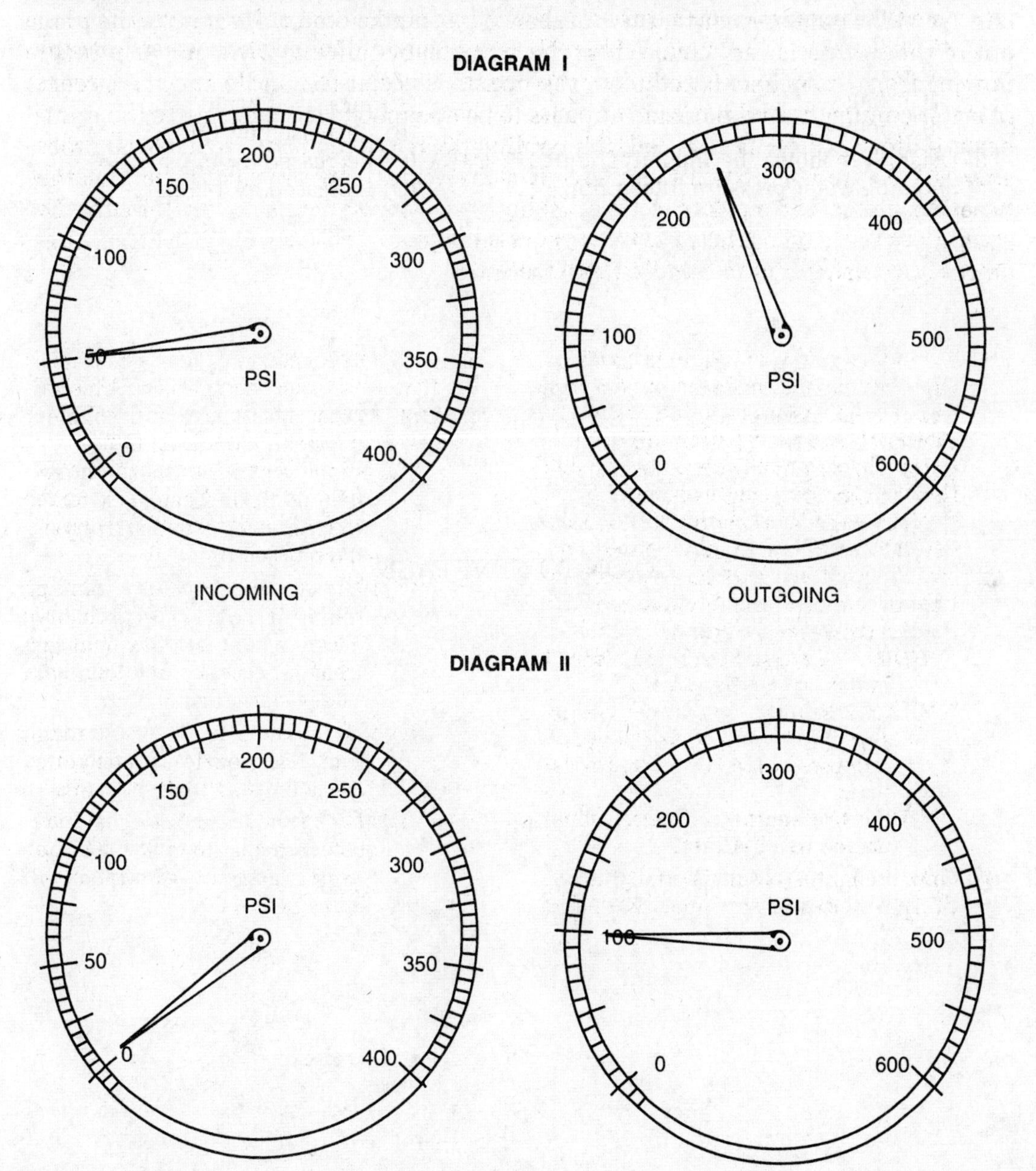

87. The firefighter looks at the gauges as shown in Diagram I and observes that the pressure in pounds per square inch (psi) of the water coming into the pumps is most nearly
    (A) 50.
    (B) 250.
    (C) 300.
    (D) 500.
88. The firefighter looks at the gauges shown in Diagram I and observes that the pressure in pounds per square inch (psi) of the water going out of the pumps is most nearly
    (A) 25.
    (B) 50.
    (C) 250.
    (D) 500.
89. Diagram II shows the incoming and outgoing water pressure fifteen minutes later. By looking at the gauges in Diagram II, the firefighter observes that the water
    (A) going out of the pumps is at 200 psi.
    (B) going out of the pumps is at 5 psi.
    (C) coming into the pumps is above 10 psi.
    (D) coming into the pumps is below 10 psi.
90. The firefighter is able to determine that, between the time of Diagram I and the time of Diagram II, the pressure of the outgoing water from the pumps
    (A) increased by 50 psi.
    (B) decreased by 150 psi.
    (C) decreased by 45 psi.
    (D) increased by 145 psi.

GO ON TO NEXT PAGE

Questions *91* and *92* are to be answered solely on the basis of the passage below.

Firefighters at times are required to work in areas where the atmosphere contains contaminated smoke. To protect the firefighter from breathing the harmful smoke, a self-contained breathing mask is worn. The mask will supply the firefighter with a limited supply of pure breathing air. This will allow the firefighter to enter the smoke-filled area. The mask is lightweight and compact which makes it less tiring and easier to move around with. The face mask is designed to give the firefighter that maximum visibility possible. The supply of breathing air is limited and the rate of air used depends upon the exertion made by the firefighter. Although the mask will protect the firefighter from some types of contaminated smoke, it gives no protection from flame, heat or heat exhaustion.

91. The rate at which the firefighter breathes the air from the mask will depend upon the
    (A) amount of energy used by the firefighter.
    (B) amount of smoke the firefighter will breathe.
    (C) color of the flames that the firefighter will enter.
    (D) color of the heat that the firefighter will enter.

92. According to the passage, the mask will protect the firefighter from some types of
    (A) flames.
    (B) smoke.
    (C) heat.
    (D) heat exhaustion.

GO ON TO NEXT PAGE

Questions *93* through *96* should be answered solely on the basis of the passage below.

In each firehouse one firefighter is always on Housewatch duty. Each 24-hour Housewatch tour begins at 9 A.M. each day, and is divided into eight 3-hour periods. The firefighter on Housewatch is responsible for the correct receipt, acknowledgment, and report of every alarm signal from any source. Firefighters on Housewatch are required to enter in the Company Journal the receipt of all alarms as well as other matters required by Department Regulations. All entries by the firefighter on Housewatch should be written in blue or black ink. Any entries made by firefighters not on Housewatch are made in red ink. Most entries, including receipt of alarms, are recorded in order, starting in the front of the Company Journal on page 1. Certain types of entries are recorded in special places in the Journal. When high level officers visit the company, those visits are recorded on page 500. Company training drills and instruction periods are recorded on page 497. The monthly meter readings of the utility companies which serve the firehouse are recorded on page 493.

93. A firefighter is asked by the company officer to find out what alarms were received the previous day, August 25, 1982, between 1 A.M. and 2 A.M. Where in the Company Journal should the firefighter look to obtain this information?
   (A) on page 493.
   (B) between page 1 and page 492, on the page for August 25, 1982.
   (C) on page 500.
   (D) between page 497 and page 500, on the page for August 25, 1982.

94. A firefighter on Housewatch is asked to find out how much electricity was used in the firehouse between the last two meter readings taken by Con Edison. On which one of the following pages of the Company Journal should the firefighter look to find the last two electrical meter readings entered?
   (A) 253
   (B) 493
   (C) 497
   (D) 500

95. A firefighter on Housewatch duty is notified by a passing civilian of a rubbish fire around the block. The company responds, extinguishes the rubbish fire and returns to the firehouse. The firefighter on Housewatch should
   (A) make no entry in the Company Journal of the receipt of the alarm because it was received orally from the civilian.
   (B) record the alarm in red ink in the Company Journal.
   (C) record the alarm in blue ink in the Company Journal.
   (D) ask the civilian to record the alarm in red ink in the Company Journal.

96. The company officer asks the firefighter on Housewatch to find out the last date on which the company had a training drill on high-rise building fire operations. On which one of the following pages of the Company Journal should the firefighter on Housewatch look to find the date of the training drill?
   (A) 36
   (B) 493
   (C) 497
   (D) 500

Answer questions *97* and *98* solely on the basis of the following passage.

Fire Department Regulations require that upon receiving an alarm while in the firehouse, the officer of the fire company directs the firefighters to take positions in front of the firehouse. The firefighters warn pedestrians and vehicles that the fire engine is leaving the firehouse. The officer directs the driver of the fire engine to move the fire engine to the front of the firehouse, then stop to check for vehicles and pedestrian traffic. While the fire engine is stopped, the firefighters will get on and the officer will signal the driver to go to the alarm location.

97. When do the firefighters who were sent to the front of the firehouse actually get on the fire engine?
    (A) as the fire engine turns into the street leaving the firehouse.
    (B) as the fire engine slows down while leaving the firehouse.
    (C) inside the firehouse, before the fire engine is moved.
    (D) after the fire engine has been moved to the front of the firehouse and stopped.

98. When responding to an alarm, why are the firefighters sent out of the firehouse before the fire engine?
    (A) To make sure that the firehouse doors are fully opened.
    (B) To go to the nearest corner to change the traffic signal.
    (C) To warn pedestrians and vehicles that the fire engine is coming out of the firehouse.
    (D) To give the firefighters time to put on their helmets and boots.

GO ON TO NEXT PAGE

99. When the engine oil drum in the firehouse is nearly empty it must be replaced by a new drum full of oil. A firefighter gives the drum a kick. It sounds empty. The firefighter then checks the written log to see how many gallons of oil have been taken out of the drum so far. Checking the written log is
   (A) unnecessary, since the oil drum sounded empty when the firefighter kicked it.
   (B) necessary, since the log should tell the firefighter exactly how much oil is left.
   (C) unnecessary, since the firefighter should avoid paperwork whenever possible.
   (D) necessary, since the firefighter should always try to keep busy with useful activity.

100. The Fire Department provides each firehouse with such basic necessities as electric light bulbs. As the items are used up, new supplies are ordered before the old ones are all gone. Of the following, the best reason for ordering more electric light bulbs before the old ones are all gone is to
   (A) decrease the amount of paperwork a firehouse company must complete.
   (B) be sure that there are always enough light bulbs on hand to replace those that burn out.
   (C) make sure that the firehouse has enough electric light bulbs to supply nearby firehouses.
   (D) decrease the cost of providing electricity to the firehouse.

END OF WRITTEN TEST

RECORDING ANSWERS—You may record your answers in the Record of Answers section of the Instruction Booklet and take the booklet with you when you leave. No additional time for recording answers in the Instruction Booklet is allowed.

ANSWER SHEET COLLECTION—When finished with the test, remain seated and summon the monitor to collect your answer sheet. You are required to hand in your test booklet and all scratch paper with your answer paper. Any candidates who leave the room with the question booklet or scratch paper in their possession will be disqualified and prosecuted to the full extent of the law.

Leave quickly and quietly.

ANSWER KEY

1. **C**	21. **D**	41. **D**	61. **B**	81. **A**
2. **B**	22. **C**	42. **D**	62. **C**	82. **A**
3. **A**	23. **B**	43. **C**	63. **A**	83. **D**
4. **A**	24. **C**	44. **D**	64. **B**	84. **B**
5. **D**	25. **A**	45. **A**	65. **D**	85. **C**
6. **A**	26. **A**	46. **B**	66. **D**	86. **B**
7. **A**	27. **B**	47. **D**	67. **A**	87. **A**
8. **D**	28. **A**	48. **C**	68. **B**	88. **C**
9. **B**	29. **D**	49. **D**	69. **A**	89. **D**
10. **A**	30. **C**	50. **A**	70. **C**	90. **B**
11. **C**	31. **B**	51. **B**	71. **C**	91. **A**
12. **B**	32. **D**	52. **C**	72. **B**	92. **B**
13. **A**	33. **B**	53. **D**	73. **D**	93. **B**
14. **D**	34. **C**	54. **A**	74. **B**	94. **B**
15. **D**	35. **B**	55. **D**	75. **C**	95. **C**
16. **B**	36. **B**	56. **B**	76. **A**	96. **C**
17. **B**	37. **B**	57. **C**	77. **B**	97. **D**
18. **A**	38. **C**	58. **A**	78. **A**	98. **C**
19. **A**	39. **A**	59. **B**	79. **C**	99. **B**
20. **D**	40. **D**	60. **D**	80. **C**	100. **B**

DIAGNOSTIC PROCEDURE

Use the following diagnostic chart to determine how well you have done and to identify your areas of weakness.

Enter your number of correct answers for each section in the appropriate box in the column headed "Your Number Correct." The column immediately to the right will indicate how well you did on each section of the test.

Below the chart you will find directions as to the chapter(s) in the guide you should review to strengthen your weak area(s).

Section Number	Question Number	Area	Your Number Correct	Scale	
One	1–10	Recall		10 right 9 right 8 right Under 8 right	Excellent Good Fair Poor
Two	11–32 46–52 55–68 80–98	Understanding Job Information		61–62 right 58–60 right 56–57 right Under 56 right	Excellent Good Fair Poor
Three	34 53–54 75–79 99–100	Judgment and Decision Making		10 right 9 right 8 right Under 8 right	Excellent Good Fair Poor
Four	36–45	Tools and Equipment		10 right 9 right 8 right Under 8 right	Excellent Good Fair Poor
Five	69–74	Mathematics and Science		6 right 5 right 4 right Under 4 right	Excellent Good Fair Poor
Six	33, 35	Dealing with People		2 right Under 2 right	Excellent Poor

1. If you are weak in Section One, concentrate on Chapter 6.
2. If you are weak in Section Two, concentrate on Chapter 7.
3. If you are weak in Section Three, concentrate on Chapter 8.
4. If you are weak in Section Four, concentrate on Chapter 9.
5. If you are weak in Section Five, concentrate on Chapter 10.

Note: Consider yourself weak in a section if you receive other than an "excellent" rating.

ANSWER EXPLANATIONS

1. **C** There is only one fire escape and only one window leading to it, the window in bedroom 3.
2. **B** The two exits from this apartment are the fire escape outside bedroom 3 and the apartment entrance door. Since the fire is in bedroom 3, the best exit route would be through the entrance hall and apartment door.
3. **A** To get to the fire escape a fire fighter must pass through both halls and bedroom 3.
4. **A** The living room, the kitchen, and the dining room cannot be closed off, but only the living room is offered as a choice.
5. **D** Every room except bedroom 2 and bathroom 2 has a window and also either a door or a doorway.
6. **A** This would permit access to the adjacent room and would provide the occupants with an escape route. Bedroom 2 (B) and the bedroom hall (D) have no windows, and the window in bedroom 3 (C) opens on the fire escape.
7. **A** The kitchen has three exits: a window, a doorway, and a door. The question asks you to select the room into which the door leads.
8. **D** The incinerator adjoins only the living room.
9. **B** Rooms that have only one means of entrance and exit should be identified and remembered when you study the plan.
10. **A** The door between the two halls divides the apartment into two parts and is the only interior entrance to the bedroom area.
11. **C** This information is found in the second sentence of the passage. Note that this statement does not state that sprinkler systems extinguish all fires or keep all fires from spreading. Choices A and B are not mentioned or implied in the passage. Choice D is incorrect; 96% refers to the rate of effectiveness.
12. **B** This information is found in the last sentence of the passage.
13. **A** The fifth sentence of the passage states this directly.
14. **D** This is the company that Fire Fighter Gold is assigned to (see the second paragraph).
15. **D** The name of the company is given in the last sentence of paragraph 2.
16. **B** The main exit, the exit to the loading dock, and the small door on the side are mentioned in the sixth paragraph.
17. **B** Paragraph 6 indicates that the side door is obstructed by cartons—a definite life-safety hazard.
18. **A** Note that the question is looking for unlocked, *not* locked, exit doors even though the locked exit would be the serious life hazard. According to paragraph 6 only the doors in choice A are unlocked.
19. **A** The third paragraph states that the exit sign over the main entrance door is unlit because of a burned out bulb. Paragraph 6 states that the other two exit signs are lighted.
20. **D** According to paragraph 7, "they [rubbish receptacles] are not covered. . . ." Since the passage does not state how many rubbish receptacles there are or how many are covered, you must assume that all of them are uncovered.
21. **D** This information is found in the last sentence of paragraph 5.
22. **C** The last sentence of paragraph 9 states that the main junction box has a cover, but the box cannot be closed because the cover is corroded.
23. **B** This information is found in paragraph 8.
24. **C** See the last sentence in paragraph 8.
25. **A** Paragraph 5 mentions that "the main control valve is sealed in the open position" (see the fifth sentence).
26. **A** This information is given in the third sentence of paragraph 4.
27. **B** The third sentence of paragraph 7 states that the rubbish receptacles "are overflowing, resulting in several piles of litter on the floor." There is no mention of the other choices.

28. **A** The name of the officer, Captain John Bailey, can be found in paragraph 2. Choice B is the fire fighter inspector. Choices C and D are incorrect mixtures of the names.
29. **D** The passage indicates that false alarms cause delays in response to actual fires and present unnecessary and additional risks to the fire fighters and citizens.
30. **C** This is stated in sentences 5 and 6.
31. **B** Sentence 3 states that ". . . the first fire fighters . . . do not know. . . ." There is no mention of any of the other choices.
32. **D** See the opening statement.
33. **B** Providing necessary first aid and summoning professional help is a primary function of fire fighters. Choices A, B, and C would all be forms of neglect of duty and might cause the elderly man to suffer serious ill effects.
34. **C** Gasoline is a volatile, flammable liquid, and all leaks must be taken care of immediately. Otherwise, a spark, a hot catalytic converter on a car, or a carelessly discarded match or cigarette may cause a violent fire. Choice A—this action would lead directly to a fire and probably serious injury to the fire fighter. Choice B—this situation is serious and must be handled immediately; leaving to chance the possibility that the attendant is aware of it is foolish and unsafe. Choice D—this would result in extreme delay and would not remove the immediate hazard.
35. **B** The question does not indicate an unusual hazard. The purpose of a fire-prevention inspection is to reduce fire hazards, and a major goal of this activity is to get the owner to maintain a fire-safe environment even when the fire department is not there to inspect. In this case the occupant has given the fire fighters a reasonable explanation of why the inspection is inconvenient at this time. The new fire fighter should *not* have been rude; the inspection should be rescheduled for the next available date. Choice A—this type of attitude does not build good rapport between the fire fighter and the owner and tends to defeat the purpose of the fire-prevention program. Choice C—this is incorrect procedure; the store is private property and if the owner refuses admittance, the fire fighter must obtain a search warrant in order to have the legal right to inspect. Note however, that this is not required where a fire condition exists. Choice D—this would represent neglect of duty.
36. **B** Illustration 7 shows a hose line and nozzle. Choice A is a portable fire extinguisher. Choice C is a portable electrical drill. Choice D is a Phillips-head screwdriver.
37. **B** Illustrations 4 and 6 show a broom and a shovel. The items in choice A are a portable straight ladder and a jimmy bar; in choice C, a first-aid kit and a hand saw; in choice D, an axe and a monkey wrench.
38. **C** Illustrations 11 and 12 show an axe and a hand saw. The items in choice A are a pipe or Stillson wrench and an engineers' pliers; in choice B, a hand drill and a slip joint pliers, in choice D, a jimmy bar and a hacksaw.
39. **A** Illustration 1 shows a portable straight ladder. Choice B is a first-aid kit. Choice C is a hacksaw. Choice D is a fire hydrant.
40. **D** A nut requires the use of a wrench, and the screw requires a screwdriver. Illustrations 16 and 18 show a double open-end wrench and a standard square-blade screwdriver. The items in choice A are a pipe wrench and engineers' pliers; in choice B, a portable electrical drill and a pipe cutter; in choice C, slip-joint pliers and a jimmy bar.
41. **D** Illustration 13 shows a monkey or adjustable wrench. Choice A is a broom. Choice B is a portable fire extinguisher. Choice C is a hand saw.
42. **D** Illustration 20 shows a fire hydrant, which is a primary source of water. Choices A (a pipe wrench), B (a fire extinguisher), and C (pliers) could not serve this purpose.

43. **C** Illustration 15 shows a hacksaw, which has fine teeth and is designed for cutting metal. Choices A (a shovel), B (a monkey wrench), and D (a screwdriver) are not used for cutting.

44. **D** The illustration indicates that the fourth rung from the bottom is missing. This can cause the fire fighter to slip from the ladder and be injured. Choice A—the coat is the proper length. Choice B—the fire fighter has a clear line of vision. Choice C—the fire fighter's feet belong on the rung.

45. **A** In this illustration the hose is at about a 45-degree angle. This allows the water to be projected upward and outward at the same time. Choice B projects the water upward only. Choice C projects the water outward but not upward. Choice D projects the water downward.

46. **B** First paragraph, second sentence states that "It [the power saw] can be dangerous . . . if it has not been inspected and tested to insure that it is in serviceable condition."

47. **D** The use of the word *foreign* in this context means "having little or no relationship to" the tool. For example, wood chips, gravel, and plaster would be foreign materials in this sense.

48. **C** The second paragraph, sentence 3, says, "The V-belt pulley, if present, must be checked . . ."; the words *if present* indicate that the item named may be lacking. Therefore some saws will work without a V-belt pulley.

49. **D** See the second paragraph, sentence 4. Choices A, B, and C occur if the belt is too tight.

50. **A** The second paragraph, sentence 2, states clearly the need for an air filter.

51. **B** Carbide-tip blades must be replaced when two or more tips are broken or missing or the tips are worn down (see paragraph 3). The other blades should be changed when cracked, badly nicked, or worn down to an 8-inch diameter or less.

52. **C** This information is given in paragraph 3, last sentence.

53. **D** To some examinees this type of question can be confusing. The best way to approach such a question is to eliminate each choice as you read it. Choice A—changed *not* every day, *only* when the indicator shows red. Choice B—replaced *not only* when the oil is changed, *but also* when the indicator shows red. Choice C—inspected *not only* when the oil is changed, *but also* each time the indicator shows red. With A, B, and C eliminated, D must be the best choice. Going back to the passage, you find that the last sentence verifies your selection.

54. **A** Fire Fighter Green *needs* 10 full air cylinders. Fire Fighter Green *has* 20 air cylinders, 5 of which are full and 15 of them empty. Since only 10 full air cylinders are needed, and 5 are already on hand, only 5 air cylinders need to be replaced.

55. **D** The next to last sentence states, "All such injuries could be eliminated if the buildings were demolished." Repairing the buildings (A) would represent only a partial solution.

56. **B** The third and fourth sentences support choice B. There is no mention of choices A, C, and D in the passage.

57. **C** See sentence 4. Choices A, B, and D are not mentioned in the passage.

58. **A** The danger of missing stairs is mentioned in the third sentence. There is no mention of items B, C, and D.

59. **B** The appropriate place to report a crime is a police station. There is only one police station, and it is located on Hemlock Street between Third Avenue and Lincoln Terrace.

60. **D** There are two public libraries. First locate Green Terrace; then trace with your fingertip the route to each library. The closest library is on Lincoln Terrace and is the Lincoln Public Library. When you have identified the correct loca-

tion, go back to the question and locate the choice that agrees with what you found. This will be the correct answer.

61. **B** Locate City Hall, write down the street it faces, and then note the cross streets. Now go to the answer choices and find the selection that agrees with what you found.

62. **C** Locate the firehouse and the high school. Using your fingertip, trace the various routes to the school until you find the most direct one. Remember to watch for one-way streets. Finally, locate the choice that agrees with the route you have selected.

63. **A** Follow the same procedure as in question 62. However, in this case (the man and his children are on foot) there is no need to consider one-way streets.

64. **B** The correct answer is found in the first sentence: ". . . must back out of a street because other fire engines have blocked the path in front of it." Choices A, C, and D are not practicable and are not mentioned in the passage.

65. **D** The fourth sentence provides the information for this answer. Choices A, B, and C are not mentioned in the passage.

66. **D** This information is found in the fifth sentence of the passage. Choices A, B, and C would be unsafe acts; they are not mentioned in the passage.

67. **A** The sixth and seventh sentences state that ". . . two fire fighters enter the intersection to direct traffic. They clear the cars and people out of the intersection. . . ."

68. **B** See the fourth sentence: "They [fire fighters] walk alongside and behind the slow moving fire engine. . . ." Fire fighter E is in front of the apparatus, the wrong position.

69. **A** Each length is 50 feet (sentence 2); therefore only one length is required.

70. **C** Each length is 50 feet.

$$\frac{\text{Distance to fire}}{\text{Single length of hose}} = \text{Number of lengths}$$

$$\frac{250 \text{ ft.}}{50 \text{ ft.}} = 5 \text{ lengths}$$

71. **C** Each length is 50 feet.

$$\frac{\text{Distance to fire}}{\text{Single length of hose}} = \text{Number of lengths}$$

$$\frac{175 \text{ ft.}}{50 \text{ ft.}} = 3\tfrac{1}{2} \text{ lengths}$$

Since you cannot use ½ length, you must round your answer to the next hightest whole number. In this case 4 = Number of lengths.

72. **B** Each length is 50 feet.

$$\frac{\text{Distance to fire}}{\text{Single length of hose}} = \text{Number of lengths}$$

$$\frac{125 \text{ ft.}}{50 \text{ ft.}} = 2\tfrac{1}{2} \text{ lengths}$$

Rounding to the next highest whole number give 3 lengths

73. **D** Each length is 50 feet.

$$\frac{\text{Distance to fire}}{\text{Single length of hose}} = \text{Number of lengths}$$

$$\frac{315 \text{ ft.}}{50 \text{ ft.}} = 6\tfrac{3}{10} \text{ lengths}$$

Rounding to the next highest whole number gives 7 lengths.

74. **B** Each length is 50 feet.

$$\frac{\text{Distance to fire}}{\text{Single length of hose}} = \text{Number of lengths}$$

$$\frac{230 \text{ ft.}}{50 \text{ ft.}} = 4\tfrac{3}{5} \text{ lengths}$$

Rounding to the next highest whole number gives 5 lengths.

75. **C** A letter regarding a heroic act is unique to the individual and would show that the fire fighter had performed outstandingly. This information should be saved and reviewed when assignments or promotions are considered. It would be kept in the individual's Personnel file.

76. **A** A ladder is a piece of equipment; therefore any information about it should be kept in the Equipment and Supplies file.

77. **B** The fire prevention schedule would be kept in the Fire Prevention file.

78. **A** As used in this passage, "property" is considered "equipment and supplies." The report should be filed under Equipment and Supplies.

79. **C** The assignment of fire fighters to units is controlled by the organization's personnel department. The roster would be kept in the Personnel file.

80. **C** Sentence 13 states, "Before actually going on the tracks, they [fire fighters] must be sure that the 600 volts of live electricity carried by the third rail is shut off." Choices A , B, and D are not mentioned or implied anywhere in the passage. Answers must be based solely on information found in the passage.

81. **A** The question could be rephrased as follows: "Why are passengers in subway cars in darkness?" Of the choices offered, the only one that will answer this question is A. Now refer to the passage to verify this choice; sentence 7 shows that you have selected the correct answer.

82. **A** To determine the correct illustration you should first mark off the outer boundaries of a "properly operating electrical system." The question states that these boundaries are 13.5 and 14.2. Mark lines through these points on each of the illustrations. Then select the illustration in which the indicating needle is within the boundaries you have marked off. Choice B is above the range, and C and D are below it.

83. **D** The indicator arrow is pointing toward a position between ¼ full and ½ full. This would tell the fire fighter that the cylinder is less than ½ full of air.

84. **B** According to the passage preceding questions 83 and 84, a full cylinder contains air at 4500 pounds per square inch. Question 84 asks you to select the gauge that indicates a cylinder that is more than ½ full. Therefore: 4500 × ½ = 2250 lb. The gauge is divided in air pressure increments of 500 pounds per square inch. A reading above 22.5 would indicate a cylinder that is more than ½ full. Choice B indicates a reading above 25; it is therefore more than ½ full. Choice A indicates 1500 psi. Choice C indicates 1000 psi. Choice D indicates 2000 psi. Each of these (A, C, and D) is below ½ full.

85. **C** Paragraph 1, fourth sentence, states that there are two causes for a partial loss of water at the nozzle: a kink in the hose, and insufficient pressure from the fire engine pump. In question 85 the fire fighter found no kinks in the hose; therefore the cause must be insufficient pressure being supplied by the fire pump. Choices A, B, and D are not mentioned in the passage, thereby elminating them from consideration.

86. **B** The second paragraph, next to last sentence, states, "If a fire fighter can physically handle the hoseline alone, the nozzle is not discharging enough water. . . ."; in other words, the fire fighting stream is probably not acceptable.

87. **A** The gauge marked "Incoming" in Diagram I shows the indicator pointing to 50. This tells the fire fighter that water is entering the pumps at a pressure of 50 psi.

88. **C** The gauge marked "Outgoing" in Diagram I shows the indicator pointing halfway between 200 and 300 psi. This tells the fire fighter that the pressure of the water leaving the pumps is 250 psi.

89. **D** The gauge marked "Incoming" in Diagram II shows the indicator

pointing just above zero and below the first indicator line. This tells the firefighter that water is entering the pumps at a pressure of less than 5 psi—therefore less than 10 psi.

90. B In the gauge marked "Outgoing" in Diagram II the indicator points to a pressure of 100 psi. In the corresponding gauge of Diagram I the pressure is 250 psi.

250 psi (original pressure) –
100 psi (current pressure) =
150 psi (decrease)

91. A The seventh sentence tells you that ". . . air used depends upon the exertion made by the firefighter"—in other words, the amount of energy used by the firefighter. Choice B is wrong because the mask protects against breathing any smoke. Choices C and D have no relationship to the mask or breathable air.

92. B This information is found in the second sentence.

93. B According to the passage, all alarms are entered in the Company Journal. If pages 493 to 500 are reserved for special purposes, then alarms must be recorded on pages 1 through 492. Choice A—page 493 is for utility meter readings. Choice C—page 500 is for visits by high level officers. Choice D—page 497 is for company drill records.

94. B Con Edison is a utility company, and meter readings taken by a utility company employee should be recorded on page 493 (see the last sentence).

95. C According to the passage, all alarms are recorded from any source, and entries by the firefighter on Housewatch should be written in blue or black ink. Choice C is the only one that meets these specifications. Choice A is incorrect; an entry must be made in the Journal. Choice B is incorrect; the entry should be made in blue or black ink. Choice D is incorrect; the entry must be made by the firefighter.

96. C According to the next to last sentence of the passage, company drills are recorded on page 497.

97. D The passage tells you that when an alarm is received the firefighters are directed to take positions in front of the firehouse, and the apparatus is driven out and stopped. Then the firefighters board the apparatus. Choice D best agrees with this procedure. Choices A, B, and C are contrary to the regulations as explained in the passage.

98. C Sentence 2 of the passage indicates that the purpose of assigning firefighters to the front of the firehouse is to caution pedestrians and vehicles.

99. B Kicking the drum will indicate only that the drum is empty or nearly empty. To know for sure how much oil is in the drum, the firefighter must know how much oil the full drum can hold and how much oil has been removed from the drum. Each time oil is taken from the drum, it is recorded in the log; a running total of how much oil has been used is also recorded. If the firefighter subtracts the total oil used from the original amount of oil that was in the drum, the firefighter will know whether any oil remains and, if so, how much.

100. B One of the strong attributes of firefighters is that they are always prepared. If the firefighters waited until the last light bulb was used, they would then be faced with the possibility that other bulbs could burn out, thereby interfering with their fast, safe response to an alarm. Obtaining replacement bulbs, parts, tools, etc., takes time and often is impossible at night. These items must be immediately available when needed.

CHAPTER 14

Practice Examination Three

In this chapter you will find the third practice examination. It has been developed by using official and sample questions from cities across the United States.

Be sure that when you take this examination you allow yourself a 3½-hour uninterrupted time period. Do not try to take the examination in parts; if you do, you will defeat its purpose. By taking the examination at one sitting you will become familiar with test conditions and with your personal idiosyncrasies about sitting and working for this period of time.

Before Taking the Examination

First, go back and review quickly the test-taking strategies outlined in Chapter 3. Then, when you begin the examination, be sure to read and follow all instructions. Time for reading the instructions has been figured into the examination. Read each question carefully, and answer only what is asked of you. Select the answer that is the one *best* choice of those provided, and then record your selection on the answer sheet.

The answer sheet precedes the examination. The Diagnostic Procedure and Answer Explanations are given at the end of the chapter.

Composite Examination

ANSWER SHEET
PRACTICE EXAMINATION THREE

Follow the instructions given in the test. Mark only your answers in the ovals below.

WARNING: Be sure that the oval you fill is in the same row as the question you are answering. Use a No. 2 pencil (soft pencil).

BE SURE YOUR PENCIL MARKS ARE HEAVY AND BLACK. ERASE COMPLETELY ANY ANSWER YOU WISH TO CHANGE.

START HERE — DO NOT make stray pencil dots, dashes or marks.

1 Ⓐ Ⓑ Ⓒ Ⓓ	2 Ⓐ Ⓑ Ⓒ Ⓓ	3 Ⓐ Ⓑ Ⓒ Ⓓ
4 Ⓐ Ⓑ Ⓒ Ⓓ	5 Ⓐ Ⓑ Ⓒ Ⓓ	6 Ⓐ Ⓑ Ⓒ Ⓓ
7 Ⓐ Ⓑ Ⓒ Ⓓ	8 Ⓐ Ⓑ Ⓒ Ⓓ	9 Ⓐ Ⓑ Ⓒ Ⓓ
10 Ⓐ Ⓑ Ⓒ Ⓓ	11 Ⓐ Ⓑ Ⓒ Ⓓ	12 Ⓐ Ⓑ Ⓒ Ⓓ
13 Ⓐ Ⓑ Ⓒ Ⓓ	14 Ⓐ Ⓑ Ⓒ Ⓓ	15 Ⓐ Ⓑ Ⓒ Ⓓ
16 Ⓐ Ⓑ Ⓒ Ⓓ	17 Ⓐ Ⓑ Ⓒ Ⓓ	18 Ⓐ Ⓑ Ⓒ Ⓓ
19 Ⓐ Ⓑ Ⓒ Ⓓ	20 Ⓐ Ⓑ Ⓒ Ⓓ	21 Ⓐ Ⓑ Ⓒ Ⓓ
22 Ⓐ Ⓑ Ⓒ Ⓓ	23 Ⓐ Ⓑ Ⓒ Ⓓ	24 Ⓐ Ⓑ Ⓒ Ⓓ
25 Ⓐ Ⓑ Ⓒ Ⓓ	26 Ⓐ Ⓑ Ⓒ Ⓓ	27 Ⓐ Ⓑ Ⓒ Ⓓ
28 Ⓐ Ⓑ Ⓒ Ⓓ	29 Ⓐ Ⓑ Ⓒ Ⓓ	30 Ⓐ Ⓑ Ⓒ Ⓓ
31 Ⓐ Ⓑ Ⓒ Ⓓ	32 Ⓐ Ⓑ Ⓒ Ⓓ	33 Ⓐ Ⓑ Ⓒ Ⓓ
34 Ⓐ Ⓑ Ⓒ Ⓓ	35 Ⓐ Ⓑ Ⓒ Ⓓ	36 Ⓐ Ⓑ Ⓒ Ⓓ
37 Ⓐ Ⓑ Ⓒ Ⓓ	38 Ⓐ Ⓑ Ⓒ Ⓓ	39 Ⓐ Ⓑ Ⓒ Ⓓ
40 Ⓐ Ⓑ Ⓒ Ⓓ	41 Ⓐ Ⓑ Ⓒ Ⓓ	42 Ⓐ Ⓑ Ⓒ Ⓓ
43 Ⓐ Ⓑ Ⓒ Ⓓ	44 Ⓐ Ⓑ Ⓒ Ⓓ	45 Ⓐ Ⓑ Ⓒ Ⓓ
46 Ⓐ Ⓑ Ⓒ Ⓓ	47 Ⓐ Ⓑ Ⓒ Ⓓ	48 Ⓐ Ⓑ Ⓒ Ⓓ
49 Ⓐ Ⓑ Ⓒ Ⓓ	50 Ⓐ Ⓑ Ⓒ Ⓓ	51 Ⓐ Ⓑ Ⓒ Ⓓ
52 Ⓐ Ⓑ Ⓒ Ⓓ	53 Ⓐ Ⓑ Ⓒ Ⓓ	54 Ⓐ Ⓑ Ⓒ Ⓓ
55 Ⓐ Ⓑ Ⓒ Ⓓ	56 Ⓐ Ⓑ Ⓒ Ⓓ	57 Ⓐ Ⓑ Ⓒ Ⓓ
58 Ⓐ Ⓑ Ⓒ Ⓓ	59 Ⓐ Ⓑ Ⓒ Ⓓ	60 Ⓐ Ⓑ Ⓒ Ⓓ
61 Ⓐ Ⓑ Ⓒ Ⓓ	62 Ⓐ Ⓑ Ⓒ Ⓓ	63 Ⓐ Ⓑ Ⓒ Ⓓ
64 Ⓐ Ⓑ Ⓒ Ⓓ	65 Ⓐ Ⓑ Ⓒ Ⓓ	66 Ⓐ Ⓑ Ⓒ Ⓓ
67 Ⓐ Ⓑ Ⓒ Ⓓ	68 Ⓐ Ⓑ Ⓒ Ⓓ	69 Ⓐ Ⓑ Ⓒ Ⓓ
70 Ⓐ Ⓑ Ⓒ Ⓓ	71 Ⓐ Ⓑ Ⓒ Ⓓ	72 Ⓐ Ⓑ Ⓒ Ⓓ
73 Ⓐ Ⓑ Ⓒ Ⓓ	74 Ⓐ Ⓑ Ⓒ Ⓓ	75 Ⓐ Ⓑ Ⓒ Ⓓ
76 Ⓐ Ⓑ Ⓒ Ⓓ	77 Ⓐ Ⓑ Ⓒ Ⓓ	78 Ⓐ Ⓑ Ⓒ Ⓓ
79 Ⓐ Ⓑ Ⓒ Ⓓ	80 Ⓐ Ⓑ Ⓒ Ⓓ	81 Ⓐ Ⓑ Ⓒ Ⓓ
82 Ⓐ Ⓑ Ⓒ Ⓓ	83 Ⓐ Ⓑ Ⓒ Ⓓ	84 Ⓐ Ⓑ Ⓒ Ⓓ
85 Ⓐ Ⓑ Ⓒ Ⓓ	86 Ⓐ Ⓑ Ⓒ Ⓓ	87 Ⓐ Ⓑ Ⓒ Ⓓ
88 Ⓐ Ⓑ Ⓒ Ⓓ	89 Ⓐ Ⓑ Ⓒ Ⓓ	90 Ⓐ Ⓑ Ⓒ Ⓓ
91 Ⓐ Ⓑ Ⓒ Ⓓ	92 Ⓐ Ⓑ Ⓒ Ⓓ	93 Ⓐ Ⓑ Ⓒ Ⓓ
94 Ⓐ Ⓑ Ⓒ Ⓓ	95 Ⓐ Ⓑ Ⓒ Ⓓ	96 Ⓐ Ⓑ Ⓒ Ⓓ
97 Ⓐ Ⓑ Ⓒ Ⓓ	98 Ⓐ Ⓑ Ⓒ Ⓓ	99 Ⓐ Ⓑ Ⓒ Ⓓ
100 Ⓐ Ⓑ Ⓒ Ⓓ		

Directions: You are given 5 minutes to study the following floor plan and to commit to memory as many details as you can. You are *not* permitted to make any written notes during the 5 minutes you are studying the illustration.

After 5 minutes, stop studying the illustration, turn the page, and answer the questions without referring to the illustration. The next time you are permitted to look at the illustration is when you have completed the test and are verifying your answers.

Now start your 5 minutes on a clock and begin.

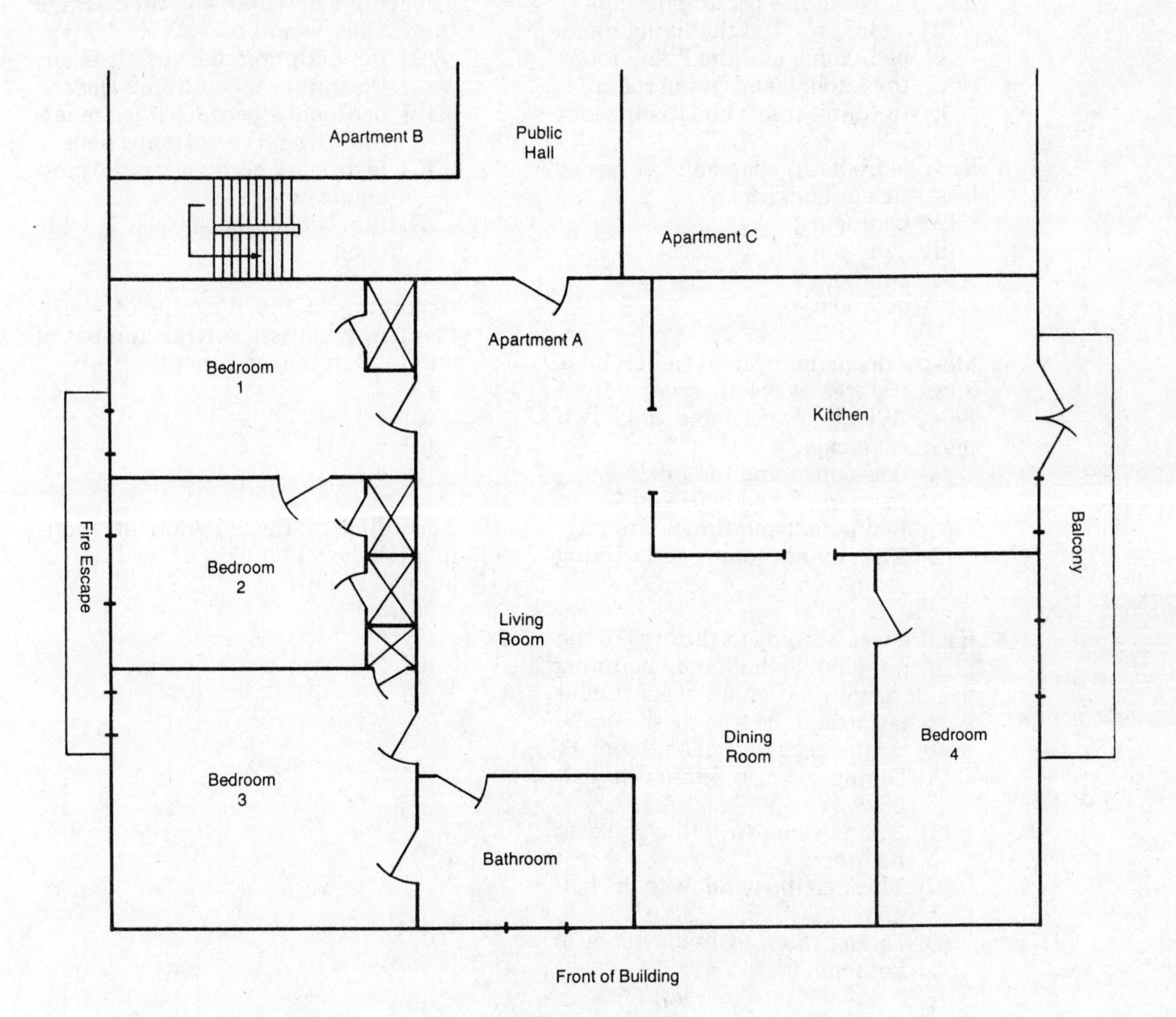

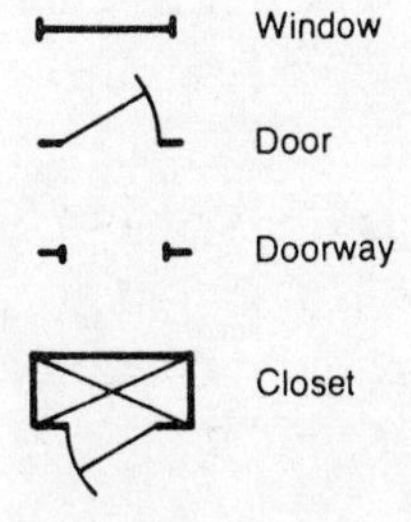

1. The occupants of apartment A could safely escape from the apartment and reach the street by how many means if no fire department ladders were used?
   (A) 1
   (B) 2
   (C) 3
   (D) 4

2. To escape from a fire in bedroom 2 of apartment A and exit via the public hall would require passing through
   (A) bedroom 1 and the living room
   (B) bedroom 3 and the living room
   (C) the kitchen and living room
   (D) the dining room and living room

3. Each bedroom in apartment A has at least one closet except
   (A) bedroom 1
   (B) bedroom 2
   (C) bedroom 3
   (D) bedroom 4

4. Most of the rooms in apartment A offer at least three ways to escape from them. Which rooms have only two means of escape?
   (A) Bedroom 1 and the kitchen.
   (B) Bedroom 2 and bedroom 4.
   (C) Bedroom 3 and the bathroom.
   (D) The living room and dining room.

5. If a fire was burning in the area of the door to the public hall and beginning to extend into bedroom 1 of apartment A, what would be the best escape route for the occupant of bedroom 4?
   (A) Dining room to kitchen to balcony.
   (B) Dining room to living room to bathroom.
   (C) Through the window to the balcony.
   (D) Dining room to living room to bedroom 3.

6. From which room in apartment A are there two ways to get directly to the outside air?
   (A) Bedroom 3.
   (B) The bathroom.
   (C) Bedroom 4.
   (D) The kitchen.

7. From which window of apartment A would it be possible for the fire department to make a rescue by using an aerial ladder?
   (A) Bedroom 4.
   (B) Dining room.
   (C) Bathroom.
   (D) Bedroom 3.

8. In a very heavy smoke condition the most difficult room in apartment A for fire fighters to search and then escape from safely would be
   (A) the bathroom because it is farthest from the entrance door
   (B) bedroom 4 because it is the farthest from the entrance door
   (C) bedroom 2 because it is the most complex
   (D) the kitchen because it is wide open

9. According to the diagram of apartment A, what is the total number of windows in the apartment?
   (A) 2
   (B) 4
   (C) 6
   (D) 8

10. According to the diagram of apartment A, how many doors lead into or out of a room?
   (A) 5
   (B) 6
   (C) 7
   (D) 8

Answer questions 11 through 15 using only the information in the following passage.

Dispensing Flammable Liquids

The inherent hazards of flammable liquids are greatest when transferring such liquids from one container to another. Heavy vapors can travel considerable distances to sources of ignition. The following precautions should be taken to ensure maximum safety when refueling the fire apparatus.

1. Prior to refueling, assure the following:
(A) Allow no visitors in quarters.
(B) Permit no smoking.
(C) Safeguard against all sources of ignition, that is,
  (1) apparatus ignition system
  (2) battery chargers
  (3) exposed battery terminals
  (4) mobile radios
(D) Close kitchen and cellar doors if present. Doors and windows on apparatus floor can provide proper ventilation.

2. Whenever possible, transfer fuel from storage directly to vehicle tank.

3. When necessary in lieu of above, use only safety-type cans and observe the following precautions:
(A) Avoid overfill or spillage.
(B) Never use a device to hold open the spring-loaded cap.

4. Immediately flush away spills with copious amounts of water. Avoid flushing spills into sewer systems. Consideration should also be given, when feasible, to sanding operations or other appropriate measures.

5. Maintain contact between fuel tank and nozzle, just prior to and during refueling operation. This method of grounding will prevent static sparks.

6. Maintain hose, nozzles, and safety cans in proper condition. Replace as necessary. Mark unserviceable safety cans appropriately and remove them from quarters.

7. Keep suitable extinguishers readily available.

11. Flammable liquids are most dangerous
(A) when being created
(B) when being used
(C) when being transferred
(D) when being stored

12. Which of the following would NOT be a viable precaution to take when handling flammable liquids?
(A) Prevent visitors from entering the refueling area.
(B) Prohibit smoking in the refueling area.
(C) Whenever possible, use a safety can to transfer fuel from storage to the vehicle tank.
(D) Keep appropriate portable fire extinguishers available.

13. When transferring gasoline from a gasoline pump to the fire apparatus tank, the nozzle should be kept inside the tank and in contact with the fill pipe. The main reason for this procedure is
(A) to prevent the loss of flammable fuel
(B) to activate the automatic nozzle shut-off
(C) to prevent the nozzle from coming out
(D) to prevent static electric sparks

14. When using a 5-gallon safety can to refuel a fire apparatus, all of the following precautions shall be adhered to EXCEPT
(A) keep all doors and windows open

(B) avoid overfill or spillage
(C) don't use a tool to prop open the spring-loaded cap
(D) do not permit any smoking

15. With respect to the dispensing of flammable liquids, it is correct to state that:
(A) The battery charger may be left plugged in.
(B) The fire apparatus radio should be turned down.
(C) Spills should be flushed with large quantities of water.
(D) Water should be flushed into the sewer.

16. It is recommended that the life net be held by not more than 14 or less than 10 fire fighters. Under certain conditions it may become necessary to use fewer fire fighters.
According to this passage, it is
(A) sometimes advisable to use more than 14 fire fighters on the net
(B) impossible to use a net unless at least 10 fire fighters are available to hold it
(C) best to use between 10 and 14 fire fighters on the net
(D) better to use 10 fire fighters on the net rather than 14

17. The flammability of gasoline fumes is amazing. In one case these fumes were carried outside a building to a lighted lamp 30 feet away. There they took fire and flashed back to the building, which was entirely destroyed.
According to this passage, gasoline
(A) fumes sometimes catch fire many feet from the liquid gasoline
(B) fumes are more dangerous then liquid gasoline
(C) flashback is an important element in building losses
(D) can be safely used if no lighted lamps are within 30 feet of the scene

18. In any fire, destruction is present in varying degrees. In addition to the destruction caused directly by the fire, there is also destruction caused in fighting the fire. If the sum of destruction to a building by fire and fire fighters is greater for one method of combating the fire than another, the method causing the lower level of destruction should be employed.
According to this passage, which of the following statements is correct?
(A) Fire fighting methods are responsible for the major destruction in most fires.
(B) Unavoidable damage by fire fighters should be ignored when choosing a fire fighting method.
(C) The aim when choosing a method of attack is to find the one that will cause the least amount of total damage.
(D) Ways of fighting fires should be chosen that are not dangerous to fire fighters.

19. Spray nozzles use considerably less water than straight-stream nozzles to achieve the same results. They are well suited, therefore, for use in situations where the water supply is limited.
According to this passage, spray nozzles
(A) should be used more often than straight-stream nozzles
(B) achieve better results than straight-stream nozzles
(C) are not effective when the water supply is large
(D) can be used where only a small supply of water is available

20. Fires cannot be fought entirely by rules or set procedures. The successful fire fighter also studies past fires and applies what he learns to future fire fighting action.
According to the above paragraph, which of the following is correct?
(A) Little can be learned about fire fighting by studying rules and procedures.
(B) The most successful fire fighter is the one with the most fire fighting experience.
(C) Studying past fires helps solve problems that may be met in future fires.
(D) Fires are successfully fought only through experience, but rules and procedures are valuable if based on a study of past fires.

Answer questions 21 and 22 using only the information in the following passage.

The most significant improvements in personnel selection procedures can be expected from a program designed to obtain more precise statements of the requirements for a particular position, to select not just those applicants who are generally best, but those whose abilities and personal characteristics provide the closest fit to the specific job requirements.

21. Better personnel selection procedures will result from
    (A) simplifying job descriptions
    (B) instituting better recruiting procedures
    (C) obtaining more detailed experience data from applicants
    (D) obtaining detailed statements of the training and skills required for various positions

22. The most desirable applicant for a position is
    (A) the person who has all the necessary training, even though he lacks the personal characteristics required
    (B) the one whose abilities and personal characteristics are of the highest order
    (C) the person who has the greatest interest in obtaining the position
    (D) the one whose qualifications are most nearly the same as the job requirement

23. A life gun is used to shoot a projectile, with a cord attached, to persons in upper stories or on roofs who cannot be reached in any other way. A life line is tied to the other end of the cord. According to this passage, use of a life gun is
    (A) usually the best way to rescue persons in upper stories or on roofs
    (B) a way to send a life line to people in upper stories or on roofs
    (C) a last resort because the projectile is dangerous
    (D) done with the lifeline attached directly to the projectile

Answer questions 24 and 25 using only the information in the following passage.

All operations at a building fire are aimed at saving life and reducing damage to property. If we are to achieve these objectives, it is almost always necessary to get inside the building quickly to locate and remove trapped persons and extinguish the fire. Often it will be necessary to ventilate the building so that search and rescue operations and fire extinguishment can be carried out. If the fire has progressed to the point at which there is danger of backdraft, or smoke explosion, we must ventilate before entering. How and where we enter and ventilate will depend on the urgency of the search and rescue problem. Since entry, ventilation, and search and rescue activities affect and are affected by each other, knowledge of each is needed before we can perform any one effectively. Coordination of these activities and the skill with which they are performed will determine their success, and to a great extent the success of the entire fire ground operation.

24. Success in saving life and reducing damage to property depends most on
    (A) proper ventilation of the building so that search and rescue can be carried out
    (B) the speed at which the building is entered so that trapped persons can be removed and the fire extinguished
    (C) how and where the building is entered and ventilated, since ventilation is vital in order to perform rescue activities
    (D) the skillful implementation of entry, ventilation, and search and rescue

25. When considering the problem of how

and where to enter a burning building, the most important aspect is
(A) the possibility of unnecessary damage to the building
(B) the number of persons in the building
(C) the urgency of the search and rescue problem
(D) the danger of backdraft and the difficulty of fire extinguishment

Answer questions 26 through 35 using only the information in the following passage.

Regulations Governing the Use and Storage of Ammunition for Powder-Actuated Tools in the Construction and Alteration of Buildings

1. The following regulations shall apply whether or not a permit for the storage of ammunition is required.

2. Powder-actuated tools using ammunition shall be of an approved type and so labeled.

3. Such tools shall not be used in an explosive atmosphere.

4. a. The main supply of ammunition shall be kept in a locked metal box interlined with ½″ asbestos or other noncombustible insulating material.
b. Storage and distribution of ammunition shall be supervised by a competent person who shall have the key to the storage box in his possession.
c. The ammunition storage box shall be kept away from heat and shall not be stored in the same compartment or shanty in which compressed gases or flammable liquids are kept.

5. The compartment or shanty in which the locked ammunition box is stored shall bear a permanent sign with the words "DANGER—AMMUNITION" in 2″ white letters on a red background.

6. There shall be provided one 2½-gallon water-type extinguisher or equivalent where ammunition is stored.

7. "No Smoking" signs shall be posted in area where ammunition is stored.

8. a. Powder-actuated tools utilizing ammunition shall be used only by a person holding a Certificate of Fitness issued by the Fire Department upon submission of evidence that said person has satisfactorily completed a training program in the safe use of such equipment, acceptable to the Fire Department.
b. No powder-actuated tool utilizing ammunition shall be used unless the Certificate of Fitness holder establishes a safe zone behind the work area by the use of ½″ steel backup plate and/or maintenance of an area clear of all people.

26. Ammunition for powder-activated tools shall be stored
(A) in a special box and in a compartment or shanty
(B) in a special box at least 5 feet from heat or compressed gases
(C) in a shanty with a black and white sign reading "DANGER —AMMUNITION"
(D) in a shanty if the special box is not locked

27. The ammunition storage box must be
(A) made of steel
(B) lined with asbestos
(C) not more than ½ inch thick
(D) equipped with a lock

28. Regulations governing the use and storage of ammunition for powder-activated tools in the construction and alteration of buildings include all of the following EXCEPT
(A) a permit is always required to store ammunition
(B) only approved and labeled equipment shall be used
(C) these tools shall be used only in a nonexplosive atmosphere
(D) the ammunition storage box

shall be kept away from flammable liquids

29. A Certificate of Fitness issued by the Fire Department is required to operate
    (A) all tools
    (B) power-activated tools
    (C) pneumatic-activated tools
    (D) powder-activated tools utilizing ammunition

30. The storage and distribution of ammunition for use with a powder-activated tool shall be supervised by
    (A) a person with a Certificate of Fitness
    (B) a person with a special permit
    (C) a competent person
    (D) any available worker

31. The shanty where the main supply of ammunition is stored should have all of the following EXCEPT
    (A) "No Smoking" signs posted in the area
    (B) two buckets of sand to cover ammunition
    (C) a 2½-gallon water-type portable fire extinguisher
    (D) a "Danger—Ammunition" sign

32. When a powder-activated tool is going to be used,
    (A) a competent person shall establish a safe zone in front of the work area
    (B) a Certificate of Fitness holder shall establish a safe zone in front of the work area
    (C) a competent person shall establish a safe zone behind the work area
    (D) a Certificate of Fitness holder shall establish a safe zone behind the work area

33. It would be correct to state that a safe zone can be established by using a
    (A) ½-inch metal backup plate
    (B) ½-inch asbestos backup plate
    (C) ½-inch P.V.C. (polyvinyl chloride) backup plate
    (D) ½-inch steel backup plate

34. At a construction site where a powder-activated tool is going to be used, no appropriate backup plate is available. In this situation
    (A) the tool may not be used
    (B) any metal backup plate may be used
    (C) no backup plate is required if the front of the work area is maintained clear of all people
    (D) no backup plate is required if the back of the work area is maintained clear of all people

35. To obtain a Certificate of Fitness as a user of powder-activated tools, a person must
    (A) be a high school graduate
    (B) pass a Fire Department examination on the operation of such tools
    (C) submit evidence that he has passed a training program approved by the Fire Department
    (D) attend the Fire Department training program and complete the course satisfactorily

Answer questions 36 through 38 using only the information in the following passage.

Flammable Wearing Apparel and Interior Furnishings

In order to actively participate and cooperate with the Federal and State authorities to achieve the intent of the Flammable Fabrics Act and the General Business Law, a report shall be transmitted by the Bureau of Fire Investigation to the Division of Fire Prevention whenever this department becomes engaged in an operation involving deaths or injuries determined to be caused by the ignition of wearing apparel, interior furnishings, sleeping bags, or portable temporary shelters which are deemed to possess extraordinary flammability characteristics. Such report shall include all information relative to the incident, and whenever possible a sample of the subject-material shall accompany the report.

36. Whenever a death occurs as a result of ignition of a person's clothing, a report must be forwarded to the
    (A) Federal Bureau of Investigation
    (B) Division of Fire Prevention
    (C) Bureau of Fire Investigation
    (D) Flammable Fabrics Commission

37. The Federal Flammable Fabrics Act controls all of the following EXCEPT
    (A) tents
    (B) kitchen appliances
    (C) sleeping bags
    (D) drapes

38. The fire investigator working at a fire in which several people were injured as a result of a flash fire of a living room couch should
    (A) send a sample of the couch material with the report
    (B) send a picture of the couch with the report
    (C) have the couch impounded and sent to the Division of Fire Prevention
    (D) send a sample of the couch and a picture of the apartment with the report

Answer questions 39 and 40 using only the information in the following passage.

Breaching Walls

Many different types of construction will be encountered in rescue operations. These include walls made of brick with lime mortar, brick with cement mortar, stone, concrete, and concrete block.

When cutting through walls or floors of large buildings, try to locate sections of the structures in which cutting can be done most quickly and safely. When cutting through walls, be sure that support beams and columns are not weakened. After a building has been subjected to bomb blast, the parts left standing may appear sound, although badly shaken and cracked. Therefore, when cutting away wall sections, especially with air hammers, care must be taken to prevent further collapse.

Openings large enough for rescue purposes usually can be made in brick walls without danger of the masonry falling. The bricks should be removed so that the opening is arch-shaped.

Concrete walls and floors, especially when they are reinforced, are difficult to cut through. Pavement breakers or other power tools will be helpful. Squad leaders should call for such equipment from the engineering services. In all walls and floors except concrete, the best method is to cut a small hole and then enlarge it. With concrete, however, it is better to cut around the edge of the section to be removed. If the concrete is reinforced, the reinforcing bars can then be cut by a hacksaw or torch, and the material removed in one piece. If a torch is used, be sure explosive gases are not present, and that flammable materials are not ignited. A fire extinguisher should be kept nearby.

39. To rescue a person trapped by collapse of a floor may require breaching a hole in a wall. When cutting a hole in a wall, a fire fighter should
    (A) make the hole as large as possible
    (B) make the hole round in shape
    (C) make sure the support beams and columns are not weakened
    (D) make sure the person to be rescued is alive

40. When breaking through a reinforced concrete wall, the steel reinforcing rods should be cut with a torch. When this is done, the fire fighter should
    (A) remove the material in a number of small pieces
    (B) cut all the reinforcing rods at the same time
    (C) make sure any flammable materials are ignited to assist with the cutting
    (D) have an appropriate portable fire extinguisher near by

Answer questions 41 and 42 using only the information in the following passage.

At every premises where gasoline, flammable liquids, and/or diesel fuel for motor vehicles are stored in underground storage systems, there shall be posted the name of the owner or person responsible for testing of such systems and the telephone number where such owner or person can be reached 24 hours/day, 7 days/week in the event of a leak or other emergency. Such sign or notice, in a form acceptable to the Fire Commissioner, shall be posted in such manner that it is easily read from the outside of the premises, and shall include, additionally, the name and phone number of the supplier or distributor who furnishes gasoline or flammable liquid product to the premises.

41. With respect to the sign described in the passage, it is correct to state that:
    (A) It shall be easily read from the outside of the premises.
    (B) It shall be posted for all underground storage systems.
    (C) It shall be made of durable materials and be white with red letters.
    (D) It shall indicate the business hours of the occupancy.

42. The sign to be posted shall contain all of the following EXCEPT
    (A) name of owner
    (B) telephone number where owner can be reached
    (C) address of owner
    (D) name and telephone number of supplier or distributor

Answer questions 43 through 47 using only the information in the following passage.

Care of Fire Hose

Fire hose is subject to injury through lack of proper care. Replacement in forward areas is often difficult or even impossible; hence the following precautions must be observed:

Protection from mechanical injury. (1) Avoid dragging hose over ground or other rough surfaces.
(2) Avoid sharp folds which may result in permanent kinks. If folds in hose on trucks and on reels are changed frequently, strain at fold points is reduced.
(3) Guard against traffic damage. Vehicles should not be permitted to drive over hose lines. Hose bridges and special guards should be provided.
(4) Eliminate friction between hose and ground when working from a pumper. Special chafing blocks, shoes, or guards may be made for this purpose.
(5) Use special care when hoisting hose over sills, parapets, or other obstructions.
(6) Avoid closing shut-off nozzles rapidly, to prevent excessive back pressure.
(7) Do not drop or drag couplings on pavements.
(8) Handle frozen hose with great care.
(9) Only in extreme emergencies is serviceable fire hose used in flushing, drainage, or other nonfire purposes.
(10) Use oldest hose or salvage hose for drill work, except when hose lines are charged.
(11) Avoid operating pumps at pressures above those recommended for the hose.

43. The fire apparatus pump operator should
    (A) avoid using the pumps in the pressure position
    (B) avoid using the pumps at pressures above the levels prescribed for the hose
    (C) avoid using the pumps in the volume position
    (D) avoid operating the pumps at

pressures above the levels recommended for the pumps.

44. A firefighter who regularly conducted dry fire-hose-handling drills used new hose so that members would "get the feel" of it. This procedure is
    (A) correct—it is best if members work with new hose because it increases their pride.
    (B) incorrect—new hose should be used only when older hose is not available.
    (C) correct—new hose is safer and more dependable.
    (D) incorrect—older hose should be used for practice and newer hose saved for fire attack.
45. A firefighter would be CORRECT if the firefighter
    (A) used a protective hose roller when hoisting hose over a parapet and onto the roof.
    (B) closed the hose nozzle rapidly to avoid excessive back pressure.
    (C) used the fire attack hose for washing the tile walls of the fire station.
    (D) allowed backed-up traffic to pass slowly over the fire hose.
46. Fire hose is subject to damage because of improper handling and care. Which of the following would NOT be considered proper care of the fire hose?
    (A) Fire apparatus is not permitted to drive over it.
    (B) Hose nozzles are shut down slowly.
    (C) Chafing blocks are used to protect against rubbing on the ground.
    (D) Folds in the fire hose are maintained in the same position.
47. Protection of fire hose from mechanical damage includes all of the following EXCEPT
    (A) not dragging fire hose over rough surfaces.
    (B) eliminating friction from the fire hose.
    (C) not dragging fire hose over the ground.
    (D) eliminating the dropping of fire hose couplings on the pavement.
48. Appropriately selected and properly installed fire detection systems can have a major effect in reducing loss of life from fire. A study of the circumstances surrounding 342 dwelling fire deaths indicated that the use of smoke detectors could result in a 41 percent saving of life; the use of thermal detectors could save 8 percent.

    According to this passage,
    (A) heat detectors have consistently proved to be important life savers.
    (B) a study found that smoke detectors resulted in the saving of 41 lives.
    (C) life safety can be improved substantially through the installation of smoke detectors.
    (D) a study indicates that 342 dwelling fires resulted in deaths.

Answer questions *49* through *52* using only the information in the following passage.

Fire Detection

The purpose of detection is twofold: it can reduce life loss and it can reduce property loss. Where human lives are at risk, time becomes an important factor. Time is needed to alert the occupants and time is needed for them to reach an area of refuge. Throughout this time period it is essential that an escape route be passable. Where the risk to life is minimal and property loss reduction is the consideration, longer detection times are often tolerated for the sake of minimizing needless alarms. Often the sprinkler system with its water flow alarm doubles as the detection system. Sprinkler systems as presently used are often slow to respond to smoldering fires—those which produce large quantities of toxic products.

49. Sprinkler systems are used to control and extinguish fires. Another use of the sprinkler system is to
    (A) act as a fire detection system
    (B) maintain the escape route in passable condition
    (C) reduce the production of toxic chemicals
    (D) limit the flow of water to needless areas of the occupancy

50. The passage places great emphasis on time. It would be LEAST correct to state that:
    (A) Time becomes important when human life is at stake.
    (B) Where risk to life is minimal, less time for fire detection is tolerated.
    (C) Time is needed to alert the occupants.
    (D) Reaching the area of refuge takes time.

51. The purpose of fire detection is
    (A) threefold
    (B) to report a fire to the fire department
    (C) to find and extinguish fires
    (D) to reduce property damage and loss of life

52. The passage states that sprinkler systems
    (A) do not react quickly to smouldering fires
    (B) produce large quantities of toxic products
    (C) will not extinguish large, smouldering fires
    (D) are not presently in use

Answer questions 53 through 57 using only the information in the following passage.

Ensuring Safety at Intersections

The crossing of intersections, particularly against traffic controls, has consistently produced some of the most severe apparatus accidents.

Many fire departments require department vehicles to *stop* before entering an intersection against a red light, stop sign, or yield sign to assure that the right-of-way is being yielded by all intersection traffic.

Slowing at an intersection without stopping offers only enough time for a quick glance in each direction. Too often this glance does not identify the vehicle that will fail to yield the right-of-way. The correct procedure, when crossing an intersection against traffic controls, is as follows: (1) Stop. (2) See that the cross traffic is yielding. (3) Proceed.

53. A speeding fire apparatus on the way to a fire should NOT
    (A) stop before entering an intersection against a red light
    (B) stop for a yield sign at an intersection
    (C) stop at all intersections
    (D) stop for a stop sign at an intersection

54. Slowing at an intersection without stopping
    (A) is permissible if the light is about to change
    (B) is not permitted
    (C) doesn't allow sufficient time to avoid a collision
    (D) is not necessary

55. The correct procedure for a fire apparatus crossing an intersection against traffic controls includes all of the following EXCEPT
    (A) stop
    (B) wait for the light to change
    (C) see the traffic stop
    (D) proceed when safe

56. Many severe fire fighting apparatus accidents occur because of
    (A) excessive speeding
    (B) failure to obey department regulations
    (C) crossing traffic against controls
    (D) the carelessness of the driver

57. As a fire fighter instructing a class of

new recruits it would be correct for you to state that:
(A) All fire departments require their vehicles to stop before entering an intersection.
(B) Crossing intersections against traffic controls causes the most accidents.
(C) Slowing at an intersection without stopping is sufficient to determine whether other traffic will yield the right-of-way.
(D) Stop signs and yield signs at intersections should be treated just like red lights.

Answer questions 58 through 60 using only the information in the following passage.

Development of an effective arson-reduction program by a community is dependent on knowing the extent and types of arson. In communities across the nation, there is an absence of reliable data on the incidence of arson. In the volunteer system, especially within unincorporated areas, there are normally no central points for the collection of arson statistics. Each fire department and fire protection district keeps its own records, with little uniformity and completeness of data among them. The state or county fire marshal's office also keeps fire investigation data. Here, classifications of fires can vary. Dollar loss estimates for arson are guesses at best. Moreover, not all arson fires are reported or even identified. The USFA National Fire Incident Reporting System (NFIRS) and the FBI Uniform Crime Reports (UCR) are improving the data situation, but on the whole, accurate, uniform arson statistics are lacking at this time.

58. The title that best fits this passage is
(A) Lack of Reliable Arson Data
(B) Statistics and Arson
(C) Arson, the Undefined Crime
(D) State and County Roles in Arson

59. The passage states that:
(A) All arson fires are investigated.
(B) The state or county fire marshal's office keeps accurate records.
(C) The classification of fires is done by the USFA.
(D) Each fire department and fire protection district keeps its own records.

60. While uniform arson statistics are not totally available, the data situation is improving because of the
(A) United States Federal Fire Service
(B) United Congressional Crime Reporting System
(C) National Fire Incident Reporting System
(D) National Fire Academy Fire Statistical Project

61. A senior fire fighter hears a newly appointed fire fighter tell a group of school children that, if their clothing begins to burn, the proper thing to do is to scream for help and beat the flames out with their hands. Knowing this is incorrect, the senior fire fighter should
(A) disagree immediately but courteously with the new fire fighter's statement, and explain the stop, drop, and roll method
(B) agree with the new fire fighter's statement but suggest a second method, the stop, drop, and roll method
(C) disagree immediately with the new fire fighter, and scold him for giving inaccurate information that could seriously endanger someone
(D) agree with the new fire fighter since he has recently finished school and is probably up to date on all the new techniques

62. A fire may be burning a block away from a fire station without the fire fighters knowing it. This shows that
(A) more fire inspections are needed

(B) the firefighters are not doing their job.
(C) firefighters should always watch hazardous locations.
(D) an alarm system is necessary.

63. If a person's clothes catch fire, one way to put the fire out is quickly to wrap a blanket around him. This method is effective mainly because
(A) it cuts off the supply of air to the fire.
(B) most blankets are fireproof.
(C) blankets are made of material that is a good heat insulator.
(D) the blanket has a cooling effect.

64. You are fighting a fire with a hose in a large storeroom and find that the situation is too dangerous for you to remain there any longer. The room is so dark and so filled with thick smoke that you cannot see even a short distance. You can probably best find your way out by
(A) lighting a match.
(B) using your recollection of where the door is located.
(C) following the hose back from the nozzle.
(D) spraying water in different directions to locate an exit by sound.

65. A firefighter working at a scene where a high-tension wire connected to a pole transformer is down should
(A) climb the pole and cut the wire close to the transformer.
(B) move the wire to a position where it will not interfere with the flow of traffic.
(C) use a rope to tie off an area around the danger zone.
(D) go to the nearest manhole or transformer vault and disconnect the power.

66. When a fire is burning in brush on the side of a hill, the fire generally covers ground fastest
(A) downhill.
(B) uphill.
(C) sideways from the fire.
(D) in the middle of the fire.

67. The knot shown in illustration 67 would be used to
(A) form a noose for hoisting tools.
(B) wrap around fire victims to lower them to the ground.
(C) shorten the effective length of a piece of rope.
(D) lash two ladders together to extend their lengths.

Illustration 67

68. While attending a local movie theater, a firefighter finds the rear exit doors secured with a chain and lock wrapped through the door handles. The best action for the firefighter to take would be to
(A) pull the interior fire alarm box to summon the local fire company to the scene to enforce the law.
(B) tell the manager to remove the lock immediately because it is unsafe and illegal.
(C) leave the theater because it is unsafe and then report the condition to the local fire company.
(D) take a seat near the open entrance door, tell an usher about the problem, and then watch the movie.

69. Two firefighters on their way to work see a fire in a store on the first floor of a four-story apartment building. Which of the following is the correct action for these firefighters to take?
(A) Both firefighters enter the building and alert the occupants to the fire.
(B) One firefighter goes to the nearest fire alarm box and waits there until the fire company arrives, while the other firefighter alerts the occupants.
(C) One firefighter goes to the fire station to report the fire and let the officer know that the other firefighter will be late in reporting for duty.

(D) One fire fighter goes to the nearest fire alarm box and awaits the arrival of the fire company, while the other continues on to the fire station to make sure the company responds and the officer is aware that the second fire fighter will be late.

70. During a serious fire a fire fighter searching an apartment house is confronted by a large and apparently ferocious dog preventing access to the apartment above the fire. The most appropriate action for this fire fighter to take is to
    (A) hold the dog at bay with an axe and yell into the apartment, "Is anybody there?"
    (B) try to find the dog's owner
    (C) report back to his officer that the apartment cannot be searched because of the dog
    (D) get help from another fire fighter to control the dog and then enter and search the apartment

71. At a fire in a large multiapartment house the officer instructed one of the fire fighters to shut off the gas supply to the entire building. After the fire a resident of an apartment three floors below the fire requested that the gas be turned back on. The fire fighter was instructed by the officer not to comply with this request. The most likely reason for this refusal is that:
    (A) The pilot lights in unoccupied apartments are out and gas will accumulate.
    (B) Fire fighters are not taught how to turn the gas supply on.
    (C) Fire fighters are permitted to turn the gas on only when requested to do so by the utility company.
    (D) The fire is probably not under complete control, and the gas would present a hazard.

72. While directing a hose stream, a fire fighter straddles a hose line. This action of the fire fighter is
    (A) safe, mainly because the fire fighter will not be hit if the hose ruptures
    (B) unsafe, because the fire fighter may become tangled in the hose if a fast escape is required
    (C) safe, because the fire fighter has complete control of the hose and nozzle
    (D) unsafe, because the fire fighter will be injured if the line ruptures

73. Every fire should be thoroughly investigated to determine how it started. The primary reason for finding out how the fire started is to
    (A) evaluate the fire fighting units
    (B) collect statistics on the number and types of fires
    (C) reduce the possibilities of future fires
    (D) prove that the fire was arson

74. Ventilation, that is, the opening of windows, doors, and the roof of a building, is a method used by fire fighters to increase the probability of rescue of occupants and rapid extinguishment of the fire. Which of the following would NOT be a benefit of ventilation?
    (A) Control of how the fire will travel.
    (B) An increase in visibility.
    (C) A slowing of the rate at which the fire burns.
    (D) A lowering of the amount of dangerous gases.

75. A problem likely to be found in a hotel fire but not usually encountered in fires involving other types of residences, such as apartment houses, is
    (A) obstructions in hallways and other passageways
    (B) large numbers of people in a small number of rooms
    (C) delay in the transmission of the fire alarm
    (D) the presence of many occupants who don't know the location of exits

76. Which tool would be best for clearing out brush to prevent the spread of a fire?

(A) 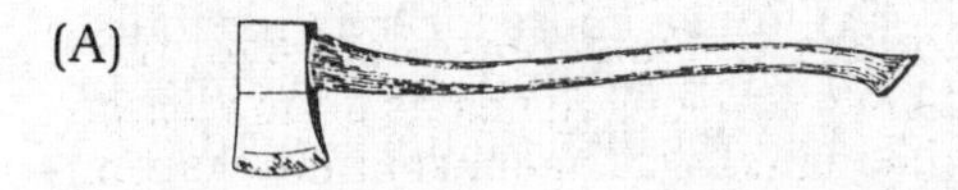

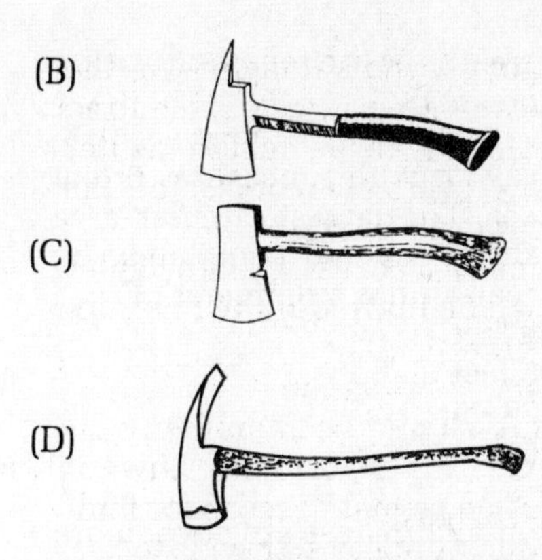

77. Which tool would be best for tightening the connections of a household water supply pipe?

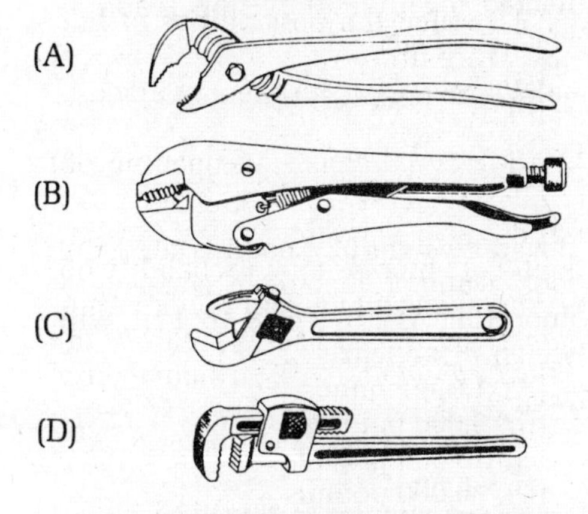

78. The long pole and hook shown in illustration 78 is called a pike pole. Fire fighters sometimes push the point and the hook through a plaster ceiling and then pull the ceiling down. Which of the following is the most likely reason for this practice?
    (A) To let heat and smoke escape from the room.
    (B) To trace defective electric wiring through the house.
    (C) To see whether hidden fire is burning above the ceiling.
    (D) To remove combustible materials that will provide fuel for the fire.

Illustration 78

79. Fire fighters must often chop holes in roof planking to let smoke escape from burning buildings. When chopping these holes, they try to place their axe cuts just alongside the heavy supporting beams underneath the planking. The most likely reason for positioning the cuts just to the sides of such beams is that:
    (A) The planking is thinnest at this point.
    (B) There are no nails in this area to dull the axe.
    (C) The axe is less likely to "bounce."
    (D) The beam provides a strong support for the fire fighters using the axe.

80. Some tools are known as all-purpose tools because they may be used for a great variety of purposes; others are called special-purpose tools because they are suitable only for a particular purpose. Generally, an all-purpose tool, as compared to a special tool for the same purpose, is
    (A) cheaper
    (B) less efficient
    (C) safer to use
    (D) simpler to operate

81. The device indicated in illustration 81 would be used to
    (A) force open center-closing elevator doors
    (B) open or close the operating valve on a fire hydrant
    (C) hold several wide boards together
    (D) jack up the bumper of an automobile

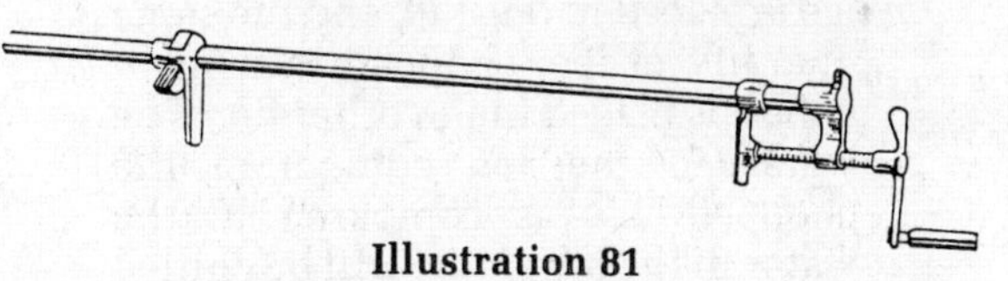

Illustration 81

82. The tool that would be used to hold the clamp together in illustration 82 until a bolt and nut could be inserted and made up would be:

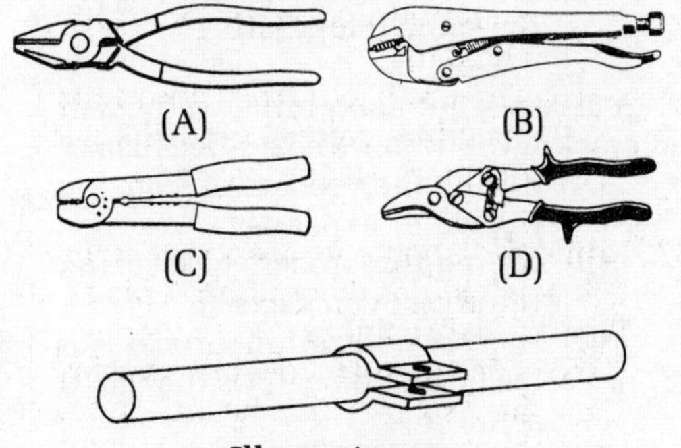

Illustration 82

83. The use of power-driven tools can be dangerous. Which of the following would NOT be a safety technique you should observe?
   (A) When possible, clamp down the object being worked on.
   (B) Use only sharp, clean, well-cared-for tools.
   (C) Approach the operator of the tool only from behind.
   (D) Plan how you will proceed with the work before you turn the tool on.

84. The tool that would be most appropriate to cut a chain securing a gate on a fence would be
   (A) an axe
   (B) a bolt cutter
   (C) a pry bar
   (D) a tin snip

85. A tool that is used to loosen or tighten a set screw is sometimes called a hex key or set-screw wrench. This tool is also known as
   (A) a crocodile wrench
   (B) a Stillson wrench
   (C) an Allen wrench
   (D) a crescent wrench

86. Assume that two identical insulated jugs are filled with equal quantities of water from a water tap. A block of ice is placed in one jug, and the same quantity of ice, chopped into small cubes, is placed in the other jug. The water in the jug containing the chopped ice, as compared to the water in the other jug, will be chilled
   (A) faster and to a substantially lower temperature
   (B) faster and to approximately the same temperature
   (C) slower but to a substantially lower temperature
   (D) slower and to approximately the same temperature

87. A clay pitcher containing water will crack if the water freezes because
   (A) during the process of freezing the water expands
   (B) during the process of freezing the water contracts
   (C) the crystallization process of ice formation causes the ice to solidify
   (D) the tension strength of clay decreases as the temperature decreases

88. A canvas tarpaulin measures 6 feet by 9 feet. The largest circular area that can be covered completely by this tarpaulin has a diameter of
   (A) 9 feet
   (B) 8 feet
   (C) 7 feet
   (D) 6 feet

89. If sections of fire hose are never more than 50 feet long and never less than 40 feet long, how many sections are needed to ensure that, when connected together, they will reach at least 300 feet?
   (A) 9
   (B) 8
   (C) 7
   (D) 6

90. If water weighs 62½ pounds per cubic foot, water 1 foot deep covering a floor that is 10 feet × 16 feet will weigh most nearly
   (A) 1,000 pounds
   (B) 6,000 pounds
   (C) 10,000 pounds
   (D) 16,000 pounds

91. In illustration 91, when part *A* revolves in the direction of arrow 1, part *B*
   (A) rotates in the direction of arrow 3
   (B) rotates in the direction of arrow 4
   (C) does not rotate
   (D) rotates first in direction 1, then in direction 2

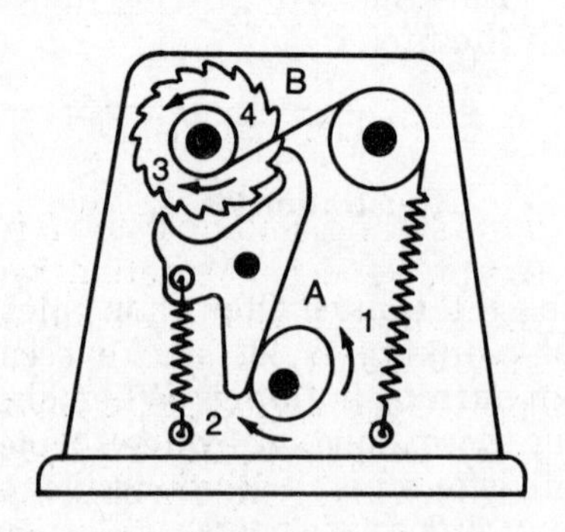

Illustration 91

92. If the pistons of both pumps in illustration 92 go up and down, which of the following statements is true?
 (A) In pump 1, water will go out at *A*.
 (B) In pump 1, water will go out at *B*.
 (C) In pump 2, water will go out at *C*.
 (D) In pump 2, water will go out at *D*.

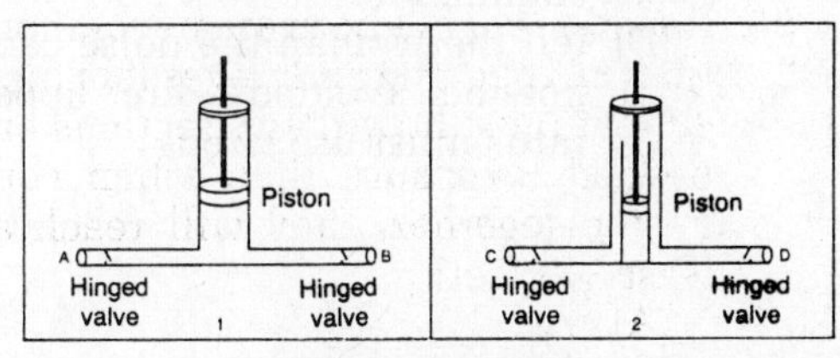

Illustration 92

93. If a piece of rope 100 feet long is cut so that one piece is 2/3 as long as the other piece, the length of the longer piece must be
 (A) 60 feet
 (B) 66⅔ feet
 (C) 70 feet
 (D) 75 feet

94. If the sliding shaft on gear *A* in illustration 94 is pushed to the left, what will happen to drum *B*?
 (A) It will turn in direction 1.
 (B) It will turn in direction 2.
 (C) It will move to the right and will not turn.
 (D) It will move to the left and will not turn.

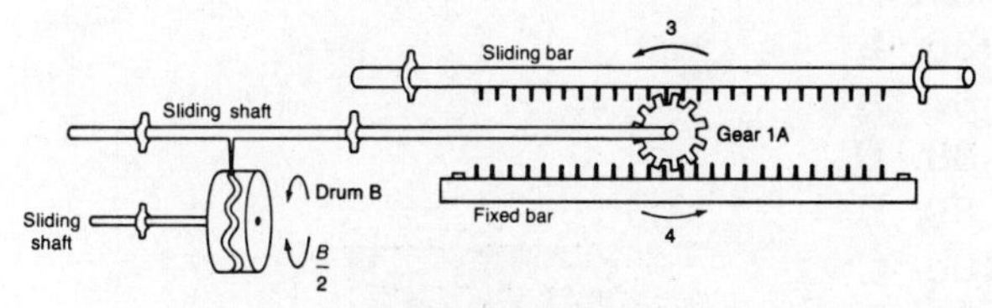

Illustration 94

95. If the sliding shaft on drum *B* in illustration 94 is made stationary, what will happen to gear *A* if drum *B* turns in direction 1?
 (A) Gear *A* will turn in direction 3.
 (B) Gear *A* will turn in direction 4.
 (C) Gear *A* will turn back and forth between directions 3 and 4.
 (D) Gear *A* will not move.

96. An elevator in a large apartment house became stuck between floors, and the fire department was called to remove the trapped passengers. Soon after arriving, the officer in command informed the passengers that the fire department was present and would start rescue operations immediately. The main reason for informing passengers of the fire fighters' arrival was to reduce the chance that:
 (A) Another agency would receive credit for the rescue.
 (B) The Fire Department would be criticized for being slow in responding.
 (C) The passengers would not cooperate with the rescuers.
 (D) The passengers would become panic-stricken if they did not know help was at hand.

97. Each year many children die in fires that they started by playing with matches. Which of the following measures would be most effective in preventing such tragedies?
 (A) Warn children of the dangers involved.
 (B) Punish parents who are found guilty of neglecting their children.
 (C) Educate adults to keep matches out of the reach of children.
 (D) Tell people to use only safety matches.

98. A senior fire fighter, after showing a probationary fire fighter how to perform a task, should
 (A) have another senior fire fighter check to see whether the probationary fire fighter has learned the job
 (B) question the probationer sharply as to whether he understands
 (C) watch the probationary fire fighter do all his tasks
 (D) let the probationary fire fighter perform the task and then check on his performance from time to time

99. Assume that you have just been transferred to a new fire station. The best action for you to take is to
 (A) advise the other members of

how you want to do your work
(B) request to be transferred out of the new station and back to your old station
(C) check the type and location of the tools on the apparatus
(D) tell the officer of the minor infractions you observe in the new station

100. A woman who lives next door to the fire station enters quarters and complains that the sirens make too much noise. The fire fighter on housewatch duty should
(A) ask the woman for suggestions as to how the company should respond with sirens
(B) explain why a fast response is needed, why traffic-warning devices are used, and that the siren is used only when necessary
(C) ask the woman to put her complaint in writing so that you can forward it to the company commander
(D) tell the woman the noise cannot be avoided—fire apparatus must use sirens

ANSWER KEY

1. **D**	21. **D**	41. **A**	61. **A**	81. **C**
2. **A**	22. **D**	42. **C**	62. **D**	82. **B**
3. **D**	23. **B**	43. **B**	63. **A**	83. **C**
4. **B**	24. **D**	44. **D**	64. **C**	84. **B**
5. **D**	25. **C**	45. **A**	65. **C**	85. **C**
6. **D**	26. **A**	46. **D**	66. **B**	86. **B**
7. **C**	27. **D**	47. **B**	67. **C**	87. **A**
8. **C**	28. **A**	48. **C**	68. **B**	88. **D**
9. **C**	29. **D**	49. **A**	69. **B**	89. **B**
10. **D**	30. **C**	50. **B**	70. **D**	90. **C**
11. **C**	31. **B**	51. **D**	71. **A**	91. **B**
12. **C**	32. **D**	52. **A**	72. **D**	92. **B**
13. **D**	33. **D**	53. **C**	73. **C**	93. **A**
14. **A**	34. **D**	54. **C**	74. **C**	94. **D**
15. **C**	35. **C**	55. **B**	75. **D**	95. **C**
16. **C**	36. **B**	56. **C**	76. **D**	96. **D**
17. **A**	37. **B**	57. **D**	77. **D**	97. **C**
18. **C**	38. **A**	58. **A**	78. **C**	98. **D**
19. **D**	39. **C**	59. **D**	79. **C**	99. **C**
20. **C**	40. **D**	60. **C**	80. **B**	100. **B**

DIAGNOSTIC PROCEDURE

Use the following diagnostic chart to determine how well you have done and to identify your areas of weakness.

Enter your number of correct answers for each section in the appropriate box in the column headed "Your Number Correct." The column immediately to the right will indicate how well you did on each section of the test.

Below the chart you will find directions as to the chapter(s) in the guide you should review to strengthen your weak areas.

Section Number	Question Number	Area	Your Number Correct	Scale	
One	1–10	Recall		10 Right	Excellent
				9 Right	Good
				8 Right	Fair
				Under 8 Right	Poor
Two	11–60	Understanding Job Information		48–50 Right	Excellent
				46–47 Right	Good
				45 Right	Fair
				Under 45 Right	Poor
Three	61–75	Judgment and Reasoning		15 Right	Excellent
				14 Right	Good
				13 Right	Fair
				Under 13 Right	Poor
Four	76–85	Tools and Equipment		10 Right	Excellent
				9 Right	Good
				8 Right	Fair
				Under 8 Right	Poor
Five	86–90	Mathematics, Machines, and Science		5 Right	Excellent
				4 Right	Good
				—	—
				Under 4 Right	Poor
Six	91–100	Dealing with People		10 Right	Excellent
				9 Right	Good
				8 Right	Fair
				Under 8 Right	Poor

1. If you are weak in Section One, then concentrate on Chapter 6.
2. If you are weak in Section Two, then concentrate on Chapter 7.
3. If you are weak in Section Three, then concentrate on Chapter 8.
4. If you are weak in Section Four, then concentrate on Chapter 9.
5. If you are weak in Section Five, then concentrate on Chapter 10.
6. If you are weak in Section Six, then concentrate on Chapter 11.

Note: Consider yourself weak in a section if you receive other than an "excellent" rating.

ANSWER EXPLANATIONS

1. **D** There are four ways to escape without using ladders: the interior public stairs and three bedroom windows to the fire escape. The balcony cannot be used to go from one level to another without the aid of a ladder.
2. **A** The door from bedroom 2 opens into bedroom 1 only. Therefore escape through the apartment door would require passing through bedroom 1 and the living room.
3. **D** Bedroom 4 has no closet. There are two closets in bedroom 2, and one closet in each of bedrooms 1 and 3.
4. **B** From bedroom 2 a person can go either out to the fire escape or into bedroom 1. From bedroom 4 a person can go into the dining room or out to the balcony and then in through the kitchen to the front door.
5. **D** This route would allow the occupants to go down the fire escape and to safety. Choices A and C result in stranding the occupant on the balcony to await help from the fire department. Choice B is incomplete and unwise. The occupant of bedroom 4 must still go through bedroom 3 to escape, and going via the bathroom would take extra time and effort.
6. **D** The kitchen has both a door and a window that open onto the balcony.
7. **C** This is the only room with a window that opens on the front of the building and is not obscured by the balcony or fire escape. Note that the dining room (B) does not have a window.
8. **C** To reach bedroom 2 requires going through one of two doors; searching involves the room and two closets. The possibility of becoming disoriented is high in this type of situation. Choices A and B (bathroom and bedroom 4) are incorrect because the distances are not excessive in this relatively small apartment. Choice D (the kitchen) would be relatively easy to search.
9. **C** There is one window in each of the four bedrooms, one in the bathroom, and one in the kitchen.
10. **D** The kitchen and each of the four bedrooms have one door apiece; there are also two doors in the bathroom and a door out to the public hall.
11. **C** The first sentence of the passage gives this information.
12. **C** Not "whenever possible," but *only* when direct transfer is not possible, should a portable can be used (see items 2 and 3 of the passage).
13. **D** See the last sentence of item 5. Placing the nozzle against the fill pipe grounds the pipe to the pump and reduces the chance of a static spark.
14. **A** Not all doors should be kept open; kitchen doors and cellar doors should be closed (item 1D) to reduce the chance of the vapors reaching the pilot light on the kitchen stove or the firehouse heating unit. The vapors of gasoline are heavier than air.
15. **C** "Large quantities" of water and "copious amounts" of water (see item 4) have the same meaning. Choice A—the battery charger should be disconnected from the apparatus and the outlet. Choice B—the radio should be turned off, not down. Choice D—spills should not be flushed into the sewer.
16. **C** This information is found in the first sentence of the passage. Choice A—the maximum number recommended is 14. Choice B—the second sentence states that ". . . it may become necessary to use fewer [than 10] fire fighters." Choice D—the exact "best" number of fire fighters is not indicated; only the range is given. Do not read into the passage facts that are not there.
17. **A** The passage indicates that gasoline vapors flashed back 30 feet. Choice B may be true, but this is not indicated in the passage. There is no mention of liquid gaso-

line. Choices C and D are incorrect and are not implied in the passage.

18. **C** This information can be found in the last sentence of the passage. Choice A—the destruction done by fire fighters can *sometimes* be greater than that caused by the fire. The passage does not indicate that this is true *most* of the time; the numbers of times are not indicated for the two causes of destruction. Choice B—the method that causes the lowest level of destruction should be chosen (last sentence). Choice D—the passage does not address the question of danger to fire fighters. The answer must be based on the information in the passage.

19. **D** See the second sentence of the passage. Choice A—the passage indicates that the spray stream can achieve the same results as the straight stream (first sentence). We can assume, therefore, that there are situations where a spray stream is preferable and other situations where the straight stream is more effective. No information is given as to which type of situation occurs more often. Choice B—spray nozzles achieve the same, not better, results. Choice C—if spray nozzles are effective with limited water supplies, they will also be effective where adequate or large-volume water supplies are available.

20. **C** This means the same as the second sentence of the passage. Choice A—the passage says that not everything can be learned from rules and procedures. However, this does not mean that only a little can be learned. Choice B—the passage tells us that the successful fire fighter is the one who studies past fires and applies what is learned; it does not mention on-job experience. Choice D—according to the passage, the application of rules, set procedures, and operations at past fires leads to successful fire fighting.

21. **D** The passage states that selection procedures can be improved by obtaining ". . . more precise statement of the requirements for a particular position." Choice A—job descriptions should be made more precise, not simplified. Choice B—"recruiting" means getting people to apply; the passage refers to choosing the best applicants from those that have already applied. Choice C—this is not mentioned or implied in the passage.

22. **D** This is another way of saying ". . . those whose abilities and personal characteristics provide the closest fit to the specific job requirements." Choice A—training is not mentioned in the passage, whereas the importance of personal characteristics is stressed. Choice B—abilities and characteristics need not be of the highest order, but should provide the best fit to the specific job requirements. Choice C—interest is an admirable quality; however, it is the person's abilities and personal characteristics that the passage stresses.

23. **B** The passage tells us that the life line is connected to the cord, the cord is attached to the projectile, and the projectile is shot to people in upper stories or on roofs. Choice A—this is not indicated in the passage; the best rescue method would be by the interior stairs. Choice C—the method is a last resort, not because the projectile is dangerous, but because there is no other way to reach the person. Choice D—the cord is attached to the projectile, and the life line is tied to the other end of the cord.

24. **D** This captures the meaning of the entire passage and can be inferred from the last sentence. Choices A, B, and C each address only one aspect each of the successful life-saving process.

25. **C** This information can be found in the fifth sentence. Choices A and B are not discussed in the passage and therefore are not relevant to answering the question. Choice D is incorrect because, according to the passage, the problem of backdraft is involved in *when* to ventilate, not in *how* and *where*.

26. **A** See Sections 4.a and 5: All ammunition must be kept in "a locked metal box" in a "compartment or shanty." Choice B—no special distance is given; the box may not be stored in a shanty with compressed gases or flammable liquids (Section 4.c). Choice C—the sign should be red and white. Choice D—the box should not be left open; it must be kept locked.
27. **D** See Sections 4.a, 4.b, and 5. Choice A—the box must be made of metal, but which metal is not stated. Choice B—the box must be lined with asbestos *or* other noncombustible insulating material (Section 4.a). Choice C—no thickness is specified for the box.
28. **A** The regulations apply whether or *not* a permit is required (see Section 1).
29. **D** See Section 8.a. Choice A—the title states specifically that the regulations apply only to "powder-activated tools." Choices B and C—these refer to other forms of energy used to activate tools; they are not discussed in the passage.
30. **C** See Section 4.b. No special requirements are indicated other than competency.
31. **B** The passage does not mention buckets of sand. Choice A is specified in Section 7; choice C, in Section 6; choice D, in Section 5.
32. **D** This answer is given in Section 8.b. Choices A, B, and C are all incorrect on the basis of the passage.
33. **D** Section 8.b specifies a steel backup plate. Choice A—it is true that steel is metal; however, not all metals are steel. Choice B—asbestos is required in the ammunition box. Choice C—P.V.C. is a plastic and hence not acceptable.
34. **D** Section 8.b states the conditions: a safe zone behind the work area, established by means of a ½-inch steel backup plate, or an area clear of all people.
35. **C** Refer to Section 8.a. Choice A is not mentioned in the passage. Choice B—the prospective user need not pass a test, but rather must submit evidence of having satisfactorily completed a training program. Choice D—the training program is not given by the Fire Department, but must be acceptable to it.
36. **B** This information is given in the first sentence of the passage. Choices A and D are not mentioned, and the report is sent *by*, not *to*, the Bureau of Fire Investigation (C).
37. **B** Kitchen appliances are made of noncombustible materials and are not considered furnishings. Choice A—tents are "portable temporary shelters." Choice C—sleeping bags are specifically mentioned in the passage. Choice D—drapes are "interior furnishings."
38. **A** See the last sentence. Choices B, C, and D (picture of the apartment) are not mentioned in the passage.
39. **C** The fourth sentence states, "When cutting through walls, be sure that support beams and columns are not weakened." Choice A—the hole should be made only as large as necessary for rescue purposes. Choice B—the shape of the hole depends on the type of wall; for brick walls the opening is arch-shaped (paragraph 3). Choice D—this is not indicated and would not be practical—the person may be unconscious and unable to respond.
40. **D** See the last sentence. Choice A—concrete should be removed "in one piece." Choice B—the torch can cut only one rod at a time. Choice C—"be sure . . . that flammable materials are not ignited" (next to last sentence).
41. **A** The second sentence states that the sign ". . . shall be posted in such a manner that it is easily read from the outside of the premises. . . ." Choice B—not all underground systems, only those storing gasoline, flammable liquids, and/or diesel oil for motor vehicles require signs. Choices C and D—while these ideas may be good, they are not mentioned in the passage and hence cannot be considered.
42. **C** The owner's address is not necessary and is not mentioned in the

passage. Choices A, B, and D are all specifically required.

43. **B** Item 11 states, "Avoid operating pumps at pressures above those recommended for the hose." Choices A and C are not mentioned in the passage. Choice D refers incorrectly to pressures recommended for the pump, not the hose.

44. **D** See item 10: "Use oldest hose or salvage hose for drill work. . . ." Choice A—this is not indicated in the passage. Furthermore, pride is not a factor to be considered for this situation. Choice B—new hose can be used if the line is to be charged, that is, filled with water (see item 10). Choice C—it may be true that new hose is more dependable and safer, but this is not the question. When hose is used dry, the dependability of the hose is immaterial. The practice stated in the item is incorrect.

45. **A** See item 5. Choice B—see item 6. Choice C—item 9. Choice D—see item 3.

46. **D** Item 2 states that frequent changing of folds reduces the strain at fold points.

47. **B** Friction cannot be eliminated from fire hose. Item 4 indicates that fire fighters should "eliminate friction between hose and ground . . . ." Choices A and C—see item 1. Choice D—see item 7.

48. **C** The second sentence states that ". . . the use of smoke detectors could result in a 41% saving of life." Choice A—the passage refers to smoke detectors, not heat detectors. Choice B—smoke detectors could result in a 41% saving of lives, not a saving of 41 lives. Choice D—the study involved 342 deaths that occurred in dwellings. The number of dwellings involved is unknown; there could have been more than one death in a single dwelling.

49. **A** The next to last sentence tells us that a sprinkler system often doubles as a detection system. Choice B—while this function is possible, the passage does not say so. The passage states that the escape route must be passable but does not tell us how to achieve this. Choice C—sprinkler systems do not control the production of chemicals; there is no mention of this in the passage. Choice D—the passage addresses the subject of needless alarms (sentence 5), but not needless water or needless areas of the occupancy.

50. **B** The fifth sentence states that ". . . longer detection times are tolerated . . ." where there is minimal risk to life, not less time.

51. **D** This information is given in the first sentence. Choice A—The purpose is twofold, not threefold (sentence 1). Choice B—this is not indicated in the passage. Choice C—in a broad, general sense this is true, but it is not specifically supported by the passage, as is Choice A.

52. **A** See the last sentence. Choice B—smouldering fires, not sprinklers, can produce large quantities of toxic products (last sentence). Choice C—this is not indicated in the passage and is not correct; responses to smouldering fires are slow but does occur. Choice D—sprinkler systems are very much in use (note "as presently used" in the last sentence).

53. **C** The second paragraph tells us specifically what a fire apparatus should stop for. Choices A, B, and D are all required, but choice C is only a "maybe." It therefore is the best choice as something that would *not* be done.

54. **C** The third paragraph explains that slowing without stopping does not provide enough time to recognize what a car or truck is going to do. There is no justification in the passage for choices A, B, and D.

55. **B** It is not necessary to wait for the light to change, but only to ensure that other vehicles have yielded the right-of-way.

56. **C** This is stated in the opening sentence in the passage. Choices A, B, and D are possible causes of severe accidents but are not mentioned in the passage and therefore cannot be considered.

57. **D** The second paragraph states that

the apparatus should stop for all three traffic control devices. Choice A—according to the passage, many, but not all, fire departments have this regulation. Choice B—not the *most* accidents, but some of the *most severe* accidents, occur when intersections are crossed against controls. Choice C—slowing is not sufficient (see paragraph 3).

58. **A** The main theme of the passage is that existing data are unreliable. Note the second sentence: "... an absence of reliable data"; the third sentence: "... no central points for the collection of arson statistics"; the fourth sentence: "... little uniformity and completeness of data...."

59. **D** This information is given in the fourth sentence. Choice A—"Moreover, not all arson fires are reported or even identified" (sentence 8). Choice B—this office keeps records, but their accuracy is questionable (sentences 6–8). Choice C—classifications can and do vary from region to region (sentence 6); the passage does not state that the USFA classifies fires.

60. **C** See the last sentence, where the FBI Uniform Crime Reports are also cited as sources of improvement. Choices A, B, and D are not mentioned in the passage.

61. **A** An incorrect statement to a young, impressionable group could result in one of them being seriously injured at some future time. Although normally it is not good practice to correct someone in public, this would be an exception (and should be courteously done) because of the potential danger involved. Choice B—the new firefighter's statement is incorrect, and the senior firefighter has the responsibility to point this out, not to agree. Choice C—the newly appointed firefighter should not be scolded; instead the firefighter should be instructed politely as to the proper action. Choice D—the senior firefighter should not assume that the recommended procedure has been changed.

62. **D** A fire alarm system allows citizens to notify the fire department directly of a fire. It tells firefighters not only that there is a fire but also where the fire is located. The fire alarm box extends the communication reach of the firefighter out into the community and permits a citizen who does not know the location of the nearest fire station to transmit a fire alarm.

63. **A** Smothering is one of the four methods of extinguishing a fire. In this case the blanket cuts off the air supply and starves the fire. Choices B and D are not true, and choice D is irrelevant to extinguishing the fire.

64. **C** The hose line serves as a guide. The hose would have been hooked up to a water supply before you advanced it into the storeroom. By following it back to the water-supply point, you can safely escape. Choice A—a match will not give off sufficient light. Choice B—in complete darkness you cannot sense direction even if you think you remember the location of an exit. Choice D—spraying water around would be a waste of time and water. Any sound or the lack of it would be misleading.

65. **C** The rope will warn people of the danger and should keep them safely away from it. Choices A and B are unsafe acts. Control of the electrical power on the pole should be left to the utility company (A); the wire should not be moved unless there is an immediate, direct hazard to life (B), in which case the wire should be moved only with great caution. Choice D is wrong because underground and aboveground wiring are generally independent of each other. Shutting off the power to the underground system should not affect the aboveground system.

66. **B** Heat rises and in doing so preheats the areas above it. As the fire grows, so does the heat output.

67. **C** Illustration 67 shows a length of rope that has been folded back on

itself several times. A loop has been made at either end to loosely bind the rope, thereby taking up the slack and reducing any hazard from the excess rope. If a force is applied to any of the center lengths, the rope will come apart rapidly. Choice A—a clove hitch and binder would be used for this purpose. Choice B—a bowline on a bight or a rescue hitch would be used for this purpose. Note: The knots in choices A and B are used for the purposes stated because they do not come apart easily. Choice D—lashing ladders together would be an unsafe act and should not be undertaken. The proper action would be to get a longer ladder.

68. **B** The manager should have the lock and chain removed at once, thus taking care of the immediate hazard. The firefighter would follow up by telephoning the fire department and notifying it of the extremely dangerous condition that had been found. Choice A—this action would activate the interior fire alarm bells and could cause a panic in the theater. It would also result in the fire apparatus responding in an unnecessarily fast and hazardous manner. Choice C—the firefighter should stay at the theater until the lock and chain are removed or the local fire company arrives. Should a fire occur meanwhile, the firefighter could probably direct the people to safety. Choice D—this would not be adequate action on the part of the firefighter.

69. **B** This makes the best use of the skills of the two firefighters. The one who remains at the fire alarm box will direct the arriving units rapidly to the fire, while the second firefighter makes the necessary notification in the apartment building. Choice A—since there are two priorities, both firefighters should not do the same thing. Choices C and D—the important function of alerting the occupants is ignored.

70. **D** The apartment immediately above the fire is one of the most seriously exposed areas and presents an extreme life hazard for the occupants. The firefighter must get into the apartment and search for overcome victims. Normally a dog can be controlled by one or more firefighters, thus allowing another firefighter to conduct the search. Choice A—yelling will not be effective if the victim is unable to hear the firefighter or has been overcome and cannot answer. Choice B—this will result in delay, and the dog's owner may meanwhile be overcome in the apartment. Choice C—this statement is untrue because, with the aid of one or more other firefighters, search is possible.

71. **A** When the main gas supply is shut down, all the gas to the building is turned off. Even though the valves on each gas stove may be shut off, there is normally a pilot light which is supposed to burn a small amount of gas continuously. If the pilot light is out and the gas goes on, an accumulation of unburnt gas can develop and can be ignited and explode. Once the gas main is shut down, only the utility company should turn it back on. Choice B—firefighters are taught how the entire system functions. Choice C—firefighters are generally not permitted to turn on the gas supply. Choice D—this contradicts the stem of the question and is incorrect.

72. **D** The fire hose under pressure can whip around when the nozzle is opened or closed and will whip violently if the line bursts or is ruptured. This can cause and has caused many serious injuries. Straddling the line puts the firefighter in a position where an injury is almost a certainty. Choices A and C are incorrect—straddling the hose line is not safe. Choice B is correct in calling this practice unsafe but gives the wrong reason.

73. **C** By determining how fires started, unsafe practices and faulty equipment can be identified and such causes can be changed, cor-

rected, or removed. Choice A—Identifying the cause of a fire would in no way help to evaluate the actions of the fire fighters. Choice B—statistics are a tool used to prove that the cause of the fire, once identified, is or is not a continuing problem. Choice D—the question refers to every fire, not just arsonous ones.

74. **C** Ventilation provides a fresh supply of oxygen, which in turn increases, not slows, the rate of burning. Choice A—heat travels upward until it is stopped by a barrier; it mushrooms out across ceilings and builds up. By providing ventilation, we allow the heat to escape and the fire to reach the fresh oxygen supply provided at the ventilation hole. Choices B and D—ventilation does improve visibility (B) and reduce the quantity of gases (D).

75. **D** A person checking into a hotel for a short stay may not take the time to locate an exit other than the one by which he entered the building. Choices A and C are problems common to all residence occupancies. Choice B is incorrect because hotels have large numbers of people but also have large numbers of rooms; also, the average hotel room is occupied by only one or two people.

76. **D** This tool has both an axe head and an adz head. The axe head could be used to cut thick roots, small trees, tree branches, etc.; the adz head, for digging up roots and debris. Choices A and B offer only an axe-head tool. Choice C is used only for digging; it has an adz head and a point head (these are not used for cutting).

77. **D** This is a pipe or Stillson wrench. Although it would be possible to use any of the four tools shown under the right conditions, the best, or most appropriate, tool for the purpose stated is choice D. Choice A is channel lock pliers; choice B, vice grip pliers; choice C, a crescent wrench.

78. **C** Fire can and will travel behind walls and ceilings for long distances. Using this tool, the fire fighter can search rapidly for signs of a traveling fire.

79. **C** At the point where the roof boards meet the roof joists, the boards are held firmly in place and have little give. As we go away from this point, the boards have increasing flexibility and room to bend. If a board is hit at a point in the center of a bay, it will deflect downward to a greater degree than if it was struck at a point near the roof joist; it will then spring back to its normal position since the fibers have not been cut. The axe is used to cut through the wood fibers. If the board is flexible, it will absorb some of the force of the axe without the fibers being cut, thus slowing the cutting operation. Choice A—planks are usually uniform in size. Choice B—this is a true statement; the nails are in the board directly above the joist, not off to the sides; however, it is not why the cut is made at this location. Choice C is a better selection. Choice D—the fire fighter stands on the side of the roof away from the cut.

80. **B** A multipurpose tool is often more convenient because it offers many different functions in one package. It replaces several special-purpose tools but is often made with a lesser degree of tolerance and hence is less efficient. Choice A—generally, all-purpose tools are more, not less, expensive than single-purpose tools. Choice C—because all-purpose tools are made to work in many different situations, they are less precise and more subject to slipping. Choice D—because all-purpose tools are not specifically designed for a single, selected task, they must be adjusted to meet the needs of a particular application, thus making them more complicated to operate.

81. **C** The tool shown in illustration 81 is a pipe clamp, used to hold wide boards together.

82. **B** This tool (illustration 82) is a combination plier-wrench. It can be used as a small, portable hand-held vice. Choice A is lineman's

pliers; choice C, an electrician's multipurpose tool; choice D, duck-bill snips.

83. **C** Do not approach the operator from the rear. He may not see or hear you and may react in such a way as to endanger you or himself.

84. **B** A bolt cutter has hardened jaws and is designed to cut steel rods, bolts, and locks, as well as chains. Choice A—an axe would not cut the chain, but the chain would damage the axe head. Choice C—a pry bar would not be effective and might result in a strain or other injury to the user. Choice D—a tin snip would not provide the cutting head or leverage needed to sever the chain links.

85. **C** An Allen wrench is a hexagonal bar that is inserted into the head of the screw and then worked in the same way as a screwdriver.

86. **B** When water changes from a solid to a liquid, it takes on heat energy. The rate of energy absorbed depends on the amount of surface area available to absorb it. When a block of ice is chopped up, more surface area becomes available to absorb heat; therefore the ice will melt faster, eliminating choices C and D. The amount of heat required to melt a given volume of ice is the same, regardless of whether the ice is in block or chopped form; therefore A is incorrect.

87. **A** Water contracts as the temperature decreases. However, as it approaches the freezing point, water ceases contracting and begins to expand. This expansion would cause the clay pitcher (or a water pipe) to crack and release the water contained in it. Choice B—although the water does contract initially as it freezes, this would not cause the pitcher to break. Choices C and D are erroneous statements.

88. **D** The diameter of a circle is a straight line that passes through the center of the circle from one side to the other. Since the tarpaulin's smallest dimension is 6 feet, this is the largest circular area that can be covered.

89. **B** If we assume the worst case, each length is 40 feet. We divide the total length needed (300 feet) by a single length (40 feet) to get the number of lengths:

Total length ÷ Single length = Number of lengths

300 feet ÷ 40 feet = x

7.5 = x

Round to the next highest whole number: x = 8.

90. **C** First determine the number of cubic feet that will be covered:

Cubic feet (volume) = L × W × H

Cubic feet = 10 ft. × 16 ft. × 1 ft.

Cubic feet = 160 ft.^{3}

Now multiply the weight of 1 cubic foot of water (62.5 lb.) by the total volume (160 ft.^{3}):

62.5 lb. × 160 ft.^{3} = Total weight

10,000 lb. = Total weight

91. **B** Part A is oval in shape, and as it turns it pushes the pawl, causing it to release the ratchet. The purpose of the pawl is to prevent the ratchet from turning in the direction in which the points of the ratchet are facing. The ratchet can turn freely in the direction opposite the ratchet points. However, it can turn in the direction of the points only when the pawl is disengaged. Therefore, when part A pushes the pawl away from the ratchet, the ratchet will turn in direction 4.

92. **B** The hinge valves at points *A* and *B* are known as check valves. They are put into the pipe to restrict the flow of water to one direction only. On the top is the swinging check valve; below it is a projection (stop) that keeps the valve from swinging in the opposite direction. The actual flow of water is in the direction in which the check valve appears to be pointing. If the flow is reversed, the check valve will be forced back against the stop and will restrict the flow in that direction. When the piston in illustration 1 is pushed down, the water in the system is pushed against the check valve at point *A*, closing it, and against the check valve at

point *B*, opening it. By raising the piston and reducing the pressure, a new supply of water will enter at point *A* and the water at point *B* will be prevented from returning into the pump. In illustration 2, on an upward stroke water will enter from points *C* and *D*, and on a downward stroke both valve *C* and valve *D* would close, preventing any discharge.

93. **A** If the smaller of the two pieces is 2/3 of the larger, then the size of the larger piece must be 3/3. The total length of the rope is then equal to the sum of the two parts (3/3 + 2/3), or 5/3. If we divide the length of the original rope (100 feet) into 5 equal parts, we get 20 feet. Now we multiply this by the length of each rope: $3 \times 20 = 60$, and $2 \times 20 = 40$. The longer rope is 60 feet long.

94. **D** As the shaft attached to gear *A* is moved, the projection from the sliding shaft into drum *B* will push the sliding shaft attached to drum *B* in the direction of the push, in this case to the left. Choices A and B are incorrect; the drum will turn only if the sliding shaft attached to it is turned. Choice C is incorrect because the drum will move in the direction of the push, *not* in the opposite direction.

95. **C** By making the shaft attached to drum *B* stationary and then turning the drum, the projection from the shaft above will ride along the grove as drum *B*. This action will result in gear *A* moving back and forth.

96. **D** When confined in a small space with no apparent way out and faced with the possibility of serious danger, many people will undertake irrational acts in an attempt to save themselves. Letting the people know that help is at hand and that something is being done will tend to reduce their fears.

97. **C** If matches are kept out of the reach of children, they cannot play with them and cannot get burned from them. Choice A—a warning serves to inform the child but does not remove the temptation to play with the matches; also, some children are too young to understand the danger. Choice B—punishing parents may not prevent the child from playing with matches again. Choice D—safety matches are more difficult to light and may self-extinguish, but a child can still play with them and light them.

98. **D** The best way for the probationary fire fighter to learn is to be shown and then to perform the task. The trainer should follow this up with corrections and appropriate questions as needed.

99. **C** This action ensures that you know what equipment the new unit has and that you will be able to get the proper tool quickly when it is needed in an emergency. Choices A, B, and D are likely to lead to the development of animosity between you and the other members of the company.

100. **B** Explaining the purpose of sirens and letting the woman know that they are not being used unnecessarily will help to resolve her complaint.

CHAPTER 15

Practice Examination Four

In this chapter you will find the fourth practice examination. It has been developed by using official and sample questions from cities across the United States.

Be sure that when you take this examination you allow yourself a 3½-hour uninterrupted time period. Do not try to take the examination in parts; if you do, you will defeat its purpose. By taking the examination at one sitting you will become familiar with test conditions and with your personal idiosyncrasies about sitting and working for this period of time.

Before Taking the Examination

First, go back and review quickly the test-taking strategies outlined in Chapter 3. Then, when you begin the examination, be sure to read and follow all instructions. Time for reading the instructions has been figured into the examination. Read each question carefully, and answer only what is asked of you. Select the answer that is the one *best* choice of those provided, and then record your selection on the answer sheet.

The answer sheet precedes the examination. The Diagnostic Procedure and Answer Explanations are given at the end of the chapter.

Composite Examination

ANSWER SHEET
PRACTICE EXAMINATION FOUR

Follow the instructions given in the test. Mark only your answers in the ovals below.
WARNING: Be sure that the oval you fill is in the same row as the question you are answering. Use a No. 2 pencil (soft pencil).
BE SURE YOUR PENCIL MARKS ARE HEAVY AND BLACK. ERASE COMPLETELY ANY ANSWER YOU WISH TO CHANGE.

START HERE DO NOT make stray pencil dots, dashes or marks.

1 Ⓐ Ⓑ Ⓒ Ⓓ	2 Ⓐ Ⓑ Ⓒ Ⓓ	3 Ⓐ Ⓑ Ⓒ Ⓓ
4 Ⓐ Ⓑ Ⓒ Ⓓ	5 Ⓐ Ⓑ Ⓒ Ⓓ	6 Ⓐ Ⓑ Ⓒ Ⓓ
7 Ⓐ Ⓑ Ⓒ Ⓓ	8 Ⓐ Ⓑ Ⓒ Ⓓ	9 Ⓐ Ⓑ Ⓒ Ⓓ
10 Ⓐ Ⓑ Ⓒ Ⓓ	11 Ⓐ Ⓑ Ⓒ Ⓓ	12 Ⓐ Ⓑ Ⓒ Ⓓ
13 Ⓐ Ⓑ Ⓒ Ⓓ	14 Ⓐ Ⓑ Ⓒ Ⓓ	15 Ⓐ Ⓑ Ⓒ Ⓓ
16 Ⓐ Ⓑ Ⓒ Ⓓ	17 Ⓐ Ⓑ Ⓒ Ⓓ	18 Ⓐ Ⓑ Ⓒ Ⓓ
19 Ⓐ Ⓑ Ⓒ Ⓓ	20 Ⓐ Ⓑ Ⓒ Ⓓ	21 Ⓐ Ⓑ Ⓒ Ⓓ
22 Ⓐ Ⓑ Ⓒ Ⓓ	23 Ⓐ Ⓑ Ⓒ Ⓓ	24 Ⓐ Ⓑ Ⓒ Ⓓ
25 Ⓐ Ⓑ Ⓒ Ⓓ	26 Ⓐ Ⓑ Ⓒ Ⓓ	27 Ⓐ Ⓑ Ⓒ Ⓓ
28 Ⓐ Ⓑ Ⓒ Ⓓ	29 Ⓐ Ⓑ Ⓒ Ⓓ	30 Ⓐ Ⓑ Ⓒ Ⓓ
31 Ⓐ Ⓑ Ⓒ Ⓓ	32 Ⓐ Ⓑ Ⓒ Ⓓ	33 Ⓐ Ⓑ Ⓒ Ⓓ
34 Ⓐ Ⓑ Ⓒ Ⓓ	35 Ⓐ Ⓑ Ⓒ Ⓓ	36 Ⓐ Ⓑ Ⓒ Ⓓ
37 Ⓐ Ⓑ Ⓒ Ⓓ	38 Ⓐ Ⓑ Ⓒ Ⓓ	39 Ⓐ Ⓑ Ⓒ Ⓓ
40 Ⓐ Ⓑ Ⓒ Ⓓ	41 Ⓐ Ⓑ Ⓒ Ⓓ	42 Ⓐ Ⓑ Ⓒ Ⓓ
43 Ⓐ Ⓑ Ⓒ Ⓓ	44 Ⓐ Ⓑ Ⓒ Ⓓ	45 Ⓐ Ⓑ Ⓒ Ⓓ
46 Ⓐ Ⓑ Ⓒ Ⓓ	47 Ⓐ Ⓑ Ⓒ Ⓓ	48 Ⓐ Ⓑ Ⓒ Ⓓ
49 Ⓐ Ⓑ Ⓒ Ⓓ	50 Ⓐ Ⓑ Ⓒ Ⓓ	51 Ⓐ Ⓑ Ⓒ Ⓓ
52 Ⓐ Ⓑ Ⓒ Ⓓ	53 Ⓐ Ⓑ Ⓒ Ⓓ	54 Ⓐ Ⓑ Ⓒ Ⓓ
55 Ⓐ Ⓑ Ⓒ Ⓓ	56 Ⓐ Ⓑ Ⓒ Ⓓ	57 Ⓐ Ⓑ Ⓒ Ⓓ
58 Ⓐ Ⓑ Ⓒ Ⓓ	59 Ⓐ Ⓑ Ⓒ Ⓓ	60 Ⓐ Ⓑ Ⓒ Ⓓ
61 Ⓐ Ⓑ Ⓒ Ⓓ	62 Ⓐ Ⓑ Ⓒ Ⓓ	63 Ⓐ Ⓑ Ⓒ Ⓓ
64 Ⓐ Ⓑ Ⓒ Ⓓ	65 Ⓐ Ⓑ Ⓒ Ⓓ	66 Ⓐ Ⓑ Ⓒ Ⓓ
67 Ⓐ Ⓑ Ⓒ Ⓓ	68 Ⓐ Ⓑ Ⓒ Ⓓ	69 Ⓐ Ⓑ Ⓒ Ⓓ
70 Ⓐ Ⓑ Ⓒ Ⓓ	71 Ⓐ Ⓑ Ⓒ Ⓓ	72 Ⓐ Ⓑ Ⓒ Ⓓ
73 Ⓐ Ⓑ Ⓒ Ⓓ	74 Ⓐ Ⓑ Ⓒ Ⓓ	75 Ⓐ Ⓑ Ⓒ Ⓓ
76 Ⓐ Ⓑ Ⓒ Ⓓ	77 Ⓐ Ⓑ Ⓒ Ⓓ	78 Ⓐ Ⓑ Ⓒ Ⓓ
79 Ⓐ Ⓑ Ⓒ Ⓓ	80 Ⓐ Ⓑ Ⓒ Ⓓ	81 Ⓐ Ⓑ Ⓒ Ⓓ
82 Ⓐ Ⓑ Ⓒ Ⓓ	83 Ⓐ Ⓑ Ⓒ Ⓓ	84 Ⓐ Ⓑ Ⓒ Ⓓ
85 Ⓐ Ⓑ Ⓒ Ⓓ	86 Ⓐ Ⓑ Ⓒ Ⓓ	87 Ⓐ Ⓑ Ⓒ Ⓓ
88 Ⓐ Ⓑ Ⓒ Ⓓ	89 Ⓐ Ⓑ Ⓒ Ⓓ	90 Ⓐ Ⓑ Ⓒ Ⓓ
91 Ⓐ Ⓑ Ⓒ Ⓓ	92 Ⓐ Ⓑ Ⓒ Ⓓ	93 Ⓐ Ⓑ Ⓒ Ⓓ
94 Ⓐ Ⓑ Ⓒ Ⓓ	95 Ⓐ Ⓑ Ⓒ Ⓓ	96 Ⓐ Ⓑ Ⓒ Ⓓ
97 Ⓐ Ⓑ Ⓒ Ⓓ	98 Ⓐ Ⓑ Ⓒ Ⓓ	99 Ⓐ Ⓑ Ⓒ Ⓓ
100 Ⓐ Ⓑ Ⓒ Ⓓ		

Directions: Questions 1 through 5 are an ORAL examination and are intended to measure your ability to hear, understand, and apply information.

At the beginning of this examination have someone READ the passage, slowly and carefully, to you. DO NOT READ THIS PASSAGE YOURSELF.

To the assistant: Read the following passage aloud. Read slowly and distinctly. Once you start, read through to the end with only normal pauses between paragraphs.

Immediately after the reading, answer questions 1 through 5 on the basis of information found in that passage.

At a recruit training session, the trainees are told they will be assigned to a building inspection team that will go into the field on Thursday. The lieutenant supervising the team is William Holt, a veteran fire officer. The team will work in the downtown area, and will be expected to visit a minimum of five buildings.

Participation in these inspections is required to familiarize the trainees with proper practices for conducting Fire Prevention Inspections. They will show typical fire hazards that may be found in each occupancy.

On the day of the inspection, the trainees are introduced to fire fighter Bill Hartman and the other members of the team—John Grasso, George Carrol, and Ray Kelly. While on route downtown, Ray tells them they will start at the Tops Hotel and that this is a continuation of the inspection they had been working on yesterday. So far, three minor violations have been found: accumulation of rubbish in the rear, a nonfunctional exit light on the fourth floor, and a recently outdated permit for the storage of fuel oil.

During the inspection, a fellow trainee finds a locked exit door and states that "this is very serious" and calls for the "immediate issuing of a summons." You are asked to act as a witness, and the lieutenant orders the other trainee to fill out and serve the summons.

The next inspection takes place in a small restaurant that is well run and in proper order. The third occupancy inspected is a retail drugstore using its basement for the storage of its combustible goods. The inspection reveals that it has sprinklers throughout the basement. The lieutenant explains, "The requirement for a sprinkler system is a recent change in the building laws designed to improve fire safety throughout the city. The law also requires the owner to have several spare sprinkler heads stored at the control valve so a quick replacement can be made if a fire occurs." A further inspection finds no spare sprinkler heads on the premises.

1. The trainees were assigned to go out into the field and to participate in inspection activities on
   (A) Monday
   (B) Tuesday
   (C) Wednesday
   (D) Thursday

2. The officer who will be in command of your inspection team is
   (A) William Holt
   (B) Ray Kelly
   (C) John Grasso
   (D) Bill Hartman

3. Which of the following occupancies was NOT inspected?
   (A) Hotel
   (B) Retail store
   (C) Cold storage warehouse
   (D) Restaurant

4. A summons was issued by
   (A) you
   (B) a fellow trainee
   (C) the supervising lieutenant
   (D) you, a fellow trainee, and the lieutenant

5. During the inspection training, which of the following violations of law were found?
   (A) An accumulation of rubbish.
   (B) A locked exit.
   (C) An out-of-date permit.
   (D) Missing spare sprinkler heads.

Directions: You are given 5 minutes to study the following floor plan and to commit to memory as many details as you can. You are not permitted to make any written notes during the 5 minutes you are studying the illustration.

After 5 minutes, stop studying the illustration, cover it, and answer the questions without referring to the illustration. The next time you are permitted to look at the illustration is when you have completed the test and are verifying your answers.

Now start your 5 minutes on a clock and begin.

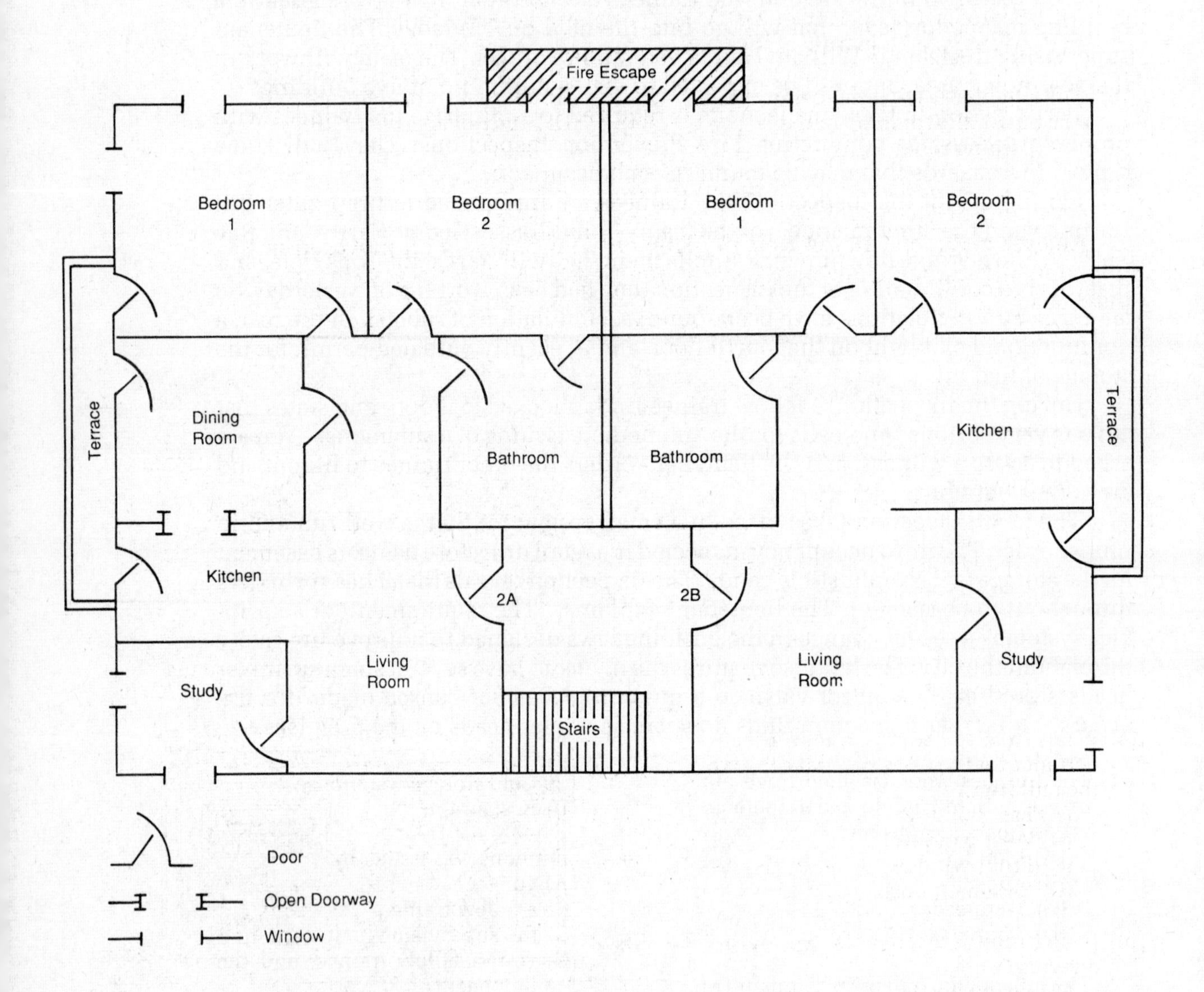

6. In apartment 2B, which room is farthest from the fire escape?
   (A) Bedroom 2.
   (B) Living room.
   (C) Study.
   (D) Foyer.

7. Which room has two doors?
   (A) Kitchen in apartment 2A.
   (B) Bathroom in apartment 2B.
   (C) Bedroom 2 in apartment 2B.
   (D) Bedroom 1 in apartment 2B.

8. Which room of apartment 2A can fire fighters reach directly from the fire escape?
   (A) Bedroom 2.
   (B) Bedroom 1.
   (C) Dining room.
   (D) Study.

9. Which room does not have egress to the exterior?
   (A) Living room.
   (B) Study in apartment 2B.
   (C) Bedroom 1 in apartment 2A.
   (D) Bedroom 2 in apartment 2B.

10. A fire fighter leaving the bathroom in apartment 2B would be in the
   (A) dining room
   (B) bedroom 2
   (C) kitchen
   (D) bedroom 1

11. Fire fighters on the terrace of apartment 2A could not directly enter the
   (A) bedroom 1 in apartment 2A
   (B) study in apartment 2A
   (C) kitchen in apartment 2A
   (D) dining room in apartment 2A

12. Which room has at least one widow on two sides of the building?
   (A) Living room of apartment 2B.
   (B) Bedroom 2 of apartment 2A.
   (C) Bedroom 1 of apartment 2B.
   (D) Study of apartment 2B.

Answer questions 13 through 16 using only the information provided in the following illustration. You may look at the illustration while answering these questions.

13. How many ways can fire fighters enter the house?
   (A) One.
   (B) Two.
   (C) Three.
   (D) Four.

14. The probable cause of the fire was
   (A) sparks from the fireplace
   (B) the overloaded electrical outlet
   (C) the combustion of flammable materials in the attic
   (D) smoking in bed

15. What would be the most direct entrance for the firemen to take to save the children?
   (A) East window.
   (B) West window.
   (C) Front window.
   (D) Basement door.

16. In which direction is the wind blowing?
   (A) To the north.
   (B) To the south.
   (C) To the east.
   (D) To the west.

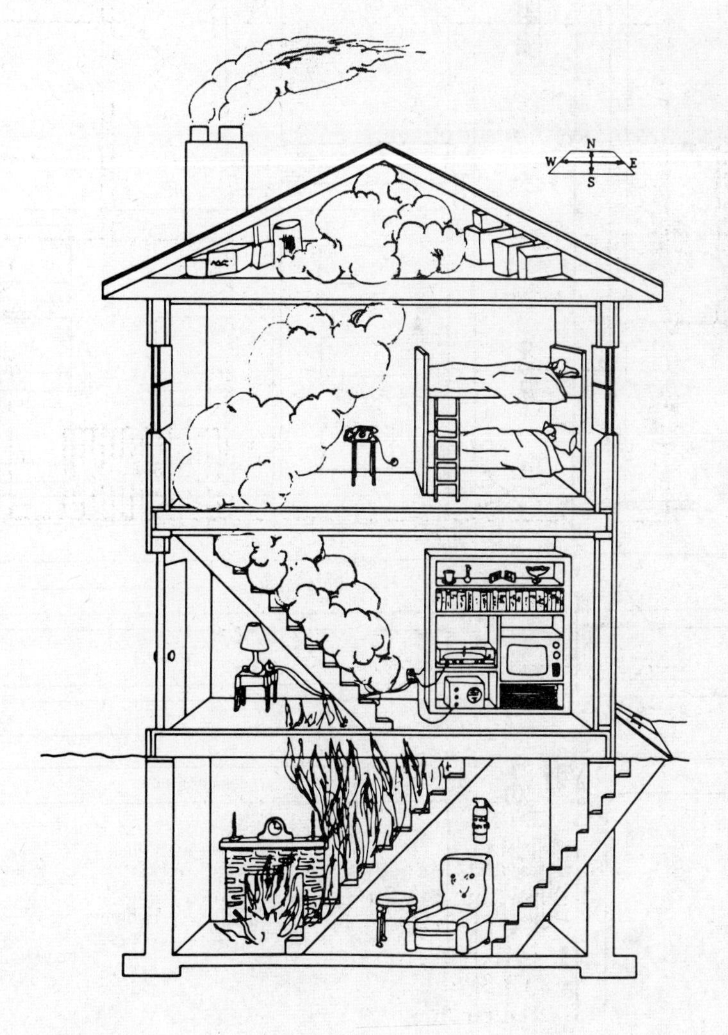

Answer questions 17 through 21 using only the information provided in the following illustration. You may look at the illustration while answering these questions.

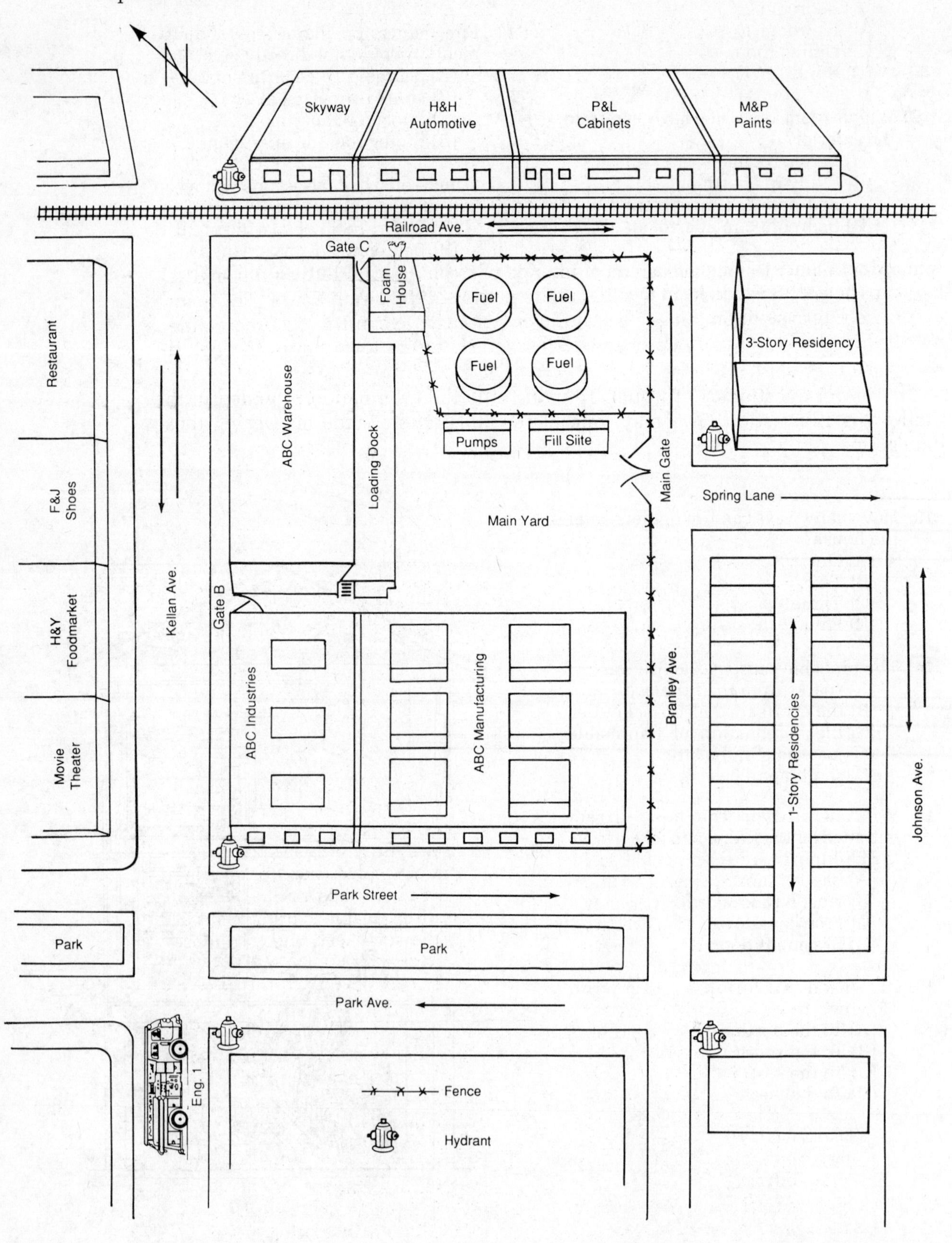

On Thursday at 10:00 hours, you and the other members of your fire unit go with the officer to perform a prefire plan at ABC Industries. The industrial park is located at 423 Kellan Ave. at Park Street. ABC Industries is part of a multinational corporation that manufactures and distributes numerous products. At the Kellan Ave. site, cabinets and outdoor furniture are produced and stored while waiting for shipment throughout the region. These items are made from plastics and wood. The finished goods are stored in the warehouse and moved onto the loading dock during the night, in preparation for loading onto trucks the next day. During an average day, 30 trucks will pick up a load of finished goods and 12 trucks will deliver raw goods. Raw materials are also delivered by train directly to the raw material storage section at the north end of the warehouse. The warehouse and the manufacturing building have sprinklers.

The ABC industrial site at Kellan Ave. is also the fuel depot for the firm's remote plants that are located on the other side of town. Fuel is delivered from the tanks to the delivery trucks at the fill site by pumps in the pump house. The foam extinguishing system on the site has sufficient capacity to control any predictable size fire in the fuel tanks. Water must be supplied to the foam house through a siamese on the exterior wall of the foam house facing Railroad Ave.

The entire premises is enclosed within an eight-foot fence. The three access gates are electrically locked at 10 P.M. There is a security force on duty that patrols the entire site once each hour.

You have been told to examine the site and become familiar with its layout and hazards. You have also been told to determine how a fire would be fought at this site.

17. Upon returning to quarters, you are called into the office and asked to make recommendations about which hydrant to connect the fire engine to so it can supply water to the siamese on the foam house. You are told that you cannot bring hose across the tracks. The route closest and most direct to get water to the foam house would be to
    (A) connect to the hydrant at Park Ave. and Kellan Ave., and bring the hose north on Kellan Ave. to Railroad Ave., then east to Gate C and then through the gate and into the foam house.
    (B) connect to the hydrant at Branley Ave. and Spring Lane and bring the hose north on Branley to Railroad Ave., then west to the siamese on the pump house.
    (C) connect to the hydrant at Park Street and Kellan Ave., then take the hose north on Kellan Ave. to Gate B, then around the warehouse, then north and directly inside the foam house.
    (D) connect to the hydrant at Branley Ave. and Spring Lane, then take the hose through the main gate and directly into the pump house.
18. The lieutenant asks you where the raw materials are stored, and if a fire occurred in this area where could fire companies park their apparatus and use their built-in heavy water streams on the apparatus to extinguish the fire. Your answer should be
    (A) in the warehouse at the north end and a pumper can be placed on Kellan Ave. or Railroad Ave.
    (B) in the warehouse at the south end and a pumper can be placed on Kellan Ave. or in the main yard.
    (C) in the manufacturing plant at the north end and a pumper can be placed on Spring Lane or in the main yard.
    (D) in the manufacturing plant at the south end and a pumper can be placed on Branley Ave. or Park Street.
19. The lieutenant asks you "If a fire develops in the fuel tanks and the wind is blowing southwest what would be in the greatest danger of catching fire?" What should you reply?
    (A) ABC Warehouse and maybe the restaurant.
    (B) fuel pumps and ABC Manufacturing plant.
    (C) Branley Ave. residences.
    (D) ABC Manufacturing.

20. If a fire occurred in the ABC Manufacturing plant, and you were with the first unit to arrive, what would be the first activity you perform?
    (A) Shut down the pumps to the fuel filling site.
    (B) Supply water to the sprinkler system in the warehouse.
    (C) Supply water to the sprinkler system in the manufacturing plant.
    (D) Supply water to the siamese on the foam house.

21. For a fire at the tank truck loading site, it would be best for engine 1 to
    (A) proceed north on Kellan Ave., west on Railroad Ave., and south on Branley Ave. to the main gate.
    (B) north on Kellan Ave., east on Park Street, north on Branley Ave. to the main gate
    (C) north on Kellan Ave., east on Railroad Ave., and south on Branley Ave. to the main gate
    (D) north on Kellan Ave., east on Park Ave., and north on Branley Ave. to the main gate

Answer questions 22 and 23 using only the information in the following passage.

The superintendent of the National Fire Academy is authorized to

1. Train fire service personnel in such skills and knowledge as may be useful to advance their ability to prevent and control fires, including but not limited to:
a. techniques of fire prevention, fire inspection, fire fighting, and arson investigation;
b. tactics and command of fire fighting for present and future fire chiefs and commanders;
c. administration and management of fire services;
d. tactical training in the specialized field of aircraft fire control and crash rescue;
e. tactical training in the specialized field of fire control and rescue aboard waterborne vessels; and
f. the training of present and future instructors in the aforementioned subjects.

2. Develop model training programs or other educational materials suitable for use at other educational institutions, and to make such materials available without charge.

3. Develop and administer a program of correspondence courses to advance the knowledge and skills of fire service personnel.

4. Develop and distribute, to appropriate officials, model questions suitable for use in conducting entrance and promotional examinations for fire service personnel.

5. Encourage the inclusion of fire prevention and detection technology and practices in the education and professional practice of architects, builders, city planners, and others engaged in design and planning affected by fire safety problems.

22. A recruit fire fighter wanting to attend a course at the National Fire Academy could expect to be able to register for all of the following courses EXCEPT
    (A) commanding the first response team
    (B) uniforms, organization, and work rules for his/her local fire departments
    (C) fire fighting tactics for ship fires
    (D) conducting arson investigations

23. It would be most accurate to state that the intent of the National Fire Academy is to train
    (A) firefighters only.
    (B) firefighters, architects, and builders.
    (C) firefighters, architects, builders, and city planners.
    (D) firefighters, architects, builders, and all private citizens for their own safety.

Answer questions *24* through *26* using only the information in the following passage.

One of the most basic and useful tools available to the fire manager is the needed fire flow for a given building. Fire flow is the amount of water needed to extinguish a fire in an occupancy. In most cases, the fire flow calculation exceeds what is really needed to control a fire. This flow is always given in gallons per minute and may be required for a prolonged period.

The needed fire flow allows the fire manager to determine the resources needed, such as manpower, water, and apparatus, before the incident occurs. If a given volume of water is required, one can calculate, rather simply, the number of firefighters required to handle the hose lines.

Consider the following example: If the needed fire flow is 250 GPM, a minimum of three firefighters are needed on each 2½ inch line or two men on each 1½ inch line. Other tactical considerations, such as the degree that the line must be maneuvered within the building, the effective reach of the 2½ inch line versus 1½ inch line, the time that will be required to get the line in service, and the ability of the stream to reach its target will be made based on local conditions and can increase the required number of firefighters for each line. When determining the number of personnel needed, the officer must also consider ventilation, search, rescue, command functions, pump operations, and similar activities.

24. The needed fire flow is
    (A) the amount of water three firefighters can handle.
    (B) not considered when figuring out the personnel required to fight a fire.
    (C) the amount of water flowing from a maneuverable effective hose line.
    (D) the amount of water required to extinguish a fire in an occupancy.
25. A minimum of two firefighters is needed
    (A) to operate each 1½ inch hose line.
    (B) to operate a 2½ inch hose line.
    (C) to be sure the needed fire flow is delivered.
    (D) for search and ventilation as well as maneuverability.
26. The needed fire flow is
    (A) determined after the fire is started.
    (B) the amount of water flowing through a 2½ inch hose line.
    (C) a rather simple calculation.
    (D) used to determine the number of firefighters required to handle the hose lines.

Directions: This part of the test measures your ability to understand directions. Each sample question describes a fire scene. You are to choose the drawing that best represents that scene.

In questions *27* through *29*, symbols are used to represent vehicles, objects, and equipment. Study the symbols before you answer the questions. The symbols and their meanings in the following box will be repeated in the test.

SYMBOLS

The folowing symbols are used in the diagrams for questions 27 through 29.

Fire truck; pumper

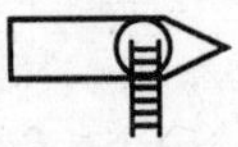

Fire truck

Other vehicles: cars, etc.

Hydrant

Hose Line (attached to hydrant)

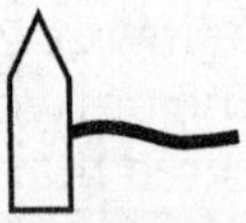

Hose line (attached to pumper)

Ladder

Building

Hose line entering at ground level

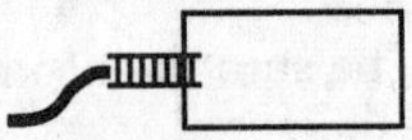

Hose line entering at upper levels

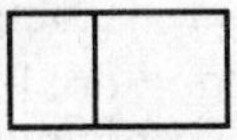

Building with a two level roof

Path of vehicle before an accident is shown with solid line and arrow

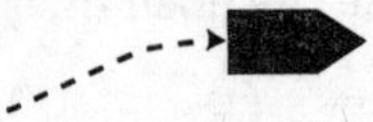

Path of vehicle after an accident is shown with dotted line and arrow

27. A fire truck facing south on a one-way street has to connect to a fire hydrant on the west side of the street to put out a fire in a building on the east side of the street.

(A)

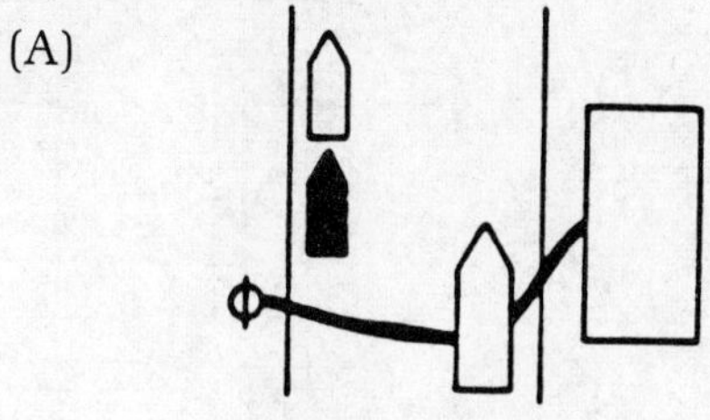

(B)

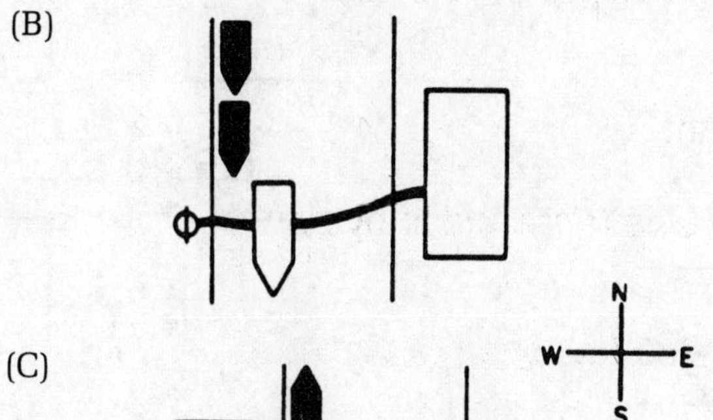

(C)

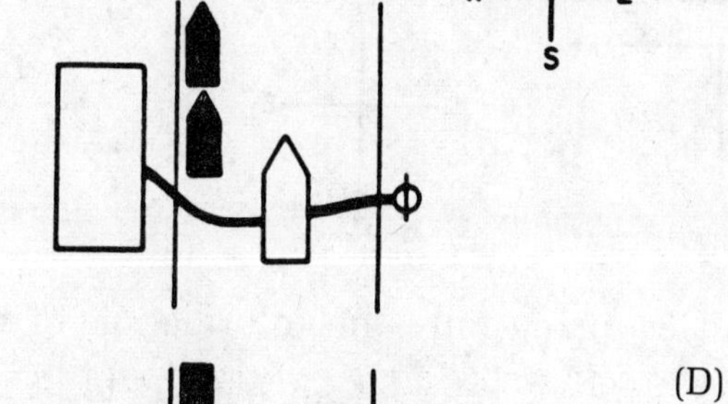

(D)

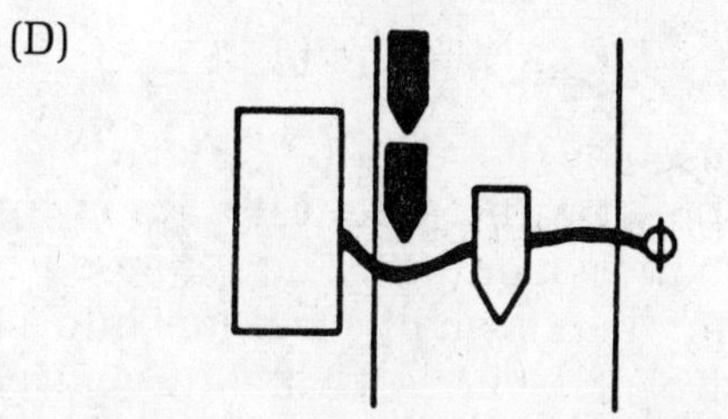

28. Two ladder trucks and one pumper arrive at a burning building. One hose from each truck is taken to the second floor.

(A)

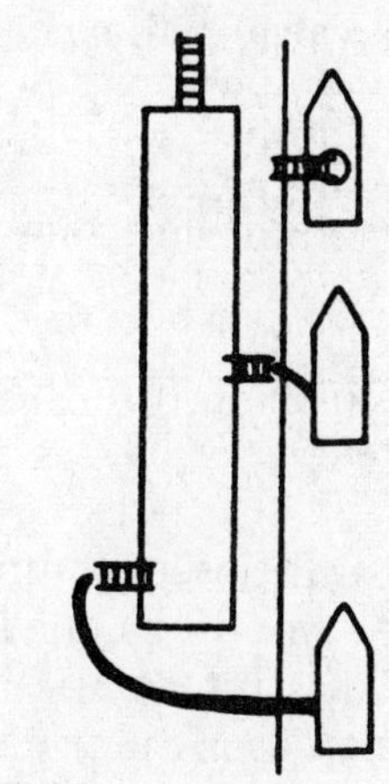

(B)

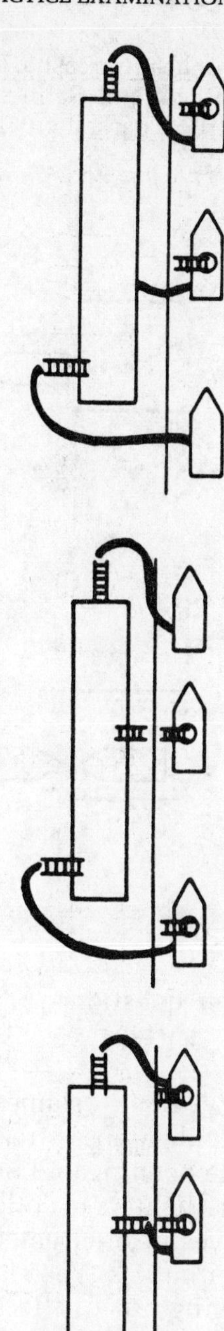

(C)

(D)

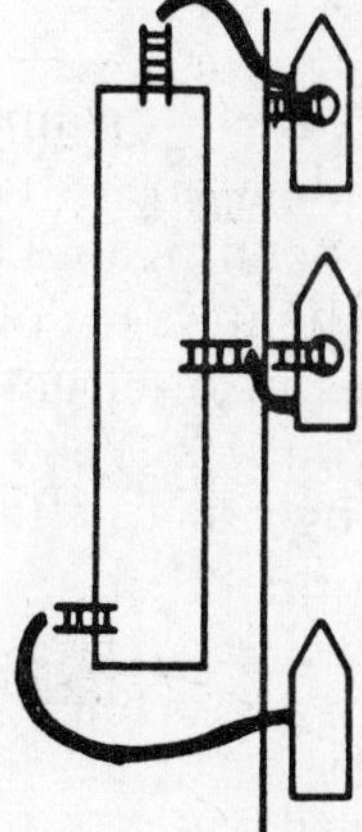

29. A pumper heading east to the intersection at Block and Rahn skids into a ladder truck parked in the opposite direction.

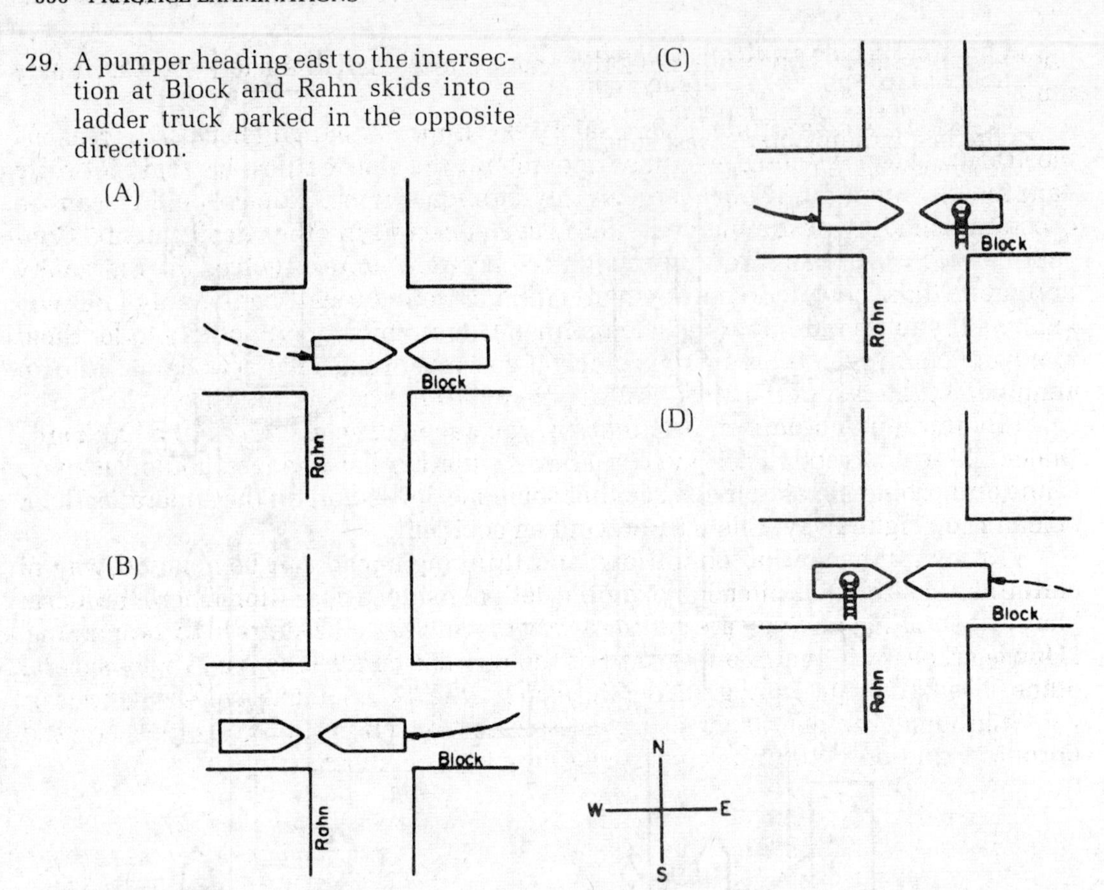

Answer questions 30 and 31 using only the information in the following passage.

The "wye" and "siamese" are often confused by the recruit. The wye is a device used to divide a hose stream into two or more streams. The water flows into the large single opening and out two or three smaller openings. The siamese is a tool used by fire fighters to collect water from a number of smaller streams into a larger stream. Wyes are often gated to allow the control of the flow of water through each of the smaller outlets. The siamese is fitted with a check valve to prevent the water from flowing out of the system and back into the smaller hose. A siamese will be found on a sprinkler and a standpipe system.

30. To reduce a hose line from 2½ inch to 1½ inch a fire fighter would use a
    (A) gated wye
    (B) gated siamese
    (C) standpipe wye
    (D) standpipe siamese

31. A "check valve" will be found on a
    (A) standpipe siamese
    (B) standpipe wye
    (C) standpipe outlet
    (D) standpipe hose line

Answer questions 32 through 34 using only the information in the following passage.

All fire fighters should be aware of their responsibilities pertaining to fire station safety. Because supervision may not be available at all times, the individual has to accept the responsibility of safety awareness. This information is gained through training and experience. Unfortunately, in many cases the fire fighters do not become safety conscious until they have a close call or a friend gets seriously hurt.

To have a good attitude toward safety, it is important to understand the reasons for safety rules. Although we may understand the safety rules, we may not fully appreciate them until they are vividly brought home; hopefully this can be accomplished by citing incidents that have occurred in other departments. Concern by all will increase as fire fighters become more involved in the safety program. Take charge of your own situation. Often you will be the only one who knows if you operate in an unsafe manner. Communicate with others to let them know of changes in their surroundings. If you must plug a battery charger into an engine, let the rest of the shift know so they will not trip on the wire.

By informing others of the situation, you can make your fire station work area safer. Taking corrective action soon after a situation has changed could save you time and problems later. If you see that someone spilled oil on the apparatus floor, clean it up right away; this will prevent an accident.

Knowing the station conditions and thinking ahead can be a sound way of eliminating problems later. For example, let's consider a condition where the alarm has just sounded and you are getting ready to run down the stairs to the apparatus. However, because you are unaware of the current conditions, you know several other fire fighters are going to take the same route. So, you take an alternate route.

Station equipment, such as power saws, drills, and grinders, should be used properly and only with the correct safety devices in place.

32. Fire fighting is a hazardous occupation and demands that the fire fighter be very aware of the safety features that can and will reduce the chance of serious injury. A fire fighter should NOT
    (A) wait to become safety conscious until a friend gets hurt
    (B) be concerned about the department's safety program
    (C) be thinking ahead and using good common sense to eliminate problems of the future
    (D) communicate with others that the conditions have changed

33. An example of a good safety attitude is expressed in which selection?
    (A) Searching the rear bedroom of a smoke filled house alone because you are only going one room in and you can easily get out.
    (B) Sharpening the axe quickly on the grinder without eye protection because it only takes a couple of seconds.
    (C) Using only a small extension cord to plug in the battery charger, which is hooked up to the apparatus battery, so nobody will trip on it.
    (D) Using different stairs than the other fire fighters when responding to an alarm.

34. All fire fighters should be aware constantly of their own safety needs and accept the responsibility for implementing them because
    (A) they are supervised at all times
    (B) it will eliminate problems later
    (C) information can only be gained through experience
    (D) we fully appreciate the value of safety

Answer questions 35 through 38 using only the information in the following passage.

Flashover and backdraft are two fire fighting problems requiring the attention of all fire fighters. These two dangerous conditions require very close coordination between hose line placement and advancement and ventilation of the building or area.

The indicators a firefighter should associate with flashover include:

1. Significant free-burning fire in a room.
2. Firefighters being forced to stay low due to the heat.
3. Increased heat build-up in the area.
4. Heavy, hot, dark smoke filling the room, with a significant increase in heat being felt.

Indicators of a potential backdraft can be seen from both the outside and the inside of a structure.

Outside the building:

1. Tightly sealed buildings with smoke being forced from cracks or small openings and then being drawn or sucked back into the building.
2. Window glass or doors hot to the touch, with little or no visible fire.
3. Tar-like, oily substance running down the inside of windows.
4. Swirling smoke within the building seen through the windows.

Inside the building:

1. A peculiar whistling sound of incoming air rushing past due to pressure differentials.
2. Smoke being drawn back into the building or area and appearing to curl around and reverse itself.
3. Heavy smoke swirling with great force.
4. Smoldering, hot fire.
5. Sickly or intermittent flame due to reduced oxygen concentration.

35. Firefighters arriving at an advanced fire late Sunday night in a downtown factory would most likely suspect that a backdraft could occur if they saw
    (A) significant free-burning fire in the factory.
    (B) a tar-like oily substance running down the inside of the windows.
    (C) windows on the upper floors open and heavy smoke billowing out.
    (D) all doors and windows on the lower floors tightly locked.

36. While advancing a hose line into the second floor, a firefighter hears a strange whistling sound and feels cool air rushing. This should indicate to the firefighter that
    (A) a flashover is imminent.
    (B) a rollover has just occurred.
    (C) it is safe to advance into the fire.
    (D) a backdraft could occur at any moment.

37. Firefighters should associate all of the following with flashover EXCEPT
    (A) a large amount of free burning fire in a room.
    (B) high heat forcing firefighters to stay low.
    (C) windows cracking from the heat, too hot to touch and with little or no visible fire showing.
    (D) no visibility because heavy hot smoke is banking down and the heat is rising rapidly.

38. Sickly or intermittent flame generally will
    (A) be seen from outside the building.
    (B) indicate heat is building up.
    (C) cause the smoke to swirl and be drawn back into the building.
    (D) indicate the fire has a reduced oxygen concentration.

Answer questions 39 through 41 using only the information in the following passage.

A fire fighter must be concerned with safety when conducting interior search and rescue. A search and escape plan should be in place before conducting the search. It must not be haphazard, uncontrolled, or uncoordinated. Search and rescue is one of the most important and probably one of the most dangerous tasks fire fighters perform. Through proper training and conscious preparation, it can be accomplished safely.

Fire fighters should be in full fire fighting protective clothing, wearing self-contained breathing apparatus (SCBA) and Personal Alert Safety System (PASS) when conducting a search. Each fire fighter should have a good working flashlight and a small strong rope, at least 25 feet long, which can be attached to the doorknob to act as a guideline for an escape route. A 2½ or 3 inch nail can be used between the inside of the door on the hinge side and the door jamb to hold the door open. Press the nail point into the door then close the door, thus pressing the head against the jamb and keeping the door open.

Another important tool fire fighters should carry is a latch strap. This can be purchased from the fire fighting supply store or may be homemade. To make your own, cut an inner tire tube into 3 inch wide strips about 12 to 15 inches long. Cut a small hole about three inches from each end. Use the rubber strip and a door with two standard door knobs to measure where to cut the holes. Make and carry several of them so you can pass through more than one door during the search.

39. During search and rescue, nails are used to
    (A) lay out a trail back to safety
    (B) prevent a door from opening before the fire fighter wants it to
    (C) prevent a door from closing
    (D) scratch a mark on the door to tell others the room has already been searched

40. A latch strap can be used to keep a door open. A fire fighter should carry
    (A) a homemade latch strap
    (B) a commercially made latch strap
    (C) a latch strap, rope, and flashlight
    (D) a rope, flashlight, and several latch straps

41. While conducting a search and rescue mission, fire fighters may NOT be
    (A) protected by SCBAs
    (B) protected by a hose line
    (C) equipped with a PASS device
    (D) concerned with safety

Answer questions 42 through 45 using only the information in the following illustration. You may look at the illustration while answering these questions.

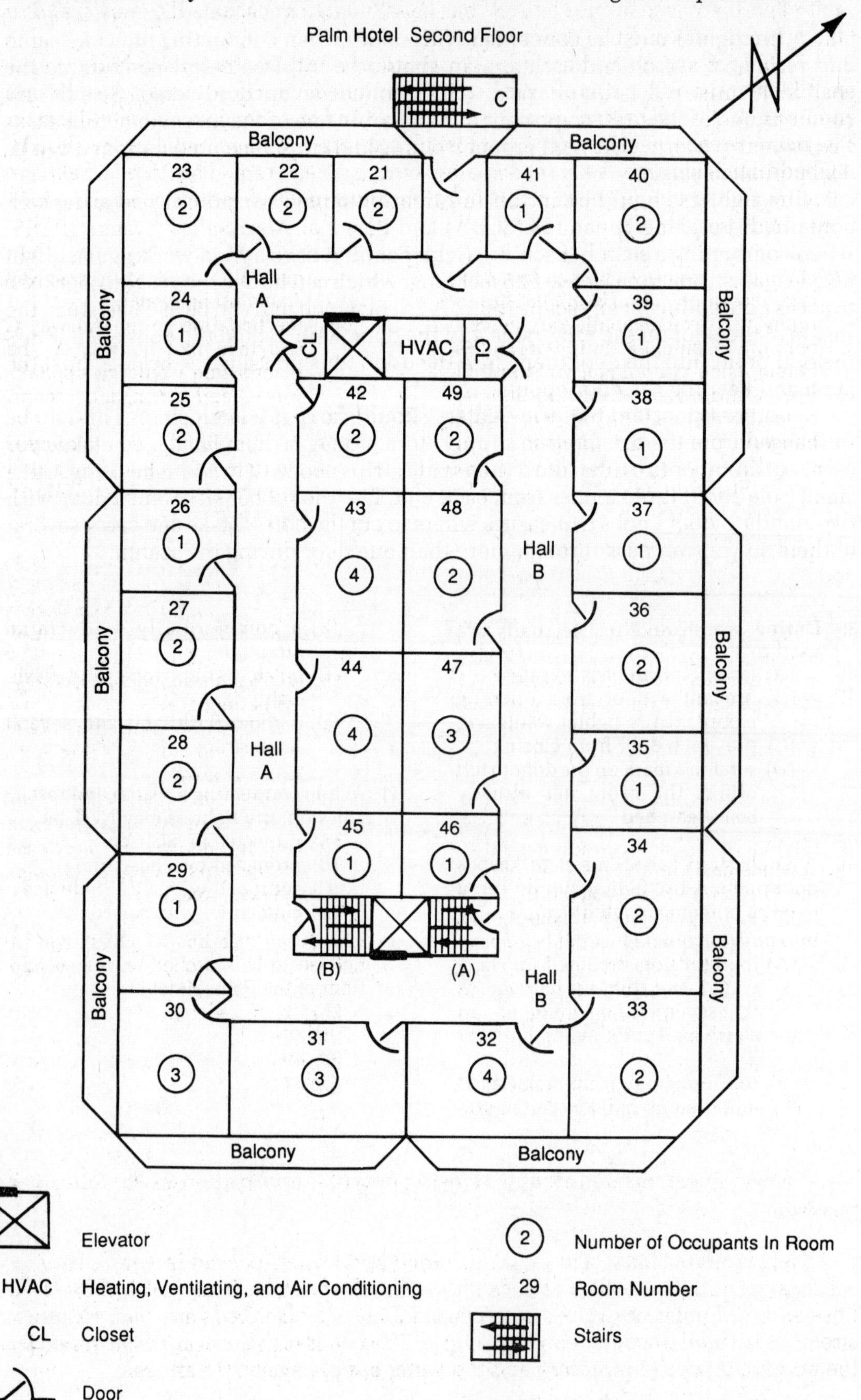

The Palm Hotel is one of the better managed hotels and has developed a good fire protection and prevention program. The manager is trained in the area of fire safety and fire evacuation. The staff has been taught to evacuate the guests, and to take precautions to limit the spread of fire.

There are smoke detectors in all the rooms and in the public hallways. A sprinkler system is planned and will be installed within a year. All stairs are enclosed by 2-hour protective walls and, like the halls, have emergency lighting. The doors to the rooms are not self-closing, but this will be remedied very shortly. All bedding is ignition resistant and slow burning; each room has drapes that have been treated to reduce the chance of burning. Until now, there have been no serious fires at this hotel.

42. On Sunday evening at 02:00, your unit is called to the Palm Hotel for a fire in room 26. As you respond, you can see a column of smoke coming from the hotel. When you arrive, you are told fire is burning out from the door of room 26 and cutting off the public hallway. From this information, you would know the people in what room are in the most danger?
    (A) Room 27.
    (B) Room 25.
    (C) Room 43.
    (D) Room 16.

43. The officer ordered two hose lines to be put into operation. One hose line will advance into the corridor to extinguish the fire. The second hose line will act as a defense barrier to stop the fire from being pushed around the hall and back onto the unit advancing the first hose line. The second line will be positioned opposite and opposing the first line. The second line will not advance into the fire; it will only stop the fire from wrapping around the hall. The best way to get the hose lines into position quickly would be to have
    (A) the first line stretched up stairway A and then up hallway A, the second line would go up stairway B and then up hallway B
    (B) the first line go up stairway B and then up hallway A, the second line would go up stairway A and up hallway B
    (C) the first line go up stairway C and then into hallway A, the second line would go up stairway C and down and around hallway B
    (D) the first line go up stairway C and into hallway A, the second line would go up stairway B and into hallway A

44. While at the command post, you hear the hotel manager tell the chief that 53 people are accounted for. You have a copy of the hotel room count and know that
    (A) this is the amount of guests staying at the hotel
    (B) this is no help to the chief because you don't know how many are missing
    (C) you should tell the chief there are four people unaccounted for
    (D) you should not get involved because this is a top level management area

45. The total number of guest rooms needed to be searched on the second floor of the Palm Hotel is
    (A) 21
    (B) 29
    (C) 49
    (D) 57

Answer questions 46 through 49 using only the information in the following passage.

The apparatus hose beds are subdivided and hose is packed into hose beds in accordance with the needs of the fire unit. The deeper hose beds are used for large hose and for longer hose stretches. The smaller, narrower beds are used for quick stretches of small diameter hand held hose. The needs of the unit are determined by the dominant type of structure and the water supply system in an area.

There are three basic hose load types:

1. The accordion load.
2. The flat load.
3. The horseshoe load.

To load a hose bed using the accordion load, the hose is folded front to rear with the flat side of the hose facing the side wall and working across the hose bed until it is full. A second, third, and fourth layer can then be added. The accordion hose load can be used when hand stretching of the line is required.

The flat load is also a back and forth, front to rear load, however the flat side of the hose is placed on the base of the hose bed. This load can also be used for hand stretching.

The horseshoe load is started similarly to the accordion load, however instead of folding it up and down, back on itself, it is wrapped around the outer edge of the hose bed and then in simulated "U"s back around the bed. This type load is effective for laying out line with the apparatus.

For all types of hose loads, the male hose butt is loaded first if the apparatus will stop at the water supply first. This is sometimes called a flying stretch. The female hose butt is loaded first if the apparatus will stop at the fire first; this is called a back stretch.

A hose load that has one end already connected to a discharge outlet on the apparatus is called a "preconnect" and is used for a rapid attack on a fire.

A hose bed that runs across the apparatus, from side to side, and usually found behind the crew cab, is called a transverse bed. It is often used to house the preconnect line.

46. The hose load that is not a basic hose load is the
    (A) preconnect load.
    (B) horseshoe load.
    (C) accordion load.
    (D) flat load.

47. The hose load that is best for laying hose from the water source to the fire building with the apparatus is
    (A) the flat load.
    (B) the horseshoe load.
    (C) the transverse bed load.
    (D) the back stretch.

48. Hose stored on fire engines is placed in compartments called hose beds. An inaccurate statement about hose beds is: "Hose beds
    (A) that are deep are used for large diameter hose."
    (B) are subdivided in accordance with the needs of the fire unit."
    (C) can run across the apparatus or run the length of the apparatus."
    (D) are all small and narrow and used for stretching hand held hose lines."

49. Firefighters have many different needs for fire hose. At a fire in a private house with people trapped on the second floor, a fast hose stretch is used. The most appropriate hose stretch would be a
    (A) preconnect stretch from a transverse bed.
    (B) back stretch from a deep bed.
    (C) flying stretch from a small bed.
    (D) a horseshoe stretch from a deep bed.

Answer question 50 using only the information in the following passage.

Arson investigation is the responsibility of specially trained personnel. They gather information from the fire scene and make a determination of the point of origin and the cause of the fire. Evidence is circumstantial or physical. At many fires the collection of physical evidence is made difficult because it has been destroyed by the fire. It may also be damaged by the water and overhauling activities of the fire fighters. Fire fighters can significantly aid in arson investigation by being observant as they respond and by preserving potential evidence at the scene.

50. It would be most correct to say that
   (A) arson investigators respond to the scene to determine if a crime has been committed and to gather evidence
   (B) fire fighters can and do destroy evidence
   (C) arson investigators arrest and prosecute people suspected of arson
   (D) specially trained fire officers are sent to the fire to aid the fire fighters in preserving the scene

Answer question 51 using only the information in the following passage.

Electricity presents a constant hazard to fire fighters. Because of the dangers, fire fighters are trained how to react when they respond and find an electrical hazard.

For fire in an underground vault, fire fighters should

1. Set up a danger zone around the vault. Request the power to be shut down and the power company to respond.
2. Place the apparatus at a safe point outside the danger zone.
3. Have a team of two fire fighters make a determination if any life hazard exists.
4. If a life is in danger, begin rescue operations using the most effective safety and rescue techniques.
5. If no life hazards exist, rope off the area and keep all people safely away.
6. Search the surrounding buildings for signs of a possible fire.
7. Await the arrival of the power company personnel and work with them to control the incident.

51. Fire fighters responding to a reported automobile fire find smoke pushing from a manhole cover in the center of the street. They notify the dispatcher to call the power company to shut down the power and respond. They then establish a danger zone, park the apparatus immediately outside the zone, and determine nobody is in immediate danger. The next thing the fire fighters should do is
   (A) search the buildings for any indication of a possible fire
   (B) wait for the power company to arrive to shut off the power
   (C) send in a team of fire fighters to see if anyone is in the manhole
   (D) rope off the area and keep the people away

Answer questions *52* and *53* using only the information in the following passage.

At a recruit training course, firefighters are taught about hose line operations. They are told that:

1. A stream that has broken into a coarse spray can be effective on small fires, but would be relatively useless on a major fire.
2. Strong winds deflect the hose stream and reduce its effective reach.
3. Damaged nozzles cause the stream to break up.
4. A 45-degree angle of attack would seem to be ideal, however, 32 degrees turns out to be best.
5. When a stream is operated close to a building, penetration on upper floors is poor.
6. From the ground, hand line steam water can be thrown approximately to the seventh floor window, but the third floor is the highest it can throw water effectively.
7. To throw water into upper floors, the nozzle must be moved away from the building and raised to a greater angle.

52. While operating a hand line stream from the sidewalk into the first floor of a four story building, you are told to put water into the third floor to stop the extending fire. Using good hose stream practices you should
(A) direct the stream upward and into the window.
(B) explain that your hand line will not be effective at that height.
(C) move back into the street and increase the angle of the nozzle.
(D) reposition the line so that the nozzle is at a 45 degree angle from the third floor window.

53. A recruit firefighter wanted to show an understanding of the rules of effective hose stream operations and told teammates that:
(1) The wind can have an adverse effect on hose stream reach;
(2) Water cannot be thrown above the third floor from a hose line on the ground;
(3) A damaged nozzle will make the stream break up; and
(4) A stream can be operated close to the building on the lower floors.

The recruit firefighter's teammates would be most correct if they accepted
(A) all but statement 1 as correct.
(B) all but statement 2 as correct.
(C) all but statement 3 as correct.
(D) all but statement 4 as correct.

Answer question *54* using only the information in the following passage.

Engine 1 responds to a reported house fire. Upon arrival, the firefighters find a distraught woman holding her baby. She tells them that a 4-inch water pipe, which supplies water to her home, broke and water is rapidly filling up her basement. The officer tells you the water is flowing at 500 gallons per minute (GPM) and that the flow will have to be stopped at the street valve. A dewatering operation is immediately started and two 250 GPM pumps are put into operation. The officer orders the water company to respond and to bring additional dewatering pumps to the scene. About ten minutes later, the woman asks a firefighter how the operation is going.

54. The firefighter should
(A) tell her that "the water is being removed as fast as it is going in and that more pumps and the water department are on their way."
(B) tell her "only the officer can give out that information."
(C) tell her "everything is OK."
(D) not answer her, just continue with assigned duties.

Answer questions *55* through *58* using only the information in the following passage.

Firefighters have many jobs to do at the scene of a fire. One of the most difficult is the position of roof ventilation. The person assigned to the task of opening up the roof must be an experienced and determined firefighter. The mission of the person assigned to roof ventilation is to open up the top of the building and release the pressure build-up of smoke and heat.

The route taken to the roof is:

1. The stairs of the adjoining building, that is, if there is an adjoining building and it is approximately the same height.
2. The aerial ladder, elevating platform, or portable ladder.
3. The fire escape, if there is one. This is the least desirable method.
4. The interior stairs of the fire building should NEVER be used except in a high rise or a large multiwing building with several enclosed remote stairs.

When arriving at the roof, the firefighter should:

1. Let the company officer know that he/she is at the roof.
2. Immediately open the door leading to the stairs of the fire building.
3. Search for any people who may have tried to use the stairs but were unable to get out. This is particularly important if the entrance door was found securely locked.
4. Remove the skylight over the stairs and hall; this will greatly aid in ventilation.
5. Remove any other skylights that could aid in ventilation.
6. Conduct a visual search of all sides of the building. From the roof, the firefighter can rapidly spot someone in trouble or someone who may have jumped before the Fire Department's arrival.

Let your officer know when the roof has been opened and you have completed your duties at that level. If you leave the roof to attempt a rescue, you must tell your officer or the officer in command of the fire.

55. The least acceptable method for getting to the roof of a small four-story residential building is the
    (A) interior stairs of the fire building.
    (B) interior stairs of the adjoining building.
    (C) aerial ladder.
    (D) the fire escape.
56. Your objective when assigned to the roof position is to
    (A) open the door to the stairs of the fire building.
    (B) remove the skylight on the fire building.
    (C) release the pressure build-up of smoke and heat.
    (D) rescue a person who may have jumped before the Fire Department's arrival.
57. After completing the roof duties, you should
    (A) notify the company officer.
    (B) notify the company officer and the officer in command of the fire.
    (C) notify the company officer or the officer in command of the fire.
    (D) come down the interior stairs to search for people who may be trapped.
58. After arriving safely at the roof, the first action you should take is to
    (A) check the perimeter of the building to see if anybody is in trouble.
    (B) get the door leading to the interior stairs open.
    (C) communicate to the company officer that you are in position on the roof.
    (D) open the skylights to relieve the trapped up smoke and heat.

Directions: Questions 59 through 63 describe a starting point and a destination. Assume you are driving in the area shown on the map accompanying the questions. Use the map as a basis for choosing the shortest way of getting from one point to another without breaking any traffic laws.

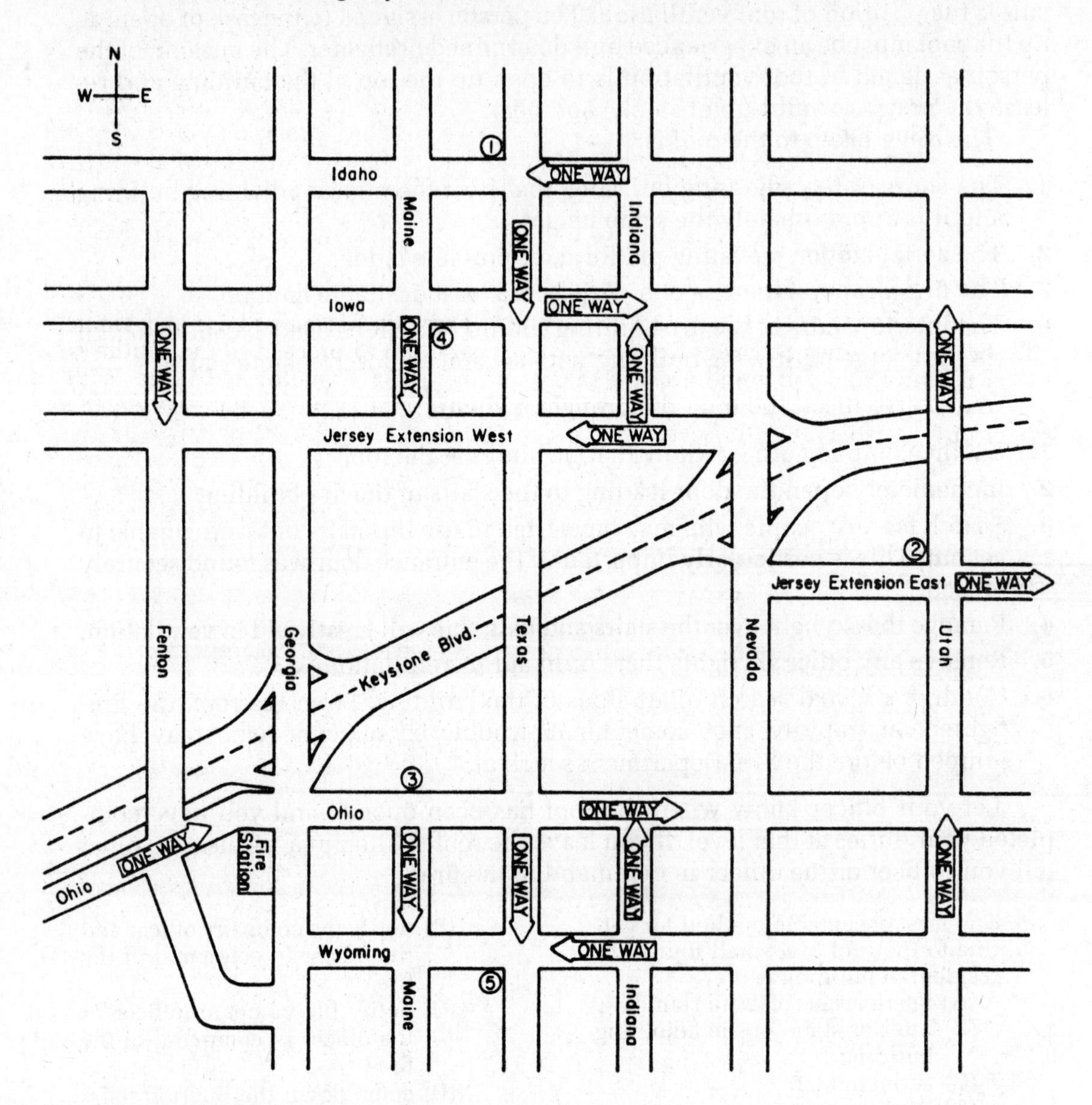

A street marked "One Way" is one way for the full length, even when there are breaks or jogs in the street. *Exception:* A street that does not have the same name over the full length.

59. From the fire station at the corner of Ohio and Georgia, you are to proceed to the northwest corner of Texas and Idaho. Without breaking any traffic laws, what is the shortest route to the corner?
    (A) East on Ohio to Utah, north to Idaho, west to the corner of Texas.
    (B) East on Ohio to Nevada, north to Idaho, west to the corner of Texas.
    (C) North on Georgia to Iowa, east to Indiana, north to Idaho, west to the corner of Texas.

60. From the corner of Idaho and Texas, you are to proceed to the northwest corner of Utah and Jersey Extension East. Without breaking any traffic laws, what is the shortest route to the corner?
    (A) South on Texas to Ohio, left to Utah, left to the corner of Jersey Extension East.
    (B) South on Texas to Ohio, left to Nevada, left to Jersey Extension East, right to the corner of Utah.
    (C) South on Texas to Iowa, left to Nevada, right to Jersey Extension East, left to the corner of Utah.

61. From the corner of Jersey Extension East and Utah, you are to proceed to the intersection of Ohio and Maine. Without breaking any traffic laws, what is the shortest route to the corner?
    (A) North on Utah to Idaho, west to Texas, south to Wyoming, west to Georgia, north to Ohio, east to the corner of Maine.
    (B) North on Utah to Idaho, west to Nevada, south to Keystone Boulevard, southwest to Georgia exit, south to Ohio, east to the corner of Maine.
    (C) North on Utah to Idaho, west to Nevada, south to Ohio, east to the corner of Maine.

62. From the corner of Ohio and Maine, you are to proceed to the southeast corner of Iowa and Maine. Without breaking any traffic laws, what is the shortest route to the corner?
    (A) South on Maine to Wyoming, west to Georgia, north to Iowa, east to the corner of Maine.
    (B) East on Ohio to Texas, north to Jersey Extension West, west to Maine, north to the corner of Iowa.
    (C) East on Ohio to Nevada, north to Idaho, west to Maine, south to the corner of Iowa.

63. From the corner of Iowa and Maine, you are to proceed to the southwest corner of Wyoming and Texas. Without breaking any traffic laws, what is the shortest route to the corner?
    (A) South on Maine to Jersey Extension West, right to Georgia, left to Wyoming, left to the corner of Texas.
    (B) East on Iowa to Texas, right to the corner of Wyoming.
    (C) East on Iowa to Nevada, right to Wyoming, right to the corner of Texas.

Answer questions 64 and 65 using only the information in the following passage.

Portable ladders are raised at building fires for many reasons. For a fire in a one-story building that has been subdivided into many stores by fire walls, ladders can be used to identify the outer limits of the subdivision. They can be used to gain access and egress from the roof. Placing ladders remotely located from each other will insure that fire fighters have a safe secondary way to get off the roof rapidly. At fires in multistory residence buildings, ladders can be used to gain entrance to lower floors or as a secondary means of egress for trapped occupants. At a fire located in either the cellar, first, or second floors, of a four-story multitenant residence, portable ladders should be raised to areas above and adjacent to the fire area even if the aerial ladder is going to be used. A heavy fire in these areas may make the interior stairs unusable and require the tenants to escape through the windows.

64. At a fire in a one-story shopping complex, portable ladders should be raised for all of the following reasons EXCEPT
    (A) to ensure that occupants can rapidly and safely exit the building.
    (B) to identify the outer subdividing walls of the fire store.
    (C) to give firefighters at least two ways, remote from each other, to get off the roof.
    (D) to gain access to the roof.

65. Arriving at a fire located in the cellar of a small multistory residence building, you are told to put up a portable ladder. You would be correct if you raised the ladder to the
    (A) roof.
    (B) window immediately above the fire.
    (C) window remote from the fire.
    (D) window adjacent and above the fire.

Answer questions *66* through *69* using only the information in the following passage.

The use of elevators during fires requires special knowledge and training. To make elevators safer, "Firefighter Service" has been installed. Firefighter service is activated by the use of a special firefighter's key. The key can be used in the lobby or in the elevator car. The three-position key switch in the lobby will (1) recall the elevators to the lobby, (2) open the car hoistway door after it has closed, and (3) return the system to normal.

To operate the car in firefighter service, insert the key in the switch and:

1. Select firefighter service.
2. Press the door close button.
3. Select a floor.
4. Once the car is moving, test the call cancel button.
5. If the car stops at the next floor, reselect your floor.
6. When you reach your floor, push and hold the door open button until the door stops.
7. To move to another floor, begin with step two.
8. When you are on your floor, the car must be returned to the lobby manually.
9. When you are finished with the elevator, the key must be returned to normal.

If water is flowing into the elevator shaft, from sprinkler systems or hose lines, do not use the elevator. Water can interfere with the electric circuits and cause the elevator to act erratically. Do not overload the elevator car. Six firefighters is the maximum number of passengers permitted. Any time the elevator acts erratically, leave the elevator, put it out of service, and tell the officer in command of the fire what you have done.

66. Upon entering the lobby of a high-rise office building, you are told by the building manager there is a fire on the twenty-third floor, and that all the occupants of that floor and the floors immediately above it have been evacuated to a position two floors below the fire. The elevators are on their way up to the twenty-first floor to bring the people down. The first action firefighters should take is to
    (A) select firefighter service.
    (B) press the door close button.
    (C) use the special key to recall the elevators.
    (D) use the special key to open the elevator hoistway door.

67. Firefighters traveling to a fire on the upper floors of a building have activated the firefighter service and are now at the desired floor landing. To exit the elevator, they should
    (A) turn the special key to normal.
    (B) push the door open button.
    (C) push and hold the door open button.
    (D) push the fire emergency release button.

68. While traveling in an elevator, firefighters discover water dripping into the car. After activating the call cancel button, the firefighters should
    (A) activate the door open button and exit the car.
    (B) let the officer in command know the elevator is not working properly.
    (C) continue to the destination floor and then put the elevator out of service.
    (D) return the car to the lobby manually.

69. A special key is used to activate firefighter service. The key switch in the lobby will NOT
    (A) open the doors.
    (B) bring the elevators down to the lobby.
    (C) put the system in regular operating mode.
    (D) close the hoistway entrance to the car.

Answer questions *70* and *71* using only the information in the illustrations.

70. Choose the floor plan that most closely demonstrates the building illustrated.
    (A) floor plan A
    (B) floor plan B
    (C) floor plan C
    (D) floor plan D

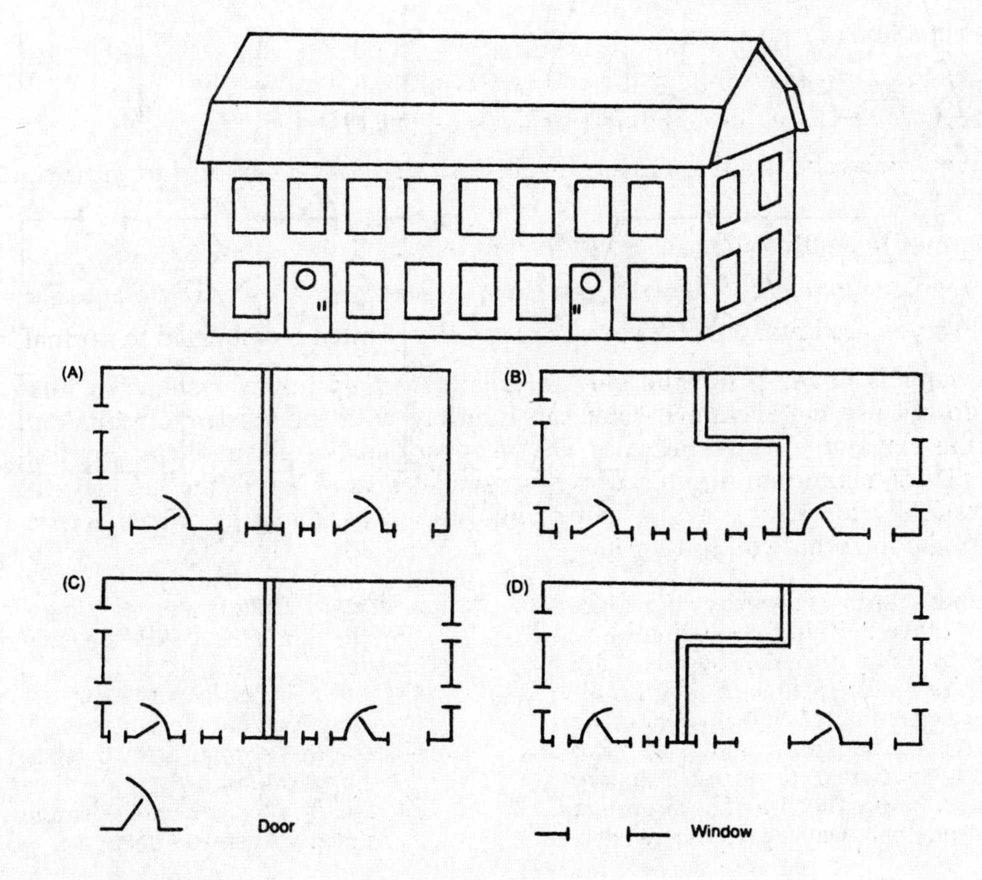

71. Choose the floor plan that most closely demonstrates the building illustrated.
    (A) floor plan A
    (B) floor plan B
    (C) floor plan C
    (D) floor plan D

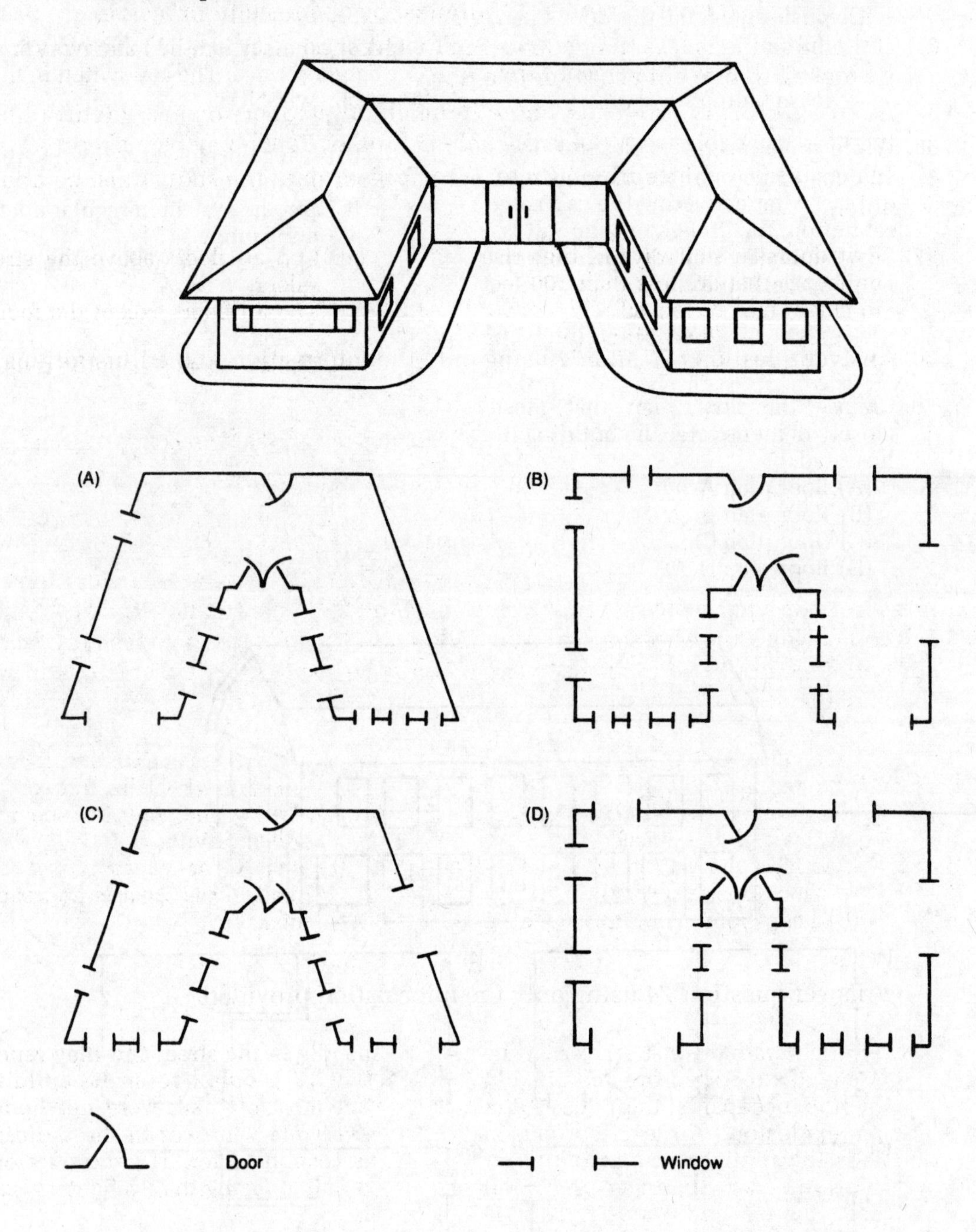

Answer question *72* using only the information in the following passage.

Building and fire prevention laws are designed to insure life safety in buildings. High-rise office buildings require that doors leading to exits must not be locked. The exceptions are:

1. Doors that lead from the stairs to the street and open directly out to the street may be locked on the street side but most be openable from inside.
2. If the building is less than 100 feet in height, then all doors in a stairway may be locked to prevent access onto a floor.
3. If the building is 100 feet or more in height, then doors on every fourth floor must be openable from the stairs side. All other doors may be locked.
4. In case of fire, all doors leading from any floor onto the stairs must be openable.

72. Exit doors in stairways of high-rise buildings that are less than 100 feet in height may be locked
    (A) on all floors including the street level.
    (B) on all floors above the street floor.
    (C) on all floors except the fourth floor.
    (D) on all floors except the roof.

Answer question *73* using only the information in the following passage.

Good driving rules are essential for your personal safety and the safety of your passengers. As a defensive driver, you should aim high while steering by making repeated glances well ahead of your driving path. Get the big picture; scan the front, sides, and rear. Anticipate and prepare when something unusual is seen. Keep your eyes moving. Do not become fixed on one target. Leave yourself room. You should prepare to stop and have enough room to do it safely. Make sure others can see you; check to see that all the car lights are working and use them to signal when you turn or change lanes.

73. A newly appointed firefighter, you are being taught to drive the fire apparatus and are told to aim high. You should take this to mean
    (A) make plans to become a certified apparatus operator.
    (B) keep your eyes moving from front to rear and side to side when driving a fire apparatus.
    (C) keep in the middle of the road when driving.
    (D) glance far ahead down the road to see what is approaching.

Answer question *74* using only the information provided.

74. Firefighters have just arrived at the scene of a reported fire and are told by the occupants there has been an explosion. An excited person tells the firefighters that the water pipes in her apartment have all been broken and water is ruining her rugs and furniture. Another person complains that the gas pipe to the meter has been damaged and is leaking. It can be seen that the roof has been damaged but does not look as if it will collapse. There are three people in the street and they report that 12 people live in the building but most of them were not home. Everyone who was in the building is accounted for. The most serious problem facing the firefighters now is
    (A) the destruction of the person's furniture and rugs.
    (B) accounting for the missing people.
    (C) shutting off the gas.
    (D) shoring up the roof.

Answer question *75* using only the information in the following passage.

As part of a six-month training program, you are detailed to work in four different fire stations and assigned to various positions. While working in station 1, you are assigned to the outside ventilation position and are told that your job is to open the windows of the fire building from the outside and to use a portable ladder or fire escape to gain access to them. When at station 2, you are assigned to the roof. While operating on the roof of a six-story building with a fire on the top floor, you observe that the outside ventilation person for unit 2 works on the roof. When at station 3, you are again assigned to the outside ventilation position and are told that if the aerial ladder is to be used, you are to assist the driver in setting it up and then to perform window ventilation from the fire escape or a portable ladder. At station 4, you are assigned to the position of forcing entry to the fire apartment. At a fire on the third floor, after forcing entry through the entrance door and while conducting a front to rear search, you find another firefighter who was assigned the outside ventilation position searching the rear bedroom.

75. During a review training session at the fire academy, you are asked to explain the duties of the outside ventilation person. You would be most correct if you said the position of outside ventilation is
    (A) flexible, and you can go wherever you think it would be most important.
    (B) clearly defined but can vary with the type and location of the fire.
    (C) to use the aerial ladder to gain access to the windows.
    (D) to gain access to the apartment by forcing the door and then searching.

Answer questions *76* through *79* using only the information in the following illustration. You may look at the illustration while answering these questions.

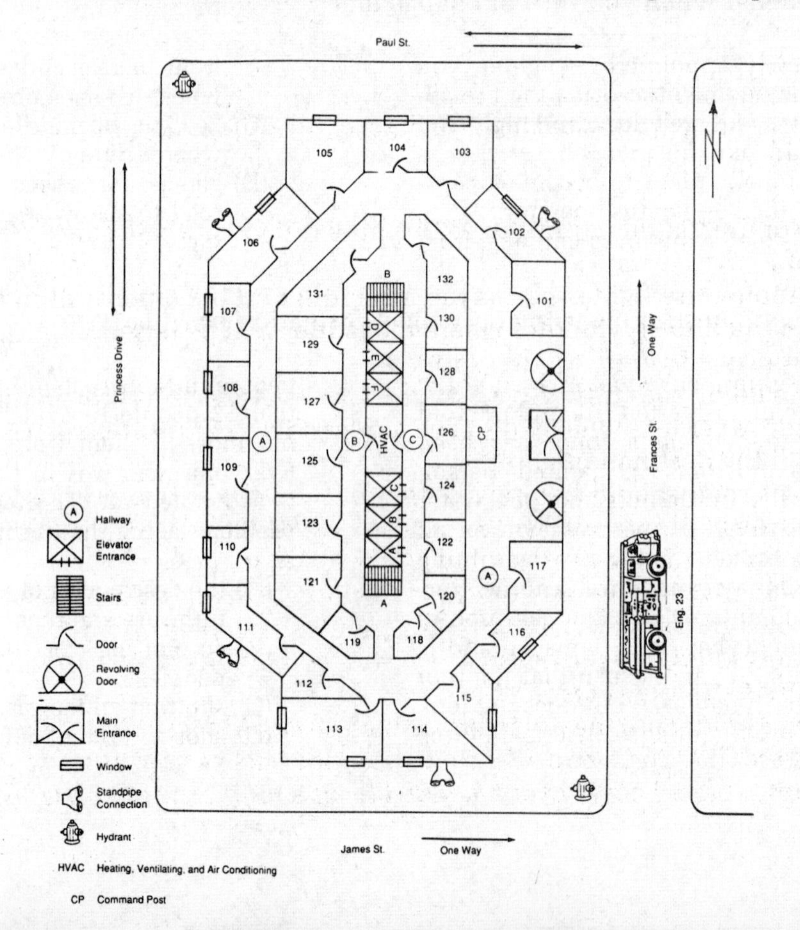

The illustration is of a typical floor of a high-rise building. The numbers designate office spaces, C.P. is the designation of the Command Post, and HVAC is the location of the Heating, Ventilation, and Air-conditioning Control Room.

76. Fire fighters reporting into the command post are directed to a smoke condition in room 121. The most direct route to this location is
    (A) north in hallway A, then southeast in hallway B
    (B) south in hallway A, then northeast in hallway C
    (C) south in hallway A, then northwest in hallway B
    (D) north in hallway A, then southwest in hallway B

77. A fire fighter directed to search all the offices with access from hallway B will be searching
    (A) rooms 101 through 105
    (B) rooms 118 through 132
    (C) rooms 117 through 112
    (D) rooms 119 through 131

78. Fire fighters responding with engine 23 have been directed to connect to a hydrant and supply water to the standpipe connection. The fire fighters see that the connection on James Street is out of service. What is the first hydrant and standpipe connection that engine 23 can drive to?
    (A) Go with the flow of traffic to the connection at the northeast portion of the building on Frances Street.
    (B) Go with the flow of traffic to the connection at the northwest portion of the building on Princess Drive.
    (C) Back up and go against traffic to the connection at James Street and Princess Drive.
    (D) Back up and go to the connection at James and Frances streets.

79. Fire fighters reporting back from conducting a search of room 106 are directed to search the rooms opposite elevators A, B, and C. The most appropriate rooms for them to search are
    (A) rooms 122 and 124
    (B) rooms 131, 129, and 127
    (C) rooms 121, 123, and 125
    (D) rooms 118 and 120

Answer questions 80 and 81 using only the information in the following passage.

At times, a portable ladder is needed on an upper floor or on the roof of a building. To raise a portable ladder to the roof, fire fighters perform the following in the order given.

1. A team of fire fighters equipped with a hose roller, rope, and safety belts goes to the roof.
2. At the roof, a fire fighter places the hose roller on the outer wall in a position between the line of windows where the ladder will be raised.
3. The rope is lowered to the street
4. A fire fighter in the street ties a bowline knot and passes it through the first two rungs above center of the ladder. The ladder should be positioned below and in line with the destination point.
5. Sufficient rope should be drawn through the ladder to allow the loop of the bowline to be slipped over both beams at the butt. The rope is then drawn taught and the ladder turned over so the knot faces the ground.
6. A signal is given to the fire fighters to haul up the ladder. The fire fighter in the street holds the butt end of the beams to guide it into position.
7. While the ladder is being raised, the fire fighter in the street shall act as a safety and keep people away from the area.
8. To lower a ladder use the procedure in steps 5 and 6, except that the knot should face toward the sky. Once on the roof, the ladder can be used as a bridge between buildings, a safety cover over a large hole, or a tool to climb to higher heights.

80. While operating at a fire in a six-story factory, you are told to help a team of firefighters bring a 20-foot portable ladder up to the roof. Your position is in the street. You would be expected to perform all of the following EXCEPT
    (A) tie a bowline knot.
    (B) put on a safety belt.
    (C) signal the roof to raise the ladder.
    (D) keep people away.

81. At a fire, firefighters at the roof lower a rope to raise a portable ladder. The first action that should be taken by the firefighter in the street is to
    (A) tie a knot in the rope.
    (B) turn the ladder over.
    (C) line the ladder up with the windows.
    (D) place a ladder roller on the wall.

Directions: Questions *82* through *85* test your ability to picture how persons or objects look from different views or after certain changes have been made in their shape or appearance. In each case, you are to study the picture and then choose the correct answer from one of the four drawings that follow.

82. On which of the four hoses below is the rope pattern the same as the one above?

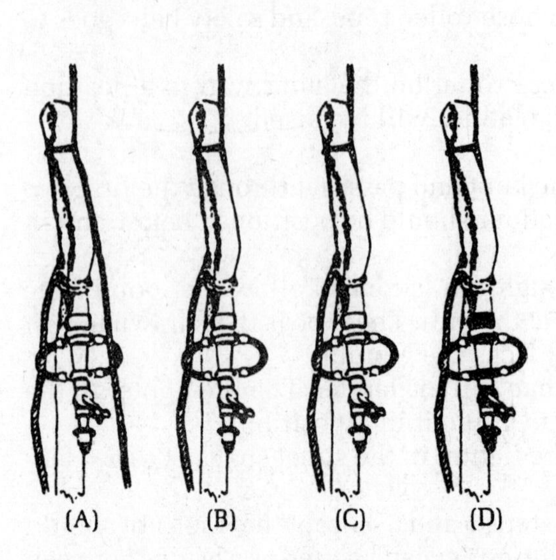

83. Which one of the four axe and rope patterns below is the same as the one above?

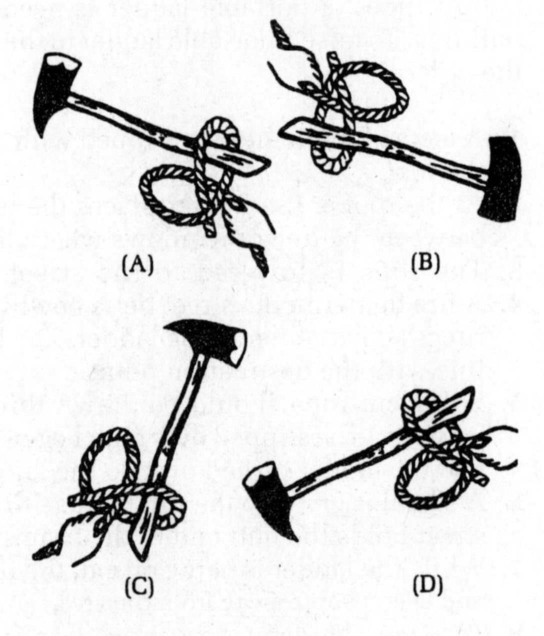

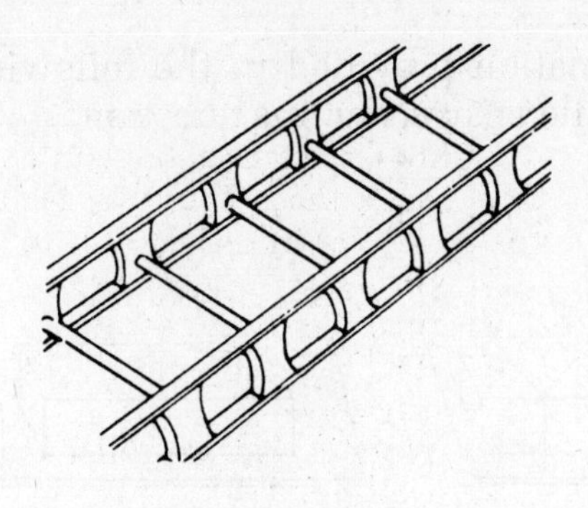

84. Which of the four ladders below is the same as the one above?

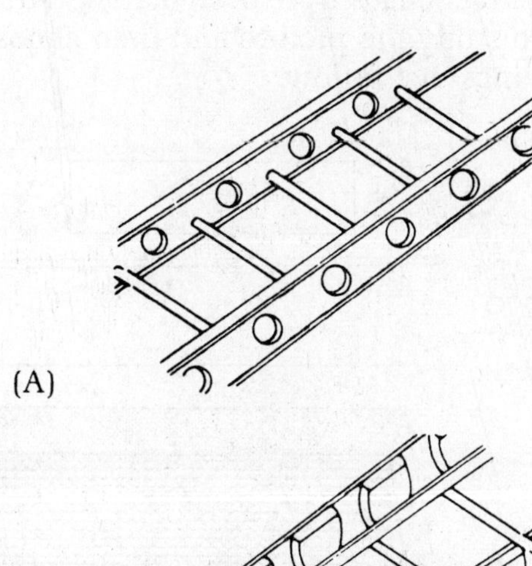

(A)

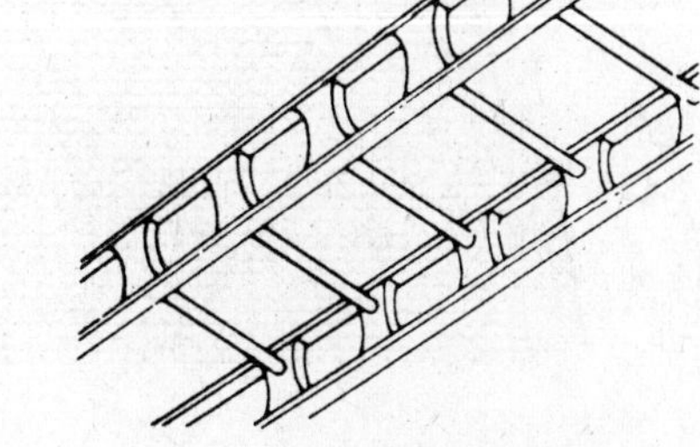

(B)

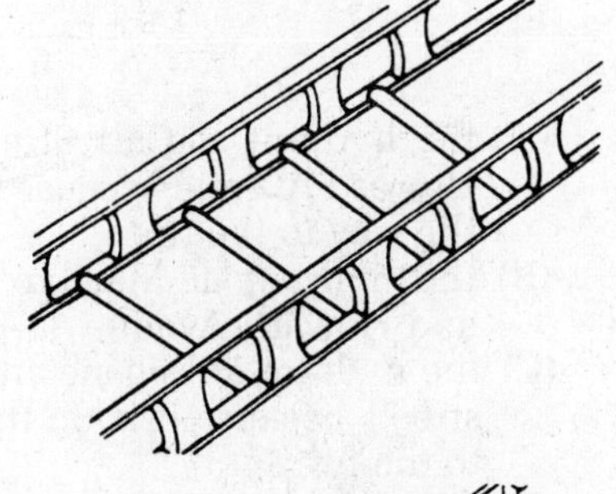

(C)

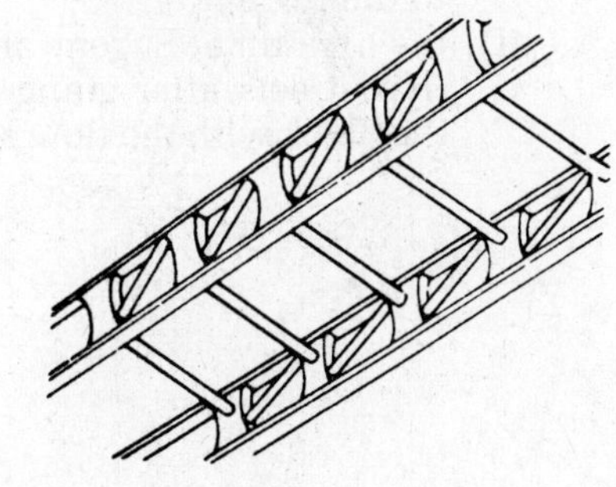

(D)

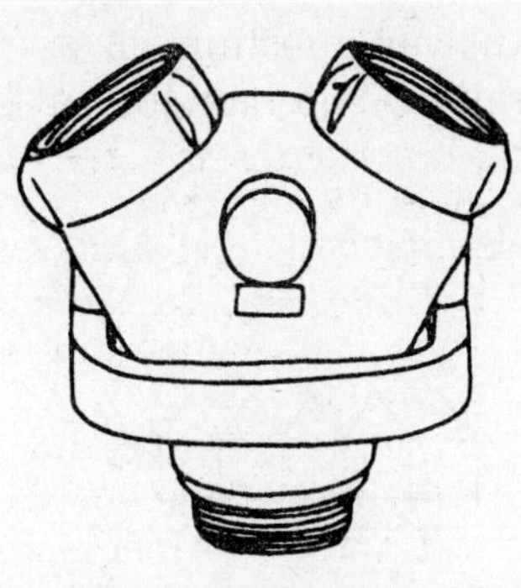

85. Which of the four couplings below is the same as the one above?

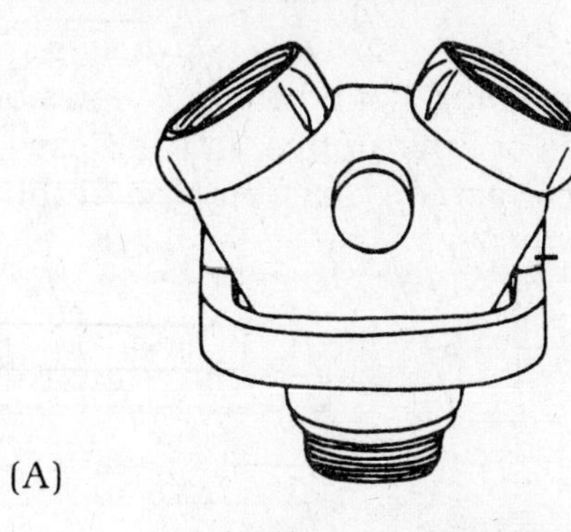

(A)

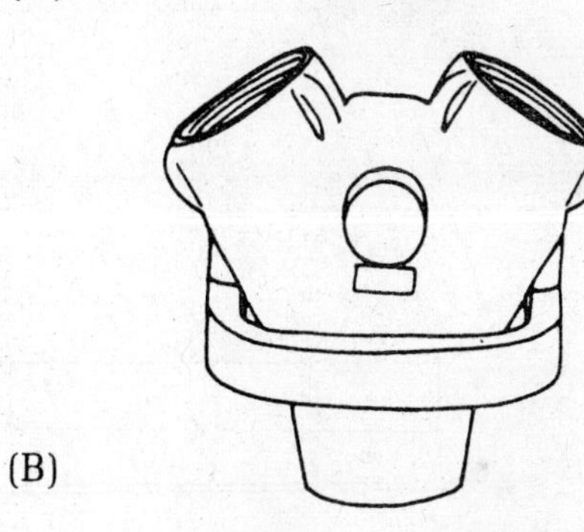

(B)

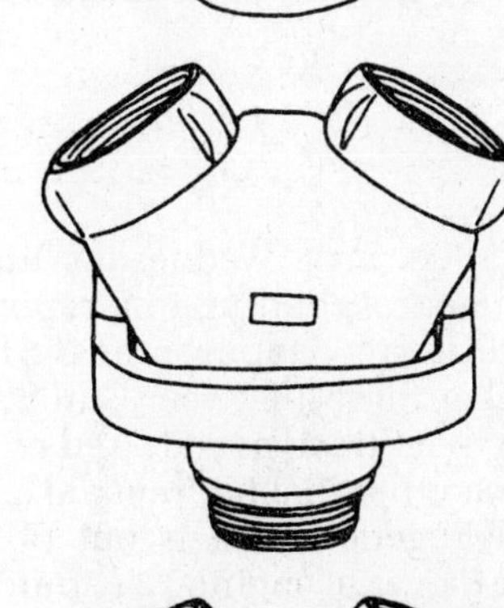

(C)

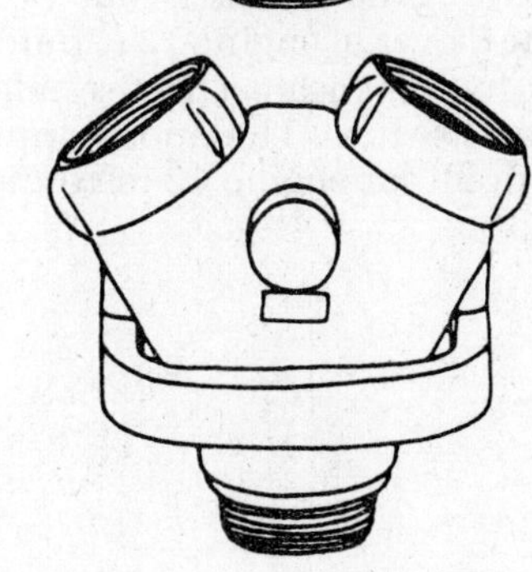

(D)

Answer question 86 using only the information provided in the following illustration. You may look at the illustration while answering this question.

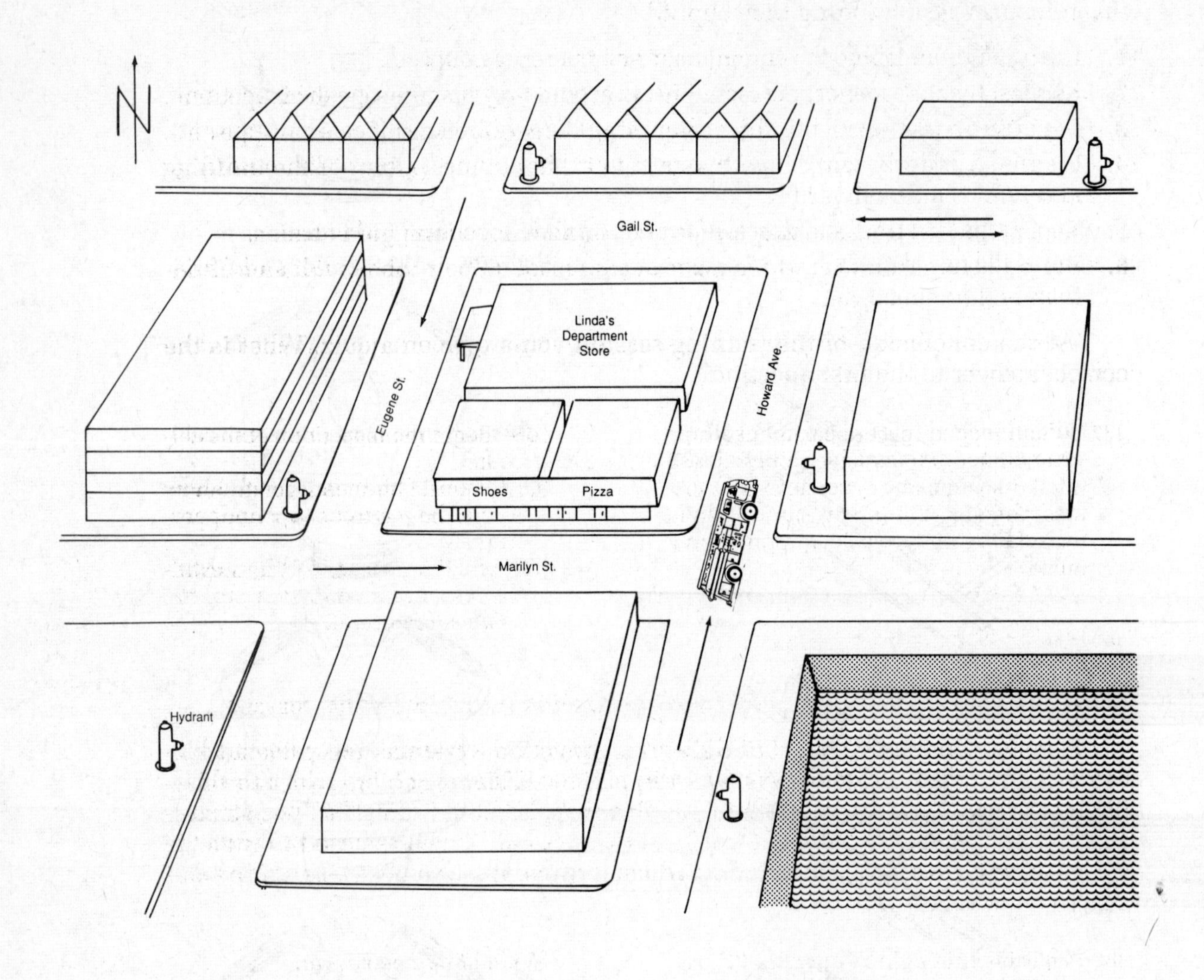

86. At 5 P.M. on a Wednesday night, fire fighters responding to a reported fire at Linda's Department Store on Eugene Street (between Gail Street and Marilyn Street) are advised by the dispatcher that the hydrant at Gail Street and Eugene Street is out of service. The driver of engine 12 should choose the hydrant nearest the fire before passing the fire. The most appropriate hydrant for engine 12 to connect to is

(A) the hydrant at Gail Street and Howard Avenue because it is on the way to the fire
(B) the hydrant at Marilyn Street and Howard Avenue
(C) the hydrant at Eugene and Gail streets because it is on the way to the fire
(D) the hydrant at Eugene and Marilyn streets after going around the block with the flow of traffic

Answer question *87* using only the information in the following passage.

During a fire prevention training session, a group of firefighters are told that when inspecting a building they should

1. Introduce themselves to the manager or his representative.
2. Request the manager or representative to accompany them during the inspection.
3. Begin the actual inspection by examining the required certificates and permit.
4. Insure an orderly work flow by starting at the topmost point of the building and working downward.
5. Record the findings, as they are found, on the inspection guide forms.
6. Before leaving, inform the manager or representative of the results and findings of the inspection.

At the conclusion of the training session you are given a quiz. What is the correct answer to the first question?

87. When inspecting a department store, you contact the manager and request that the manager accompany the team during the inspection. What would be the correct action to do next?
    (A) Begin the inspection at the roof.
    (B) Begin the inspection at the cellar.
    (C) Request the manager to show you the certificates and permits.
    (D) Ask the owner if the certificates and permits are up to date and current.

Answer question *88* using only the information in the following passage.

Siamese connections for building fire protection systems are protected by covers. The covers of these systems are painted different colors to let the firefighters know what type of system it is. Covers painted red indicate a "wet standpipe system," green indicates a "wet sprinkler system," silver is used to indicate a "dry sprinkler system," and yellow indicates a combination "wet standpipe and sprinkler system."

88. The driver of Engine Company 21 arriving at a fire in a building equipped with a standpipe system would expect to find a siamese with a
    (A) red or green cover.
    (B) yellow or silver cover.
    (C) green or yellow cover.
    (D) red or yellow cover.

Answer question *89* using only the information in the following passage.

Handling and using hand tools require an understanding of how the tool works and the proper method for using it. When using a wrench, it is better to pull than push. If you must push, keep your palm open to prevent your knuckles from damage. When using a hammer, choose the right type for the job. A carpenter's hammer is used for nails and a maul is used for heavy-duty work. Hammers should not be used for forcing a screw into something hard; use the correct type and size screwdriver.

89. Hand tools usage forces the user to
    (A) push, not pull the tool.
    (B) make use of the maul for forcing screws into hard objects.
    (C) work intelligently.
    (D) understand only the purpose of the tool.

Answer questions *90* and *91* using only the information in the following passage.

When a person is severely injured, immediate care and aid are very important. The loss of body heat can send the person into shock. This is a strong possibility in cases involving serious injuries. An essential and often overlooked action that should be taken when encountering a seriously injured person is properly blanketing the person. Many times a blanket is put over the person; however, little thought is ever given to the loss of body heat from the back of the person. Proper protection for an injured person lying on the street and awaiting transportation to a hospital should be wrapping the individual with blankets. To accomplish this, with minimum movement of the person, take a blanket and place it lengthwise alongside the person. Tuck about two thirds of the blanket snugly against the person. Having completed this, gently roll the person about an eighth of a turn away from the blanket and push the tucked blanket as far under the person as possible. Now roll the person back onto the blanket and then pull the blanket out the other side and wrap it around the top of the body.

90. Arriving at the scene of an automobile accident, you find a seriously injured young woman lying on the street. She is not bleeding, but in severe pain. It would be best if you put a blanket
    (A) over her to keep her warm.
    (B) under her to keep her warm.
    (C) around her to keep her warm.
    (D) around her injury to keep her warm.

91. The first thing to do after pushing the blanket under the person is to
    (A) roll the person toward the blanket.
    (B) roll the person away from the blanket.
    (C) pull the blanket over the person.
    (D) pull the blanket under the person.

Answer question *92* using only the information in the following passage.

When an obstructed airway emergency occurs, there is little time to take the proper actions. If the item caught in the throat cannot be dislodged rapidly, then the abdominal thrust should be used to save the person's life. The abdominal thrust is more commonly known as the Heimlich maneuver. To administer an abdominal thrust, stand behind the person and wrap your arms around the person in trouble. Make a fist and put the thumb side of the fist against the front of the upper body, slightly above the navel but well below the center of the chest bone. Now grasp your fist with the other hand and give a quick inward and upward pull or thrust. This may have to be repeated a number of times to dislodge the object.

92. To execute the Heimlich maneuver, you should
    (A) stand in back of the person, put your fist in the stomach area, and pull up and in quickly.
    (B) face the person, wrap your arms around the person, place your fist slightly above the navel, and thrust up and in rapidly.
    (C) place your fist on the chest cavity and with a quick inward and upward pull force the air out of the lungs to make the object pop out.
    (D) displace the object by giving several abdominal punches in quick succession in an upward and inward direction.

Answer questions *93* through *95* using only the information in the following passage.

Firefighters will often encounter dogs when arriving at the scene of a fire or emergency. Dogs come in many varieties and have several purposes for being on the premises.

Three major categories are used to subdivide the many reasons for having dogs. They are: security dogs, house dogs, and household pets.

Security dogs are trained specially for this purpose. They look vicious and often are. They react to their training. This training may be to prevent intruders from entering or leaving an occupancy. They generally will not leave the area they protect. Security dogs can be subclassified into attack dogs, watch dogs, and command dogs.

Attack dogs are taught to attack anyone except their trainer.

Watch dogs are taught to frighten off the intruder by barking but not to attack.

Command dogs are taught to attack at the command of the master.

House dogs are trained to protect the owner's property. They will attack anyone who enters except the owner. During a fire, they may abandon the premises but will return when the conditions improve.

Household pets normally are untrained and usually will not attack the firefighter.

Dogs can and will be unpredictable; great caution should be exercised when encountering them. Wear all your safety clothing and do not assume they are friendly. Seek professional help. Do not try to hit the dog. An injured dog is the most dangerous. If you can, get the dog to bite a tool, broom, or chair, then pull the tool away with only enough force to provide resistance. This will keep the dog safely at bay, while help is on its way.

93. The most dangerous dog a firefighter may encounter is
    (A) an attack dog.
    (B) a household pet.
    (C) a trained Doberman pinscher.
    (D) an injured dog.

94. If you encounter a barking dog that appears to be dangerous, it would be best if you
    (A) call for help from the police department or the trainer.
    (B) use a tool to hold the dog off while another firefighter searches the occupancy.
    (C) hit the dog with the tool to knock it out.
    (D) yell at the dog with loud commands to intimidate it.

95. During a discussion of problems that happen at fires, the subject of dogs comes up and there is a great amount of conflict among the firefighters in the group. Which of the following firefighters is correct?
    (A) FF Abe Jones reports he has had great success with attack dogs by giving them plenty of room and letting them get out of the building.
    (B) FF Bob Robinson claims household pets are more dangerous than attack dogs because you never know what they will do.
    (C) FF Jim Thompson said, "Any dog can be controlled with a broom handle."
    (D) FF Tom Whyith explained that, as it relates to the firefighter, command dogs generally are safer than house dogs.

96. Choose the tool that best goes with the device in the first illustration.

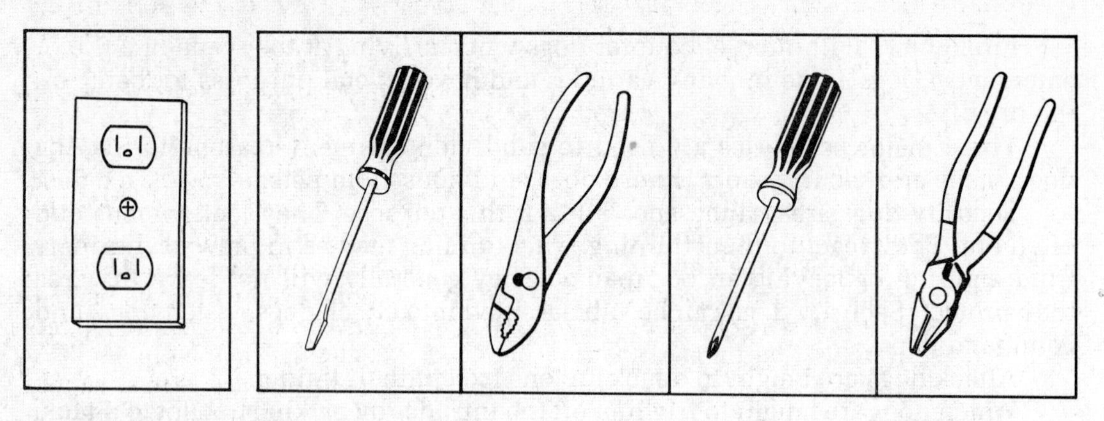

97. Choose the tool that best goes with the device in the first illustration.

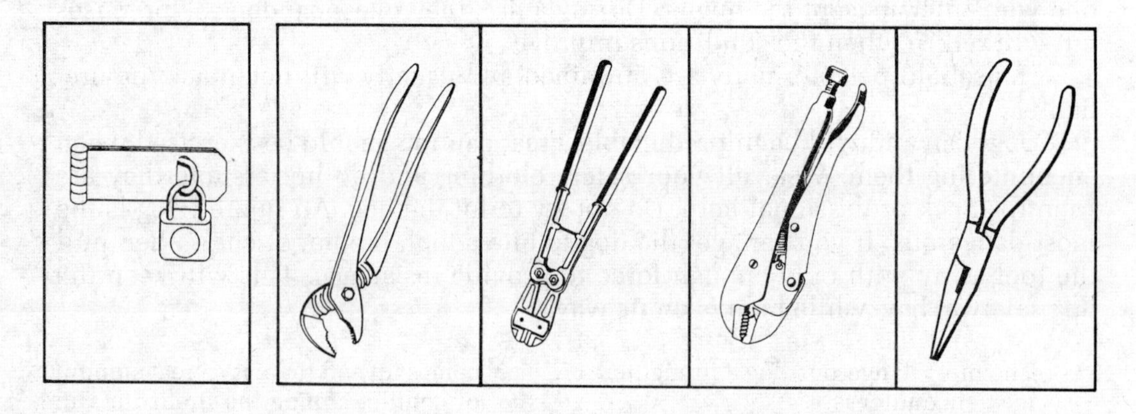

Answer questions 98 and 99 using only the information in the following passage.

Oils and liquids, such as gasoline, present a unique and difficult fire problem to fire fighters. Because oils are lighter than water, they rise to the surface when mixed with water. Several additives, known as foams, are available to help fire fighters with this problem.

All foams extinguish fires in four ways:

1. They smother the fire by floating over the surface and prevent the fire from mixing with the vapors.
2. They suppress the generation of flammable vapors.
3. They separate the flames from the fuel vapors.
4. They cool the fuel's surface and the adjacent surfaces.

Foams are made from protein and synthetic materials. The foam concentrate is mixed with water to produce a soaplike mixture. The soaplike mixture is then gently applied to the surface of the fuel. Foams come in low expansion varieties, which produce a thick substance, or in high expansion types, which are much like a bubble bath. Special foams are used for fire in alcohol based fuels. For fires during low temperature conditions, a fluoroprotein cold foam could be used.

98. On a night shift during a bitter cold spell, fire fighters respond to a fire in a large fuel oil storage tank. They set up hose lines and prepare to fight the fire. To extinguish this fire, fire fighters should
    (A) pour water directly on the fire
    (B) pour water on the exposed tank and let the fire burn
    (C) mix the water with a foam concentrate and then pour it on the surface of the fire
    (D) mix the special alcohol foam with the low expansion foam and pour it on the surface of the fuel.

99. Foams extinguish a fire by all of the following EXCEPT
    (A) smothering
    (B) vapor suppression
    (C) flame-fuel separation
    (D) dilution

100. At a difficult and smokey fire on the top floor of a building the lieutenant orders the roof of the building cut open and the ceiling immediately below it pushed down. Which of the following is the most correct reason for this order?
    (A) It provides fresh air to the fire fighters entering the top floor.
    (B) The heated fire gases will rise up and out through the hole.
    (C) It provides a clear observation point from which the lieutenant can direct the fire fighters to the seat of the fire.
    (D) It provides a large fresh supply of oxygen, which will help to extinguish the fire rapidly.

ANSWER KEY

1. **D**	21. **B**	41. **B**	61. **B**	81. **A**
2. **A**	22. **B**	42. **C**	62. **A**	82. **D**
3. **C**	23. **C**	43. **D**	63. **B**	83. **D**
4. **B**	24. **D**	44. **C**	64. **A**	84. **B**
5. **D**	25. **A**	45. **B**	65. **D**	85. **D**
6. **C**	26. **D**	46. **A**	66. **C**	86. **B**
7. **C**	27. **B**	47. **B**	67. **C**	87. **C**
8. **A**	28. **D**	48. **D**	68. **A**	88. **D**
9. **A**	29. **C**	49. **A**	69. **D**	89. **C**
10. **C**	30. **A**	50. **B**	70. **C**	90. **C**
11. **B**	31. **A**	51. **D**	71. **A**	91. **A**
12. **D**	32. **A**	52. **C**	72. **B**	92. **A**
13. **D**	33. **D**	53. **B**	73. **D**	93. **D**
14. **A**	34. **B**	54. **A**	74. **C**	94. **A**
15. **A**	35. **B**	55. **D**	75. **B**	95. **D**
16. **C**	36. **D**	56. **C**	76. **C**	96. **C**
17. **B**	37. **C**	57. **A**	77. **D**	97. **B**
18. **A**	38. **D**	58. **C**	78. **B**	98. **C**
19. **B**	39. **C**	59. **C**	79. **A**	99. **D**
20. **C**	40. **D**	60. **C**	80. **B**	100. **B**

DIAGNOSTIC PROCEDURE

Use the following diagnostic chart to determine how well you have done and to identify your areas of weakness.

Enter your number of correct answers for each section in the appropriate box in the column headed "Your Number Correct." The column immediately to the right will indicate how well you did on each section of the test.

Below the chart you will find directions as to the chapter(s) in the guide you should review to strengthen your weak areas.

Section Number	Question Number	Area	Your Number Correct	Scale	
One	1–12	Recall		12 Right 10 Right 9 Right Under 9 Right	Excellent Good Fair Poor
Two	22–26 30–41 46–50 55–58 64–69 86 98–100	Understanding Job Information		36 Right 34–35 Right 32–33 Right Under 32 Right	Excellent Good Fair Poor
Three	13–21 27–29 42–43 51 59–63 70–79 87 90–95	Judgment and Reasoning		36–37 Right 34–35 Right 32–33 Right Under 32 Right	Excellent Good Good Poor
Four	52–53 80–85 88–89 96–97	Tools and Equipment		12 Right 10 Right 9 Right Under 9 Right	Excellent Good Fair Poor
Five	44–45	Mathematics, Machines, and Science		2 Right Under 2 Right	Excellent Poor
Six	54	Dealing with People		1 Right	Excellent

1. If you are weak in Section One, then concentrate on Chapter 6.
2. If you are weak in Section Two, then concentrate on Chapter 7.
3. If you are weak in Section Three, then concentrate on Chapter 8.
4. If you are weak in Section Four, then concentrate on Chapter 9.
5. If you are weak in Section Five, then concentrate on Chapter 10.
6. If you are weak in Section Six, then concentrate on Chapter 11.

Note: Consider yourself weak in a section if you receive other than an "excellent" rating.

ANSWER EXPLANATIONS

1. **D** This information is found in the first sentence: "At a recruit training session, the trainees are told they will be assigned to a building inspection team that will go into the field on Thursday."
2. **A** The name of the officer is found in the second sentence: "The lieutenant supervising the team is William Holt, a veteran fire officer."
3. **C** During the inspection, you viewed a hotel, a small restaurant, and a (retail) drugstore. You did not inspect the storage warehouse. A cold storage warehouse is a building without heat, used exclusively for the long-term storage of people's goods.
4. **B** This information is found in the second sentence of the fourth paragraph: "You are asked to act as a witness, and the lieutenant orders the other trainee to fill out and serve the summons."
5. **D** This is explained in the last sentence of the passage: "A further inspection finds no spare sprinkler heads on the premises."
6. **C** To get from the fire escape to the study, a person would have to pass through bedroom 1, the kitchen, and the living room.
7. **C** Of the selections offered, only Choice C has two doors. Choice A is incorrect because the kitchen has a door and a doorway.
8. **A** In 2A, bedroom 2 is the only room directly accessible from the fire escape.
9. **A** The lving room has an access to a hall and stairs, but no direct egress to the exterior..
10. **C** In 2B, the bathroom door opens directly into the kitchen space.
11. **B** In 2A, there are three doors from the terrace. They lead to the kitchen, dining room, and bedroom. To get to the study from the terrace, you must pass through the kitchen and the living room.
12. **D** Of the choices offered, only the study in apartment 2B has access to two sides of the building.
13. **D** There are two windows, a front door and cellar doors in the rear.
14. **A** The main body of fire is from the fireplace and up the stairs. Choice B is incorrect; there is an overloaded electrical outlet on the first floor, but there is no indication of fire near. (Note: fire burns upward, until blocked, then horizontally, until it can find a way to go up again.) Choice C is incorrect; these materials are remote from the fire. Choice D is incorrect; there is no indication of anyone smoking in bed.
15. **A** The children are sleeping on the second floor next to the window. Based on the direction indicator at the top of the illustration, this is the east side of the house.
16. **C** The direction of the wind is indicated by the bending of the smoke coming out of the chimney.
17. **B** The object is to get water, through a hose, from the hydrant to the foam house. You must use the nearest available hydrant with the most direct route. The hydrant at Spring Lane and Branley Ave. is the nearest available hydrant and the directions in Choice B are correct. Choice A is incorrect; this is not the nearest hydrant. Choice C is incorrect; this is not the most direct route. Choice D is incorrect; this is direct but does not take you to the siamese on the foam house; it is located outside the foam house on Railroad Ave.
18. **A** The passage tells you the raw material storage is in the north end of the warehouse. A look at the map shows that apparatus could be placed on Kellan Ave., Railroad Ave., and in the main yard. Choices B and D are incorrect; raw material is stored at the north end of the warehouse. Choice C is incorrect; the placement of the apparatus on Spring Lane would not be effective.
19. **B** When working with a map or floor plan, find the arrow that indicates north. If necessary, turn the page so north is at the top (pointing away from you), east is on your right side, south is pointing toward you, and west is on your

left side. For this question you must find southwest—a direction directly between and equidistant from south and west. Choice A is incorrect; this would be in the west direction only. Choice C is incorrect; this would be south. Choice D is incorrect because it omits the first hazard, the fuel pumps.

20. **C** The sprinkler is designed to put water directly on the fire automatically. Of the four selections given to you, the best is to supply water to the system where the fire is. Choice A is incorrect; this may need to be done at some later time, however upon arrival, the danger is in the manufacturing plant and this is remote from the fuel filling site. Choices B and D are incorrect; these activities would be done after water was supplied to the manufacturing plant.

21. **B** This would be the most direct route and would put engine 1 in the best position to gain access to the site and fight the fire.

22. **B** The emphasis of this passage is on training at the National Fire Academy and refers only to training in the areas of fire fighting, command, and future trainers. Local Fire Department personnel problems and rules are not mentioned.

23. **C** This choice is the most accurate of all, although incomplete. Choice D is incorrect; the National Fire Academy is *not* for *all citizens.*

24. **D** This information is the central theme of the paragraph and is found in the second sentence. Choice A is incorrect; three fire fighters can handle a 2½ inch line, however more than one 2½ inch line may be required to extinguish the fire. Choice B is incorrect; the needed fire flow is considered. Choice C is incorrect; the needed fire flow is the amount of water flowing from one hose line but not necessarily the amount of water required to put the fire out.

25. **A** This information is found in the beginning of the third paragraph. Choice B is incorrect; three fire fighters are needed for this size line. Choices C and D are incorrect; the passage does not tell us how many fire fighters are required for these activities.

26. **D** This information is found in the second paragraph. By knowing the amount of water required, we can quickly determine a minimum number of fire fighters required. Choice A is incorrect; resource needs are determined *before* the fire. Choice B is incorrect; 250 is the fire flow for one 2½ inch hose line. Choice C is incorrect; the manpower calculation is simple, not the needed fire flow. We are not told in the passage how difficult it is.

27. **B** Only choice B has the hydrant on the west side and the fire engine truck south. Choices A and C are incorrect, the fire truck faces north. Choice D is incorrect; the hydrant is on the east side.

28. **D** According to the symbols chart, a hose line entering at an upper level must have a ladder and a hose touching the building. The question tells you that all three fire units took hose to the second floor. Only Illustration D has the lines going to the second floor.

29. **C** The pumper is heading east on Block across Rahn and headed for the ladder truck on Block. Choices A and B are incorrect; both units involved are pumpers. Choice D is incorrect; the pumper is headed west.

30. **A** Sentences two and five in the passage will lead you to this answer. Choices B and D are incorrect; a siamese is used to increase the flow. Choice C is incorrect; there is no mention of the use of a wye on a standpipe.

31. **A** The sixth sentence of the passage makes this statement. The check valve would be found on the input side of a device. Choices B, C, and D are discharge devices.

32. **A** This information is found in the last sentence of the first paragraph. Choices B, C, and D are incorrect; you *should* do these activities.

33. **D** This example is found in the fourth paragraph. Choice A is incorrect; because the house is smoke filled, you may not be able to see your way out, and because you are alone, you will have no one

to help you. Choice B is incorrect; safety eye protection should always be worn when working with power tools. Choice C is incorrect; use the proper size cord and tell the other fire fighters of the hazard.

34. **B** This is the central theme of the passage and is stated in the first sentence of paragraph four. Choice A is incorrect; it should read "not supervised." Choice C is incorrect; information can also be learned through education and training. Choice D is incorrect; as indicated in the last sentence of the first paragraph, we often do not appreciate safety until something goes wrong.

35. **B** This information is in item 3 under "Outside the building." Choice A is incorrect; it is an indicator of flashover. Choice C is incorrect; this is not mentioned in the passage, however, it is a good sign that the fire is ventilated. Choice D is incorrect; it would be normal to find the doors and windows in this condition late Sunday night.

36. **D** This information is in item 1 under "Inside the building."

37. **C** This information is an "Outside the building" indicator of a backdraft.

38. **D** This information is in item 5 under "Inside the building."

39. **C** The second paragraph, third sentence tells us this. Choices A, B, and D are incorrect; they are not mentioned in the passage.

40. **D** When answering a progression question, you must select the most correct and most complete answer. Choices A, B, and C are accurate answers but less accurate than Choice D. Therefore, only Choice D is correct.

41. **B** There is no mention of a hose line in the passage. Choices A, C, and D are incorrect; fire fighters should be protected by these devices and concerned with safety.

42. **C** This room is directly across from the fire and has no other way out. Choices A and B are separated from the fire room and have a balcony as a second means of escape. Choice D is incorrect; there is no room 16 shown on the floor plan.

43. **D** The objective of this hose line placement is to get the hose into position using the most direct route while providing the most safety to the fire fighters as they attack and extinguish the fire. As stated in Choice D, the first line should go up stairway C and the second line up stairway B. Both lines should operate in the A hallway. Choices A, B, and C are incorrect; each of these would result in a longer hose stretch.

44. **C** A quick count of the room occupants shows there were 57 guests staying at the hotel—57 minus 53 equals 4. Choice A is incorrect; it is not the amount of guests staying at the hotel, only the amount of guests accounted for. Choice B is incorrect; by knowing how many people were staying at the hotel and subtracting the number accounted for, we can determine if any are missing. Choice D is incorrect; if you have information that can be of critical help, you are required to notify those in command.

45. **B** This number can be found by counting the guest rooms on the map. Choice A is incorrect; there are 21 rooms on the outer periphery and eight rooms in the center. Choice C is incorrect; 49 is the highest room number. Choice D is incorrect; this is the number of guests staying on the floor.

46. **A** The preconnect load means that it is one of the three basic loads, with one end of the hose already connected to the pumper. See the next to last paragraph of the passage.

47. **B** The horseshoe load is found in the fifth paragraph. Choice A is incorrect; based only on the information given, this load is used for hand stretching. Choice C is not a hose load; it is a hose storage location—a hose bed. Choice D refers to the method for stretching hose from the building to the water source.

48. **D** The word *all* makes this an incorrect statement. You should be very careful of choices that use such terms as *all, never, always, must.*

49. **A** This information is explained in the last two paragraphs of the passage.

50. **B** You will find this information in the fifth sentence. The destruction of evidence is not done with malicious intent or to cover up, but does occur during the process of fire extinguishment. Choice A is incorrect; arson investigators respond to determine the point of origin and to gather evidence. Choices C and D are incorrect; this may be true in real life, but the passage does not provide this information and you are told to use only the information in the passage.
51. **D** This is item 5 in the passage. You have been told that items 1 through 4 already have been complied with. Choice A is incorrect; this is done after the area has been roped off. Choice B is incorrect; there are actions that have not been done but should be while waiting for the power company's arrival. Choice C is incorrect; there is no need to do this. The cover is still on the manhole and the firefighters have determined no life hazard exists.
52. **C** Items 6 and 7 explain this. Choice A is incorrect; from your position on the sidewalk, the stream will not penetrate the window. Choice B is incorrect; a hand line will be effective into the third floor. Choice D is incorrect; 45 degrees is the wrong angle.
53. **B** Statement 2 is incorrect. Item 6 in the passage explains that the third floor is the highest water can be thrown *effectively,* but water can be thrown above this level.
54. **A** Two times 250 GPM equals 500 GPM; the flow in is equal to the flow out; the operations are effective but will require more pumps and the water to be shut down. Giving accurate and proper information at an incident is an important part of a firefighter's role. Choice B is incorrect; if you can help the person, you should. Choice C is incorrect; everything is not "OK," the water has not been stopped and the basement still has water in it—do not give out incorrect information. Choice D is incorrect; do not insult or ignore the person.
55. **D** This information is in item 4 under "route taken to the roof." Choice A is incorrect; item 5 tells you this NEVER should be done. Choice B is incorrect; this is the most acceptable method. Choice C is incorrect; it is the second most acceptable method.
56. **C** This information is found in the last sentence of the first paragraph. Choices A and B are incorrect; these are methods for accomplishing Choice C. Choice D is incorrect; one of the roof person's duties is to act as a spotter and to locate any jumper.
57. **A** This information is found in the first sentence of the last paragraph. Choices B and C are incorrect; the officer in command does not need to be notified that you have completed your duties. Choice D is incorrect; this is one of the duties of the roof person and should be completed before notifying the company officer.
58. **C** This is found in item 1 under "When arriving at the roof."
59. **C** This is two blocks shorter than Choice B and 4 blocks shorter than Choice A. Choices A and B are incorrect because the driver must go beyond Indiana to Nevada or Utah and then return the same distance.
60. **C** This is the shortest route. Choices A and B take the driver past Jersey Extension East and then require the driver to come back.
61. **B** This is the shortest route. A is incorrect; the driver must go past Ohio to Wyoming and then return the same distance. Choice C is incorrect; the driver cannot go east on a one-way west-bound street.
62. **A** This is the shortest route. Choice B is incorrect; the driver cannot go north on Maine. Choice C is incorrect; it is several blocks longer.
63. **B** This is the shortest route. Choice A would require proceeding against the flow of traffic on Wyoming; Choice C is several blocks longer.
64. **A** At a one-story shopping complex, the occupants will not be found on the roof.
65. **D** This information is found in the sixth sentence of the passage.
66. **C** This information is found in the opening paragraph of the passage.

You have been told the elevators are traveling to the upper floors. To use the elevators, you must recall them. You are also told the people are now safe two floors below the fire.

67. **C** This information is item 6. Choice A is incorrect; this would disable the special safety features. Choices B and D are incorrect; these buttons normally are not found in elevator cars.

68. **A** This requires knowledge from two different parts of the passage. The last paragraph directs you to leave the car immediately; to do this you must push the door open button.

69. **D** You are looking for an incorrect action. Choices A, B, and C are correct actions as explained in the first paragraph of the passage. There is no mention of how the hoist doors are closed.

70. **C** For this illustration, use the first floor as a guide. The first floor shows two entrance doors, one picture window, and a total of seven standard size windows. Compare the floor plan layout with the three-dimensional representation. Choice C most accurately portrays the floor layout.

71. **A** The three-dimensional representation shows a U-shaped entrance with two doors, one picture window, and seven standard-size windows. Compare the floor plan layout with the three-dimensional representation. Choice A most accurately portrays the floor layout.

72. **B** Item 1 in the passage tells the reader that the street-level door must be openable from the inside. Item 2 tells the reader that in a building less than 100 feet in height, doors opening onto a floor of the building may be locked. This means that all floors above the street floor may be locked in buildings less than 100 feet high.

73. **D** This is explained in the second sentence. Choice A is incorrect; it is a bogus statement. Choice B is incorrect; this is "getting the big picture." Choice C is incorrect; it is not mentioned in the passage. Driving safely requires that you stay in the center of your lane, not in the center of the road.

74. **C** Gas leaking from a pipe in an apartment house creates a very dangerous condition. If it should ignite, an explosion and fire will occur. Choices A and D are incorrect; they require attention, but not until after the gas leak has been corrected. Choice B is incorrect; you were told all the people are accounted for.

75. **B** From the passage, you know the position of outside ventilation is responsible for opening the windows from the outside. You are also told that this person may have to perform other activities, such as assisting in raising the aerial ladder and that this is done before opening the windows. In addition, you are told that the outside ventilation person is seen in positions other than at the windows. There is no indication that anyone has done anything wrong or is at any time out of position; from this you can conclude that Choice B is the most correct. Choice A is incorrect; in the passage, you have been told, twice, your specific duties and the order in which to do them. Choice C is incorrect; you are told to help get the aerial ladder into position, and then to open the windows from the fire escape or portable ladders. Choice D is incorrect; the outside ventilation person is found in the rear bedroom. The firefighter did not come through the locked entrance door that you forced open; therefore, the firefighter must have come in through the window that the firefighter had opened.

76. **C** This is the most direct route. Choices A and D are incorrect; they are significantly longer. Choice B takes the firefighter in the wrong direction.

77. **D** Odd-numbered rooms 119 through 131 open onto hallway B.

78. **B** Going with the flow of traffic is the best method; the nearest hydrant would be at the northwest corner of Princess Street.

79. **A** Room 122 and 124 are directly opposite Elevator A.

80. **B** There would be no need for a safety belt in the street. Choices A, C, and D, are required actions.

81. **A** This is item 4. Choice B is incor-

rect; the ladder is turned over after the knot has been tied and the rope is secured to the ladder. Choice C is incorrect; the roller is lined up between the windows. Choice D is incorrect; the roller is put into position on the roof before the rope is lowered.

82. **D** When working with this type question, try to eliminate the wrong choices first. After eliminating the incorrect choices, prove to yourself that the remaining choice is correct. Choice A has an extra rope. Choices B and C have the hose butts and nozzle clear. (In the illustration on the left, these parts are blackened in.) Choice D is the same pattern as the illustration on the left.

83. **D** See the first two sentences of Answer 82. Choice A shows the rope behind the loop passing around the axe; the illustration at the top of the page shows it going through this loop. Choice B shows the outside the rope loop. Choice C shows the rope passing over the loop beyond the axe; the illustration shows it passing through the loop. Choice D is the same as the illustration. Hint: rotate the paper to get the object into the same viewing position.

84. **B** See the first two sentences of Answer 82. Choice A has round holes in the beams of the ladder. Choice C has the rungs of the ladder attached to the cut out portion of the beams. Choice D shows an additional rod in the cut out of the beams. Choice B is the same construction as the illustration on the top of the page.

85. **D** See the first two sentences of Answer 82. Choice A lacks the small rectangle below the round object in the center of the siamese. (*Note:* the object portrayed is a portable siamese.) Choice B lacks the thread on the bottom of the siamese. Choice C lacks the round object above the rectangle in the center. Choice D is the same construction as the illustration on the left.

86. **B** Connect to the hydrant and then drive with the flow of traffic to the entrance of the building. A is incorrect; there is no hydrant at this intersection. C is incorrect; you are told that this hydrant is out of service. D is incorrect; the driver would have to pass the entrance.

87. **C** According to the information given in the question, you have completed step 2. The next thing to do is examine the certificates and the permits.

88. **D** The passage tells you that the cover of a standpipe siamese can be red, and a combination standpipe and sprinkler siamese is yellow.

89. **C** The passage tells the reader there are different types in the same class of tools. It also explains the right tool should be selected for each job. Choice A is incorrect; the tool operator is not obligated to push or pull a tool. The choice depends on the situation. Choice B is incorrect; a maul would not be used for this. The word "only" makes Choice D incorrect.

90. **C** This is the theme of the passage. Choices A, B, and D are incorrect; these are examples of incomplete blanketing.

91. **A** This information is found in the ninth sentence of the passage. Choice B is incorrect; this should be done before you push the blanket under the person. Choices C and D are incorrect; these are done after Choice A.

92. **A** Choice B is incorrect; stand in back of the person. Choice C is incorrect; place the first in the abdominal area. Choice D is incorrect; you would not punch the person.

93. **D** This information is found in the fifth sentence of the last paragraph.

94. **A** This information is found in the third sentence of the last paragraph. Choice B is incorrect; this would be a last resort activity. Choice C is incorrect; hitting the dog will make conditions worse. Choice D is incorrect; yelling is not mentioned in the passage.

95. **D** As explained in the passage, house dogs attack anyone except their master. Attack dogs respond to commands from their master. Choice A is incorrect; dogs will return when conditions improve. Choice B is incorrect; the passage tells you that household pets usually will not attack the firefighter.

Choice C is incorrect; the passage does not say *all* or *any*, but tells you a broom handle may be helpful if done correctly and if you can get the dog to bite it; this means it may not always be possible.

96. **C** The Phillips head screw driver would be the most effective tool to remove the screw.

97. **B** The bolt cutter is the most effective tool for cutting the hasp or lock.

98. **C** Foam is made by mixing the concentrate with water. To extinguish the fire, the foam is gently applied to the surface of the burning fuel. Choice A is incorrect; the water will sink into the fuel and make conditions worse. Choice B is incorrect; you have been asked to extinguish the fire. Choice D is incorrect; there is no mention of mixing the foams.

99. **D** Dilution is not mentioned in the passage and is not one of the ways foam extinguishes a fire. Choices A, B, and C are equivalent to items 1, 2, and 3.

100. **B** Heat rises. Gases when heated expand and rise rapidly. Cutting a hole over the fire lets the heated gases and smoke rapidly vent to the outside. A is incorrect; the rising heated gases pushing through the hole would prevent the air from entering. C is incorrect; the rising smoke and heated gases would make it very difficult to see into the top floor. D is incorrect; a large supply of oxygen would not extinguish the fire, it would cause it to grow rapidly.

PART FIVE

A Final Word

CHAPTER 16

What to Do Immediately Before, During, and After the Examination

The Week Before the Examination

The time draws near. What should be done in the days immediately before the examination?

Do not be fearful. If you have prepared well, you should do well. Remember that you have helped yourself pass the examination every time you worked with this text. In a sense, therefore, you will be collecting your reward when you take the examination.

But what actions should be taken on the days just before the examination?

SEVEN DAYS BEFORE

Review the Diagnostic Examination in Chapter 4. Identify again the areas in which you have the greatest difficulty.

Some time during this week, you should ensure that you know the exact location of the examination site and the best way to get there. If you are going on public transportation, take a "dry run." Note exactly where the bus or train stops.

Chances are that the examination will be given on a weekend, so ask about the transportation schedule for that day. Do not be left waiting for a bus or train that does not run on the weekend. You have greatly increased your chances of successfully competing in the examination by following the instruction given in this book, but you have to be there to take the test.

The same applies if you are driving. Make sure the car you are using is in mechanical order. It is not a bad idea to have someone drive you, if you can. In this way, should any difficulties with the vehicle arise, you can continue on your own. Be certain that you know the route to the examination and a safe, legal parking area. Taking the examination while you are preoccupied about your car is not a good idea. Give yourself every advantage.

SIX AND FIVE DAYS BEFORE

Zero in on the chapter or chapters that deal with your greatest weakness. For example, if your scores on the Diagnostic Exam or the three practice examinations indicate a problem in understanding mechanics, concentrate on Chapter 10. Spend these two days on this topic.

FOUR, THREE, AND TWO DAYS BEFORE

On each of these three days, review one of the sample firefighter examinations given in Chapters 12 through 14. This will reacquaint you with what to expect when you take the actual examination. Review the tests under simulated examination conditions, as follows.

Sit in a room at a desk or table. Time yourself, making sure that you stay within the time allotted. You will then be less uneasy about this during the actual examination. If you are a smoker, do not smoke; the chances are that you will not be able to do so during the actual examination. Use the test techniques explained in Chapter 3.

Use the same kind of pencil you will use on examination day. Usually, you will be asked to bring a #2 lead pencil. The idea is to make conditions duplicate as closely as possible those you will be experiencing on examination day.

ONE DAY BEFORE

Reread Chapter 3, which deals with test taking. That's it—by now you should be prepared. Do not try to cram on this last day. Your test preparation is over. Begin to relax.

Eat a normal dinner on the evening before the examination. Get your regular amount of sleep the night before. Although you do not need more rest, you should not get less. Be careful of taking sedatives to help you sleep. You may feel drowsy the next day. This is generally not a good night to go to a party.

The theme is to keep yourself as you would at any other time. Remember: Tomorrow is just another day. You have prepared well for the examination; nothing more need be done.

The Day of the Examination

On the day of the examination, wake up with enough time to dress, have a good breakfast, and check over whatever test-taking equipment you will be taking with you. While an alarm clock may get you up, a call from a friend is sometimes helpful.

In regard to test-taking equipment, follow whatever instructions you have received from the testing agency (for example, about pencils or a watch). Bring an extra sweater. If it's not needed to keep warm, you can always take it off. (You may want to sit on it; a desk seat can become very uncomfortable after an hour or so.)

What else? Well, a pencil sharpener is helpful, as are a good eraser and a working pen to sign in ink if required.

Some candidates have found that a chocolate bar is a source of quick energy during the examination. If you wear glasses, make sure you have an extra pair, if possible, and something with which to clean them.

Before you arrive at the examination, charge up your brain and body with a little stretching exercise. Do just enough to stimulate your heart and breathing; avoid overdoing it and becoming tired.

When you reach the examination site, go to your assigned room and follow the instructions of the proctors. Inspect your seat and report any problems to your proctor immediately. Follow instructions to the letter.

DEALING WITH ANXIETY AND PANIC

During the examination you may begin to experience sweating palms, heart palpitations, queasiness, inability to concentrate, or a general feeling of discomfort. These common signs of stress, anxiety, and fear can be overcome. Stop taking the examination for a few moments. During this short break use a predampened handkerchief to wipe your face, hands, and the back of your neck. Breathe slowly and deeply, and concentrate on something positive, such as a previous success that you have had.

Believe in your ability. Many others just like you have passed firefighter examinations. You can too!

After You Have Finished the Examination

You must prepare for the end of the examination; you cannot just get up and walk out. You must have a plan for what you will do during the last 15 minutes, and you must know what to do after the examination is over.

Before the examination begins, you will be told, either in the examination booklet or by the proctor, whether you will be permitted to take the examination booklet home with you. This information is very important. You must have a copy of the answer choices you selected so that you can compare them with the official answer key and find out how you did. You will also need a record of your choices if you plan to protest an answer.

If you can take the test booklet home, the answers that you marked in the booklet will serve as your answer list; otherwise, you must make a copy of your answer choices to take with you, following the procedure outlined below.

THE LAST 15 MINUTES

The last 15 minutes of the examination is the time to double check that you have recorded all your selections properly. This is your last chance to check for recording errors during the test. You may have made a simple error of marking the wrong box or line for a question. Now is the time to uncover that error and correct it. Once you are sure you have properly recorded your choices, make a copy of the list to take with you. Record, on the test booklet or the answer sheet, the number of your test booklet and any other identifying symbols that are available. If you have been following the procedures taught in this guide, you will have ample time at the end of the examination to do this.

THE END OF THE EXAMINATION

The end of the examination will be signaled by a bell, horn, or verbal notification. You must stop working at that point; if you don't, you may be disqualified. Remain seated, and follow the directions of the proctor. Finally, make sure that all parts of your test are properly collected and checked.

A Word About the Future

Once you have become a firefighter, you must direct your efforts toward professionalism and advancement. Many fire science programs are offered in our nation's colleges, and the National Fire Academy makes available excellent courses that can be taken either on or off campus. In addition, there is an abundant supply of job-related texts and guides for you to study.

You'll be back to the beginning, striving to master new material and reach a new goal with new rewards.

Welcome to the fire service!